Positive Linear Systems

PURE AND APPLIED MATHEMATICS

A Wiley-Interscience Series of Texts, Monographs, and Tracts

Founded by RICHARD COURANT
Editors Emeriti: PETER HILTON and HARRY HOCHSTADT
Editors: MYRON B. ALLEN III, DAVID A. COX, PETER LAX,
    JOHN TOLAND

A complete list of the titles in this series appears at the end of this volume.

Positive Linear Systems
Theory and Applications

LORENZO FARINA
SERGIO RINALDI

A Wiley-Interscience Publication
JOHN WILEY & SONS, INC.
New York / Chichester / Weinheim / Brisbane / Singapore / Toronto

This text is printed on acid-free paper. ⊗

Copyright © 2000 by John Wiley & Sons, Inc. All rights reserved.

Published simultaneously in Canada.

No part of this publication may be reproduced, stored in a retrieval system or transmitted in any form or by any means, electronic, mechanical, photocopying, recording, scanning or otherwise, except as permitted under Sections 107 or 108 of the 1976 United States Copyright Act, without either the prior written permission of the Publisher, or authorization through payment of the appropriate per-copy fee to the Copyright Clearance Center, 222 Rosewood Drive, Danvers, MA 01923, (978) 750-8400, fax (978) 750-4744. Requests to the Publisher for permission should be addressed to the Permissions Department, John Wiley & Sons, Inc., 605 Third Avenue, New York, NY 10158-0012, (212) 850-6011, fax (212) 850-6008, E-Mail: PERMREQ @ WILEY.COM.

For ordering and customer service, call 1-800-CALL-WILEY.

Library of Congress Cataloging-in-Publication Data:

Farina, Lorenzo, 1963–
    Positive linear systems : theory and applications / Lorenzo Farina, Sergio Rinaldi.
        p. cm. — (Pure and applied mathematics)
    Includes bibliographical references and index.
    ISBN 0-471-38456-9 (cloth : alk. paper)
    1. Nonnegative matrices. 2. Linear systems. I. Rinaldi, S. (Sergio), 1940– II. Title. III. Pure and applied mathematics (John Wiley & Sons : Unnumbered)
QA188.F37 2000
512.9'434—dc21   00-021312

Printed in the United States of America

10 9 8 7 6 5 4 3 2 1

Contents

Preface	ix
PART I DEFINITIONS	1
1 Introduction	3
2 Definitions and Conditions of Positivity	7
3 Influence Graphs	17
4 Irreducibility, Excitability, and Transparency	23
PART II PROPERTIES	33
5 Stability	35
6 Spectral Characterization of Irreducible Systems	49

7	Positivity of Equilibria	57
8	Reachability and Observability	65
9	Realization	81
10	Minimum Phase	91
11	Interconnected Systems	101

PART III	APPLICATIONS	107
12	Input–Output Analysis	109
13	Age-Structured Population Models	117
14	Markov Chains	131
15	Compartmental Systems	145
16	Queueing Systems	155
Conclusions		167
Annotated Bibliography		169
Bibliography		177

Appendix A: Elements of Linear Algebra and Matrix Theory 187
    A.1 Real Vectors and Matrices   187
    A.2 Vector Spaces   189
    A.3 Dimension of a Vector Space   193
    A.4 Change of Basis   195
    A.5 Linear Transformations and Matrices   196
    A.6 Image and Null Space   198
    A.7 Invariant Subspaces, Eigenvectors, and Eigenvalues   201
    A.8 Jordan Canonical Form   207

A.9	Annihilating Polynomial and Minimal Polynomial	210
A.10	Normed Spaces	212
A.11	Scalar Product and Orthogonality	216
A.12	Adjoint Transformations	221

Appendix B: Elements of Linear Systems Theory — 225

B.1	Definition of Linear Systems	225
B.2	ARMA Model and Transfer Function	228
B.3	Computation of Transfer Functions and Realization	231
B.4	Interconnected Subsystems and Mason's Formula	234
B.5	Change of Coordinates and Equivalent Systems	237
B.6	Motion, Trajectory, and Equilibrium	238
B.7	Lagrange's Formula and Transition Matrix	241
B.8	Reversibility	244
B.9	Sampled-Data Systems	244
B.10	Internal Stability: Definitions	248
B.11	Eigenvalues and Stability	248
B.12	Tests of Asymptotic Stability	251
B.13	Energy and Stability	256
B.14	Dominant Eigenvalue and Eigenvector	259
B.15	Reachability and Control Law	260
B.16	Observability and State Reconstruction	264
B.17	Decomposition Theorem	268
B.18	Determination of the ARMA Models	272
B.19	Poles and Zeros of the Transfer Function	279
B.20	Poles and Zeros of Interconnected Systems	282
B.21	Impulse Response	286
B.22	Frequency Response	288
B.23	Fourier Transform	293
B.24	Laplace Transform	296
B.25	Z–Transform	298
B.26	Laplace and Z–Transforms and Transfer Functions	300

Index — 303

Preface

The aim of this book is to introduce the reader to the world of positive linear systems, a particular, but important and fascinating, class of linear systems. We have made absolutely no effort to hide our enthusiasm for the topics presented in the hope that this will be enough of an excuse for being "informal" at times. We have divided the subject into three parts. The first part contains the definitions and the basic properties of positive linear systems. In the second part, the main theoretical results are reported. The third part is devoted to the study of some classes of positive linear systems relevant in applications. The reader familiar with linear algebra and linear systems theory should appreciate the way the arguments are treated and the subject is presented. A number of excellent books on these topics are available; nevertheless we have included two appendixes for making the book (reasonably) self-contained. The exposition of all the topics is supported with a number of examples and problems. To facilitate the reader, the theoretical (T) or applicative (A) character of each problem is explicitly pointed out, together with its level of difficulty (I, II, or III). We would like to express gratitude to our colleagues Luca Benvenuti, Luca Ghezzi, Salvatore Monaco, Simona Muratori, and Carlo Piccardi for their support, suggestions and ideas provided during our research work on positive systems and to Ms. Patrizia Valentini for artful LaTeXing.

L.F.

S.R.

Rome, May 2000

Part I
Definitions

1
Introduction

This book is concerned with *positive linear systems*, a remarkable class of linear systems. For the sake of simplicity we will only refer to the case of time-invariant, finite-dimensional single input, single output systems, described by state equations of the form

$$\dot{x}(t) = Ax(t) + bu(t)$$

or

$$x(t+1) = Ax(t) + bu(t)$$

which correspond, respectively, to continuous-time (t is defined over the reals) or discrete-time (t is defined over the integers) systems, and by an output transformation of the form

$$y(t) = c^T x(t)$$

Such systems are identified by the triple (A, b, c^T), which has peculiar features imposed by the positivity conditions highlighted in Chapter 2.

Positive systems are, by definition, systems whose state variables take only nonnegative values. From a general point of view, they should be viewed as very particular. To see this, consider a random generation of the elements a_{ij} of the matrix A and of the elements b_i and c_i of the vectors b and c^T. The probability of obtaining a positive system in this way is very low [$2^{-(n^2+2n)}$ for a n-dimensional discrete-time system]. From a practical point of view, however, such systems are anything but particular since positive systems are often encountered in applications.

Positive systems are, for instance, networks of reservoirs, industrial processes involving chemical reactors, heat exchangers and distillation columns, storage systems (memories, warehouses, ...), hierarchical systems, compartmental systems (frequently used when modeling transport and accumulation phenomena of substances in the human bodies), water and atmospheric pollution models, stochastic models where state variables must be nonnegative since they represent probabilities, and many other models commonly used in economy and sociology. One is tempted to assert that positive systems are the most often encountered systems in almost all areas of science and technology, except electro mechanics, where the variables (voltages, currents, forces, positions, velocities) may assume either positive and negative values. However, the existence of positive systems in an electrical or mechanical context cannot be excluded. Consider, as an example, a simple mechanical system composed of a point mass driven along a straight line by an external force. Position and velocity of the mass cannot become negative provided their initial values are nonnegative and the force is unidirectional: This is a positive system. On the other hand, even the simplest electrical circuit, namely, the $R - C$ circuit, is a positive system since the voltage on the capacitor remains nonnegative if initially such.

Positive linear systems, as any other linear system, satisfy the superposition principle and also a peculiar one, that of *comparative dynamics*. Such a principle can be expressed by saying that *"positive perturbations of inputs, states, and parameters cannot produce a decrease of the state and output at any instant of time following the perturbation"*. This rule can be quite useful whenever one is interested in a qualitative analysis of the influence of some design parameter (or input) on the system.

Among a number of properties holding for positive systems, the one concerning a dominant mode undoubtedly stands out. It often allows one to dramatically simplify the stability analysis. This property is expressed through a series of results known as the *Frobenius–Perron theorems*, holding for matrices with positive entries. But, even more important is the fact that a number of properties rely only on the structure of the system, that is on the structure of existing influences among all the input, state, and output variables. In other words, it often suffices to know *"who influences who"* in order to give a complete answer to fundamental questions. In fact, if the influence of one variable on another is always positive, the compensation among different paths of influence will not be allowed. Due to this property, *the influence graph*, which shows the direct influences among the variables, becomes a valuable tool (*structural model*) of analysis. For this reason, after the definition of positive systems, we will introduce the notion of the influence graph and will systematically highlight which properties rely on the topology of the graph and which on the *"level of influence"*.

We will first discuss the classical properties of dynamical systems, that is, stability, reachability, observability, input–output maps, and minimum phase. Obviously, other properties of peculiar interest for positive systems, such as cyclicity, primitivity, excitability, and transparency will also be considered. These properties will enable us to give a better physical interpretation of the various results presented in the book. Following the exposition of the theory, we will consider a number of applications tied to models widely used by researchers and professionals during the last decades. We will discuss, in particular, the Leontief model used by economists for predicting productions and prices; the Leslie model used by demographers to study age-structured populations; the Markov chains; the compartmental models; and the birth and death processes, relevant to the analysis of queueing systems.

At the end of this book, we will present a detailed guided bibliography and two appendixes concerning linear algebra and linear systems theory in order to make, if needed, the reader familiar with the mathematics used throughout the book.

2
Definitions and Conditions of Positivity

In this chapter, we give two definitions of positivity for linear systems (called *internal* and *external*) and we derive the corresponding necessary and sufficient conditions. Some of these conditions require the notion of positivity of matrices, vectors, and functions.

We will say that a matrix F is *strictly greater* than a matrix G (having the same number of rows and columns) and denote this by $F \gg G$, if and only if *all* the elements f_{ij} of F are *greater* than the corresponding elements g_{ij} of G. If *all* the elements of F are *greater than or equal to* the corresponding elements of G, but *at least one* of the f_{ij} is *greater* than g_{ij}, we will say that F is *greater* than G and denote this by $F > G$. Obviously, both $F \gg G$ and $F > G$ imply $F \neq G$. The notation $F \geq G$, which should be read F *greater than or equal to* G, will mean that *all* the elements of F are *greater than or equal to* the corresponding elements of G. Thus $F \geq G$ is satisfied also when $F = G$. These definitions justify the use of the following notation and terminology:

strictly positive matrix $F : F \gg 0 \quad (f_{ij} > 0 \quad \forall(i,j))$
positive matrix $F : F > 0 \quad (f_{ij} \geq 0 \forall(i,j), \exists(i,j) : f_{ij} > 0)$
nonnegative matrix $F : F \geq 0 \quad (f_{ij} \geq 0 \quad \forall(i,j))$
strictly positive diagonal matrix $F : F \gg 0 \quad (f_{ij} = 0 \quad \forall i \neq j, f_{ii} > 0 \quad \forall i)$

The last notation, though not rigorous, will be used in the sequel for the sake of simplicity. Analogous definitions and notations can be given also for n-dimensional vectors with $n \geq 2$. When dealing with scalars, however, strict positivity ($a \gg 0$)

coincides with positivity ($a > 0$). It is also worth noting that the following rules hold:

$$F \gg 0, G > 0 \Longrightarrow FG > 0$$
$$F > 0, G \gg 0 \Longrightarrow FG > 0$$
$$F \gg 0, x > 0 \Longrightarrow Fx \gg 0$$
$$F > 0, x \gg 0 \Longrightarrow Fx > 0$$
$$x \gg 0, y > 0 \Longrightarrow x^T y = y^T x > 0$$

Finally, it is important to note that there is no trivial link between positive matrices (in the above mentioned sense) and positive definite matrices.

Analogous to what was previously stated for matrices and vectors, we will say that a real function $u(\cdot)$ of a real variable is strictly positive in an interval and denote this by $u(\cdot) \gg 0$ provided that $u(\xi) > 0$ at every point in the interval. Similarly, we will say that a real function is positive, provided that $u(\xi) \geq 0$ at every point in the interval and $u(\xi) > 0$ at least at one point. Finally, we will call nonnegative the functions $u(\cdot)$ for which $u(\xi)$ is nonnegative at every point in the interval. Obviously, in the case of real functions of integer variables strict positivity (or positivity, or nonnegativity) of the function $u(\cdot)$ in an interval $[0, t-1]$ coincides with strict positivity (or positivity, or nonnegativity) of the t-dimensional vector

$$u_0^{t-1} = (u(t-1)u(t-2)\cdots u(0))^T$$

We are now able to give the first definition of positivity of a linear system.

DEFINITION 1 *(externally positive linear system)*

A linear system (A, b, c^T) is said to be *externally positive* if and only if its forced output (*i.e.*, the output corresponding to a zero initial state) is nonnegative for every nonnegative input function.

External positivity is a property that is often easy to check, since, in most cases, for physical reasons input and output variables are necessarily positive. For example, a hydrological system composed of a series of lakes in which the input is the inflow into the upstream lake and the output is the outflow from the downstream lake is an externally positive system. Other obvious examples are chemical systems composed of a set of reactors in which the input is the feed concentration and the output is the product concentration and educational systems in which input and output are the annual number of freshmen and graduates.

Before we proceed, it is worth making two remarks. The first is that positivity is not independent of the basis used for representing inputs and outputs. For example, in the case of the hydrological system we could decide to assign a positive sign to the inflow into the upstream lake and a negative sign to the outflows of the downstream lake. Thus, the system would have positive inputs and negative outputs and would not be an externally positive system. It is clear, however, that this problem could be easily avoided by giving a slightly more general definition of external positivity. This extension has not been made here since in most relevant applications the natural choice of the basis is also the correct one.

The second, and more important, remark is that there exist externally positive systems in which the input and output variables can also take negative values. For example, in the electrical network depicted in *Fig. 2.1*, which is externally positive (see *Problem 1*) for appropriate values of the electrical parameters, the current u and the voltage y could also be negative. In such cases, it is not possible to easily deduce the external positivity of the system.

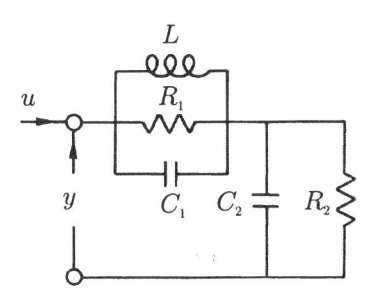

Figure 2.1 An electrical network that is externally positive for appropriate values of the parameters.

The following theorem is therefore of interest:

THEOREM 1 *(condition for external positivity)*

A linear system is externally positive if and only if its impulse response is nonnegative.

Proof. Consider a continuous-time linear system with zero initial state. The output is the convolution integral of the input and the impulse response, namely,

$$y(t) = \int_0^t h(t-\xi)u(\xi)d\xi$$

with

$$h(t) = c^T e^{At} b \quad t \geq 0$$

Therefore, if the impulse response $h(t)$ is nonnegative, the output $y(t)$ is nonnegative for every nonnegative input $u(t)$, so that the system is externally positive. On the other hand, if the system is externally positive, then $h(t)$ must be nonnegative. In fact, if this were not the case, $h(t)$ would be negative at least at one point and by continuity in a whole interval $[t_1, t_2]$. Thus, the output $y(t)$ would be negative for $t > t_2$ for every input function $u(t)$, which is strictly positive in $[(t-t_2),(t-t_1)]$

10 DEFINITIONS AND CONDITIONS OF POSITIVITY

and zero elsewhere. This would contradict the external positivity of the system. A similar proof can be given for discrete-time systems.

◇

EXAMPLE 1 (externally positive electrical network)
Consider the electrical network in *Fig. 2.2* with positive parameters R, L, and C.

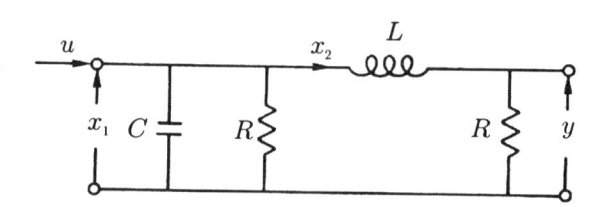

Figure 2.2 An electrical network that is externally positive for $L/R + RC \geq \sqrt{8LC}$.

The network is described by the triple

$$A = \begin{pmatrix} -1/RC & -1/C \\ 1/L & -R/L \end{pmatrix} \quad b = \begin{pmatrix} 1/C \\ 0 \end{pmatrix}$$

$$c^T = \begin{pmatrix} 0 & R \end{pmatrix}$$

and its eigenvalues $\lambda_{1,2}$ have negative real parts since the system is asymptotically stable. Moreover, it is easy to verify that they are a complex conjugate pair ($\lambda_{1,2} = a \pm ib$) if $L/R + RC < \sqrt{8LC}$ and real otherwise. In the first case, the system cannot be externally positive since its impulse response is oscillatory. In the second case, the impulse response is the sum of two exponentials

$$h(t) = \alpha e^{\lambda_1 t} + \beta e^{\lambda_2, t}$$

Since $h(0) = c^T b = 0$, it follows that $\alpha = -\beta$, and then

$$\dot{h}(t) = \alpha(\lambda_1 e^{\lambda_1, t} - \lambda_2 e^{\lambda_2 t})$$

The impulse response is therefore stationary (maximal or minimal) as a function of time only once at

$$t^* = \frac{\log(\lambda_1) - \log(-\lambda_2)}{\lambda_2 - \lambda_1}$$

This means that the impulse response is positive if and only if $\dot{h}(0) > (0)$. Since

$$\dot{h}(0) = c^T A b = \frac{R}{LC}$$

we can conclude that the system is externally positive if and only if $L/R + RC \geq \sqrt{8LC}$.

♣

The step response of an externally positive system (starting from a zero state) is non-decreasing since it is the integral of the impulse response that is nonnegative. Therefore, when a constant input is applied to the system, its output tends toward an equilibrium without *overshooting* it. This is true also when there are oscillations in the impulse response of the system.

Both *Definition 1* and *Theorem 1* make clear that external positivity of a system is a property of its input–output relationship for a zero initial state. This implies that the knowledge of the *ARMA model* of a discrete-time system

$$y(t) = \sum_{i=1}^{n}(-\alpha_i)y(t-i) + \sum_{i=1}^{n}\beta_i u(t-i)$$

or that of a continuous-time system

$$y^{(n)}(t) = \sum_{i=1}^{n}(-\alpha_i)y^{(n-i)}(t) + \sum_{i=1}^{n}\beta_i u^{(n-i)}(t)$$

is sufficient for checking external positivity of the system. In other words, given the $2n$ parameters $\alpha_i, \beta_i, i = 1, 2, \ldots, n$, it must be possible to derive whether the system is externally positive or not. Unfortunately, a simple algorithm that performs this task is not yet known. There exist, however, useful sufficient conditions for external positivity. A trivial one is the following

$$\alpha_i \leq 0 \quad \beta_i \geq 0 \quad i = 1, 2, \ldots, n$$

which can be easily proved for discrete-time systems. The proof of this property for a continuous-time system is proposed to the reader as an exercise in *Problem 3*. In some cases, the check of the external positivity seems to be possible only by a brute force approach, namely, by a numerical computation of the impulse response. Practically, this means that the state equations have to be solved for $x = (0) = b$ and that nonnegativity of the impulse response $h(t)$ must be checked for a long interval of time in order to be reasonably sure that $h(t)$ also continues to be nonnegative outside of such an interval.

PROBLEM 1 *(A-III) Determine the values of the parameters R_1, R_2, L, C_1, C_2 for which the electrical network in Fig. 2.1 is externally positive. If this problem is too difficult, prove (using arguments related with the dynamics of the system with slow and fast components) that the network is externally positive for $R_1 = L = C_1 = C_2 = 1$ and $R_2 = 10$ and that it is not externally positive for $R_1 = R_2 = L = C_1 = C_2 = 1$. In case this problem is still too difficult, give an empirical proof with the aid of a computer.*

12 DEFINITIONS AND CONDITIONS OF POSITIVITY

Now we could give the second definition of positivity, which could be called *internal*. We do not do this for the sake of brevity and for consistency with current terminology.

DEFINITION 2 *(positive linear system)*

> A linear system (A, b, c^T) is said to be *positive* if and only if for every nonnegative initial state and for every nonnegative input its state and output are nonnegative.

This definition says that all trajectories emanating from any point in the positive orthant $\mathbb{R}^n_+$ (boundary included) of the state space $\mathbb{R}^n$ obtained by applying a nonnegative input to the system remain in the positive orthant and yield a nonnegative output. *Figure 2.3* shows three trajectories obtained by applying nonnegative inputs to a second-order continuous-time positive system. As expected, the trajectories a and b which start in the first quadrant, remain confined in it. The trajectory denoted by c is not contradictory since it enters the first quadrant but does not leave it.

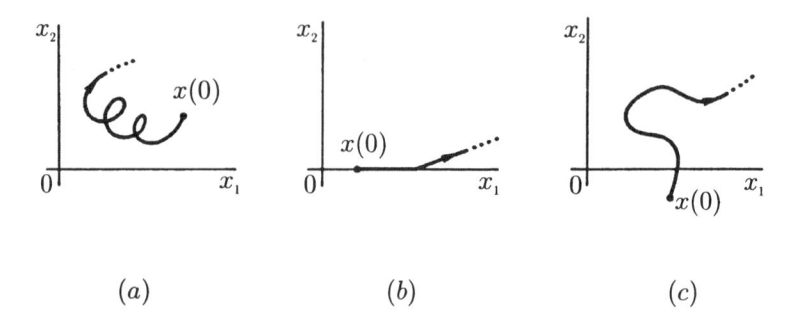

(a)   (b)   (c)

Figure 2.3 Three trajectories $(a), (b)$, and (c) obtained by applying a nonnegative input to a second-order continuous-time system.

The positive orthant is therefore a "trap" in which the trajectories generated by nonnegative inputs can possibily enter but from which they cannot get out. Formally, this can be expressed by saying that the positive orthant $\mathbb{R}^n_+$ is a *nonnegative invariant set*. This is the reason some authors define positive systems by requiring the existence of an invariant set without requiring, however, that such an invariant set be the positive orthant $\mathbb{R}^n_+$.

It is important to note that positivity, besides depending on the basis of the input and output spaces, depends also on the basis of the state space. For example, the hydrological system of *Fig. 2.4(a)*, which is obviously positive from the external point of view, is also positive if the state variables are the two storages x_1 and x_2, but is not such if the state variables are the sum z_1 and the difference z_2 of the two storages. In fact, in the first case [*Fig. 2.4(b)*], the invariant set is the first quadrant while, in the second case [*Fig. 2.4(c)*], the invariant set is not the first quadrant.

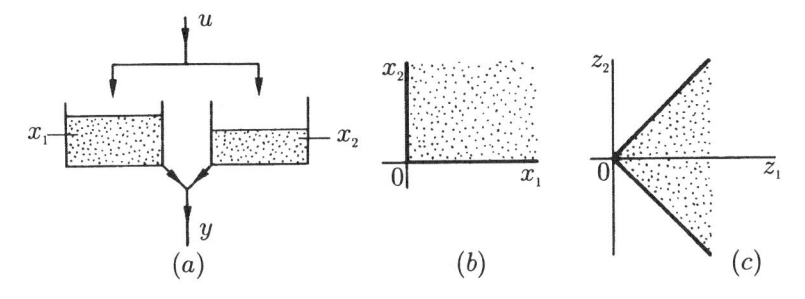

Figure 2.4 A positive system composed of two reservoirs (a) and its invariant sets (b) and (c).

In general, it is not easy to check if there exists a basis for which a given linear system is positive. Nevertheless, in the majority of the applications when such a basis exists, it is the natural one (*i.e.*, the state variables are reservoir storages, concentrations of chemical compounds, voltages on capacitors, etc.). This means that it is often easy to know whether a system admits a basis for which it is positive.

The definition of positivity (*Definition 2*) requires the nonnegativity of the output for every nonnegative initial state $x(0)$, rather than only for $x(0) = 0$ as for external positivity (*Definition 1*). This means that positivity implies external positivity while the opposite is not true, namely an externally positive system can be not positive. Moreover, it can be said that there exist systems that are externally positive and cannot be made positive through any change of basis of the state space. To see this, consider a continuous-time third-order system. Suppose that such a system is asymptotically stable and has a complex conjugate pair of eigenvalues. This implies that in the state space there exists a plane X (corresponding to the complex eigenvalues) on which the trajectories for $u = 0$ spiral toward the origin and a straight line r (corresponding to the real eigenvalue), which coincides with a trajectory tending toward the origin. *Figure 2.5(a)* depicts this situation. If the modulus of the real part of the complex eigenvalues is smaller than the modulus of the real eigenvalue (*i.e.*, if the complex eigenvalues are dominant), the trajectory representing the impulse response starts from the point corresponding to the vector b, rapidly approaches the plane X, and then slowly spirals around the origin, as shown in *Figure 2.5(b)*. The trajectory is on one side of the plane X and is tangent to it as $t \to \infty$. If the vector c, which identifies the output transformation $y(t) = c^T x(t)$, is orthogonal to X the impulse response is positive and, in view of *Theorem 1*, the system is externally positive. It is clear, however, that it is not possible to draw three different straight lines through the origin so that the trajectory lies entirely in the positive orthant determined by these three straight lines. In conclusion, the system cannot be made positive by means of any choice of the basis of the state space.

14 DEFINITIONS AND CONDITIONS OF POSITIVITY

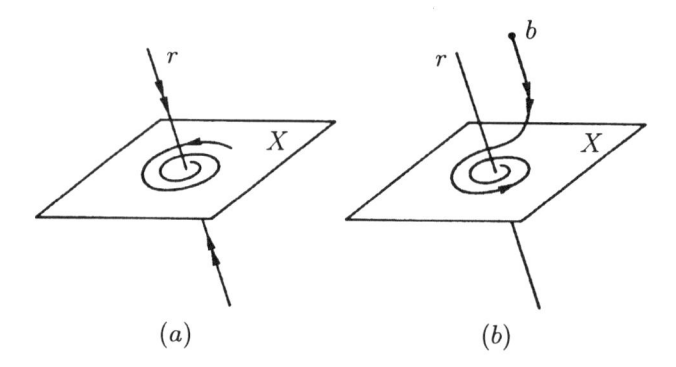

Figure 2.5 The free motion of a third-order system with complex dominant eigenvalues: (a) invariant subspaces; (b) the impulse response (the double arrow indicates fast motion).

We show now how it is formally possible to determine if a given linear system is positive.

THEOREM 2 *(condition for positivity)*

> A discrete-time linear system (A, b, c^T) is positive if and only if $A \geq 0, b \geq 0, c^T \geq 0^T$. A continuous-time linear system (A, b, c^T) is positive if and only if the matrix A is a Metzler matrix, that is, its nondiagonal elements are nonnegative $[a_{ij} \geq 0, \forall (i,j), i \neq j)]$ and $b \geq 0, c^T \geq 0^T$.

Proof. We report here the proof for continuous-time linear systems, leaving to the reader its extension to the case of discrete-time systems.
Necessity. Letting $x(0) = 0$, positivity implies $\dot{x}(0) = bu(0) \geq 0$ for every $u(0) \geq 0$, that is, $b \geq 0$. Moreover, $y(0) = c^T x(0)$ so that positivity [$y(0) \geq 0$ for every $x(0) \geq 0$] implies $c^T \geq 0^T$. Finally, letting $x(0) = e_j$ (unit vector of the x_j axis) it follows $\dot{x}(0) = Ae_j = j$-th column of A. But the trajectory of a positive system cannot leave the positive orthant $\mathbb{R}^n_+$, so that $\dot{x}_i(0) \geq \forall i \neq j$. Therefore, the elements of A that are not on the diagonal must be positive or zero, that is, the matrix A must be a Metzler matrix.
Sufficiency. It is clear that $c^T \geq 0^T$ and $x(t) \geq 0$ imply $y(t) = c^T x(t) \geq 0$. On the other hand, in order to prove that $x(t) \geq 0$, it is sufficient to check that the vector $\dot{x}(t)$ does not point toward the outside of $\mathbb{R}^n_+$ whenever $x(t)$ is on the boundary of $\mathbb{R}^n_+$. This is equivalent to verifying that the components of the vector $\dot{x}(t) = Ax(t) + bu(t)$ corresponding to the zero components of $x(t) \geq 0$ are nonnegative. Denoting by I the set of indices of such components, [*i.e.* $x_i(t) = 0$ for $i \in I$], we can write

$$\dot{x}_i(t) = \sum_{j \notin I} a_{ij} x_j(t) + b_i u(t) \qquad i \in I$$

so that, from the nonnegativity of the a_{ij} with $i \neq j$ it follows that $\dot{x}_i(t) \geq 0$.

◊

Thus, given a system (A, b, c^T), it is straightforward to know whether the system is positive or not. It is sufficient to verify that there are no negative elements in A, b and c^T, with the exception of the elements on the diagonal of matrix A in the case of continuous-time systems (in the case of systems with multiple inputs and multiple outputs, conditions $b \geq 0$ and $c^T \geq 0^T$ become $B \geq 0$ and $C \geq 0$).

As an exercise on the use of positive matrices and Metzler matrices, the reader is asked to solve the two following problems. The first one is concerned with the *law of comparative dynamics* mentioned in Chapter 1.

PROBLEM 2 *(T-II) Let $x(t) \geq 0$ and $y(t) \geq 0$ be the state and the output corresponding to an initial state $x(0) \geq 0$ and to an input $u(t) \geq 0$ of a positive linear system (A, b, c^T). Prove that by positively perturbing the initial state, the input and the elements of the matrix A and of the vectors b and c^T, one obtains a perturbed state $x'(t) \geq x(t)$ and a perturbed output $y'(t) \geq y(t)$.*

PROBLEM 3 *(T-I) Prove that a continuous-time linear system with coefficients of the ARMA model such that $\alpha_i \leq 0$ and $\beta_i \geq 0, i = 1, 2, \ldots, n$ admits a positive realization (A, b, c^T).*

3
Influence Graphs

In this chapter, we introduce the notion of the *influence graph* and show how it is useful when dealing with positive systems.

The influence graph of a continuous-time dynamic system (even nonpositive and nonlinear)

$$\dot{x}_i(t) = f_i(x_1(t), \ldots, x_n(t), u(t)) \quad i = 1, 2, \ldots, n$$
$$y(t) = g(x_1(t), \ldots, x_n(t), u(t))$$

or discrete-time dynamic system

$$x_i(t+1) = f_i(x_1(t), \ldots, x_n(t), u(t)) \quad i = 1, 2, \ldots, n$$
$$y(t) = g(x_1(t), \ldots, x_n(t), u(t))$$

is a graph G with $n + 2$ nodes $0, 1, 2, \ldots, n, n+1$. The first (0) and last $(n+1)$ nodes are associated with the input u and the output y of the system and the remaining nodes $(i = 1, 2, \ldots, n)$ with the state variables $(x_1, x_2, \ldots, x_n)$. The (oriented) arcs of the graph point out the "direct influences" among input, state, and output variables. Therefore, the presence of an arc $(0, j), j = 1, 2, \ldots, n$, means that the input $u(t)$ influences $\dot{x}_j(t)$ or $x_j(t+1)$ (according to whether the system is continuous or discrete-time). Similarly, the arc (i, j) $i, j = 1, 2, \ldots, n$ appears in the graph G if and only if the function f_j depends on the variable x_i. The arcs $(i, n+1)$ refer, instead, to the output transformation $g(\cdot)$ and reveal the influence of $x_i(t)$ on $y(t)$. These arcs have, however, a different nature when compared to the previous ones since they represent the influence of a variable (x_i) on another variable (y) at the same time t. Instead, a delay is associated with the arcs (i,j), $j \neq n+1$ since $u(t)$ and $x_i(t)$ influence $\dot{x}_j(t)$ or $x_j(t+1)$. In the sequel, in order to facilitate the reader, the arcs $(i, n+1)$ will be drawn with a dashed line and the input and output nodes will be characterized with a square instead of a circle.

EXAMPLE 2 (forced pendulum)
Consider the pendulum depicted in *Fig. 3.1(a)* where $U(t)$ is an external force and $x_1(t)$ and $x_2(t)$ are the angular position and velocity. If the output $y(t)$ is the horizontal position of mass m, the state equations and the output transformation are

$$\dot{x}_1(t) = x_2(t)$$
$$\dot{x}_2(t) = \tfrac{1}{mL^2}(u(t)L\cos x_1(t) - mgL\sin x_1(t) - hx_2(t))$$
$$y(t) = L\sin x_1(t)$$

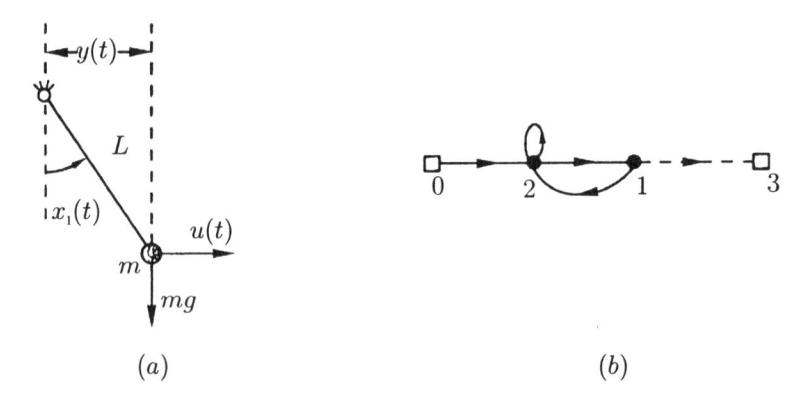

Figure 3.1 A forced pendulum (a) and its influence graph (b).

The influence graph G of the pendulum is therefore the graph depicted in *Fig. 3.1(b)* in which the self-loop of node 2 is absent if the friction coefficient h is zero.
♣

In example 2, the influence graph G has been obtained from the state equations $f_i(\cdot)$ and from the output transformation $g(\cdot)$ of the system. In many cases, however, it is natural to describe a system without specifying its state equations and output transformation. This is possible, obviously, whenever it is clear *a priori* what the direct influences existing among all the variables of the system are. The following is an example of such a possibility:

EXAMPLE 3 (chemostat)
The *chemostat* [see *Fig. 3.2(a)*] is a laboratory apparatus, used by biologists when studying the influence of a nutrient on an aquatic ecosystem. It is composed

of a reservoir (endowed with a feeding and drainage system) in which the substances are suitably mixed so that their concentrations, though varying in time, are constant in space.

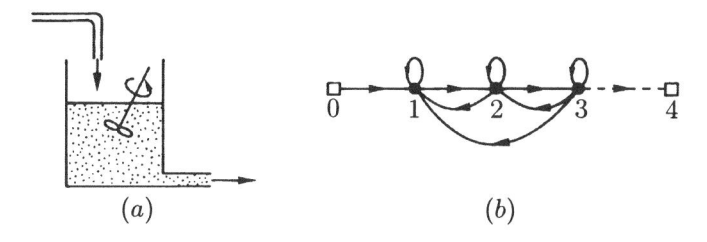

Figure 3.2 A chemostat (a) and its influence graph (b).

Suppose that some experiments on the transparency y of the water (due to the presence of suspended phytophagous algae) have been performed by varying the concentration u of the nutrient contained in the inflow. In order to explain the results of the experiments, it is then useful to make use of a model of the chemostat. The structure of this model, namely its influence graph shown in *Fig. 3.2(b)*, can be *a priori* determined on the basis of the following simple considerations. First, we can argue that the model must have three state variables: the concentrations x_1 of the nutrient, x_2 of the algae, and x_3 of the phytophagous algae, which feed on the previous ones. The input variable u influences the first state variable, [arc $(0, 1)$ in the graph], while the output y is influenced by the third state variable [arc $(3, 4)$]. In order to complete the graph, the interactions among the three variables must be identified. To do this, note that the derivative $\dot{x}_i$ of the concentration of each component i is equal to the difference between the growth and removal rate (conservation of mass). Hence, nodes $1, 2$, and 3 must have self-loops since among the removal rates of each component i there is the loss rate due to the drainage which is, obviously, proportional to x_i. Moreover, the nodes 1 and 2, must be linked together in both directions since the nutrient uptake of algae (depending on both x_1 and x_2) is responsible for their growth. The same thing holds for the two species of algae that are in a predator–prey relationship. Finally, the arc $(3, 1)$ is present if the increase of nutrient due to the decomposition (performed by bacteria) of dead phytophagous algae is not negligible.

♣

PROBLEM 4 *(A-I) Determine the influence graph G of a model of a population of Pacific salmon, taking into account that their life cycle lasts 5 years and that they breed once in their lifetime (Hint: Use a discrete-time model and denote by $x_i(t)$ the number of salmon of age i in year t).*

20 INFLUENCE GRAPHS

Any influence graph G is identified by a triple $(A^\sharp, b^\sharp, {c^\sharp}^T)$, where the $(n \times n)$ matrix $A^\sharp$ and n-dimensional vectors $b^\sharp$ and $c^\sharp$ are exclusively composed of *zeros* and *ones*. The triple $(A^\sharp, b^\sharp, {c^\sharp}^T)$ is defined in the following manner:

$$a_{ij}^\sharp = \begin{cases} 1 \Leftrightarrow \exists(i,j) \text{ in } G \\ 0 \Leftrightarrow \not\exists(i,j) \text{ in } G \end{cases} \quad b_i^\sharp = \begin{cases} 1 \Leftrightarrow \exists(0,i) \text{ in } G \\ 0 \Leftrightarrow \not\exists(0,i) \text{ in } G \end{cases} \quad c_i^\sharp = \begin{cases} 1 \Leftrightarrow \exists(i,n+1) \text{ in } G \\ 0 \Leftrightarrow \not\exists(i,n+1) \text{ in } G \end{cases}$$

For example, the influence graph G in *Fig. 3.2(b)* is identified by the triple

$$A^\sharp = \begin{pmatrix} 1 & 1 & 0 \\ 1 & 1 & 1 \\ 1 & 1 & 1 \end{pmatrix} \quad b^\sharp = \begin{pmatrix} 1 \\ 0 \\ 0 \end{pmatrix}$$

$${c^\sharp}^T = (0 \ 0 \ 1)$$

When dealing with linear systems, the influence graph G is readily inferred from the triple (A, b, c^T) since every nonzero entry of $A, b,$ and c^T corresponds to an arc of G.

However, the arc (i,j) $i, j = 1, 2, \ldots, n$ exists in G if and only if $a_{ji} \neq 0$ (note the exchange of the indices i and j). Thus, we can conclude that in a linear system $A^\sharp, b^\sharp,$ and ${c^\sharp}^T$ can be obtained from $A^T, b,$ and c^T by replacing all the nonzero entries (both positive and negative) with a 1.

PROBLEM 5 *(T-I) Determine the triple $(A^\sharp, b^\sharp, {c^\sharp}^T)$ and draw the influence graph of a linear system in Markov canonical form*

$$A = \begin{pmatrix} 0 & 0 & \ldots & 0 & -\alpha_n \\ 1 & 0 & \ldots & 0 & -\alpha_{n-1} \\ 0 & 1 & \ldots & 0 & -\alpha_{n-2} \\ \vdots & \vdots & & \vdots & \vdots \\ 0 & 0 & \ldots & 1 & -\alpha_1 \end{pmatrix} \quad b = \begin{pmatrix} 1 \\ 0 \\ 0 \\ \vdots \\ 0 \end{pmatrix}$$

$$c^T = (g_1 \ g_2 \ \ldots \ g_{n-1} \ g_n)$$

where the α_i's are the coefficients of the characteristic polynomial of A and the g_i's are the so-called Markov parameters (coinciding with the first n values of the impulse response in the case of discrete-time systems).

Whenever a property of a linear system depends solely on the matrix A [or on the couple (A, b) or (A, c^T)], the system Σ is often briefly denoted by A [or (A, b) or (A, c^T)]. In terms of influence graphs, this corresponds to a subgraph G_x defined by $A^\sharp$ and obtained by eliminating the node 0 from G and every outgoing arc and the node $n+1$ and its incoming arcs. Similarly, the subsystem (A, b) is described by an influence graph G_{ux} [identified by $(A^\sharp, b^\sharp)$] with no output and the subsystem

(A, c^T) is described by an influence graph G_{xy} [identified by $(A^\sharp, c^{T\sharp})$] with no input.

An arc (i, j) in G, represents the direct influence of x_i on x_j. But, a variable x_i indirectly influences another variable x_j if there exists in the graph G_x a path composed of $k \geq 2$ arcs between i and j. In some cases, it is interesting to know if two nodes of an influence graph G_x are linked by means of a path of length k. To do this, one can refer to the following result (and use specific algorithms to generate automatically the paths of length k).

THEOREM 3 *(number of paths in the influence graph)*

> The number of paths of length k between any pair (i, j) of nodes in G_x is equal to the element (i, j) of the matrix $A^{\sharp^k}$. The number of paths of length k between node 0 and node i in G_{ux} is equal to the ith element of the vector $A^{\sharp^{k-1}}b^\sharp$. The number of paths of length k between node i and node $n+1$ in G_{xy} is equal to the ith element of the vector $A^{\sharp^{k-1}}c^\sharp$. Finally, the number of paths of length k between node 0 and node $n+1$ is given by $c^{\sharp T}A^{\sharp^{k-2}}b^\sharp$.

Proof. For the sake of brevity, we will prove only the first statement of the theorem. Note that the statement holds for $k = 1$. We will show that if the statement is true for k, it is also true for $k+1$. Suppose, then, that the element (i, j) of $A^{\sharp^k}$, denoted by $a_{ij}^{\sharp(k)}$, coincides with the number $n_{ij}{}^k$ of paths of length k between i and j in G_x. Thus, the number $n_{ij}{}^{k+1}$ of paths of length $k+1$ between i and j can be evaluated taking into account that such paths are the concatenation of an arc (i, h) with a path of length k from h to j. Consequently,

$$n_{ij}{}^{k+1} = \sum_{h=1}^{n} a_{ih}{}^\sharp n_{hi}{}^k = \sum_{h=1}^{n} a_{ih}{}^\sharp a_{hj}{}^{\sharp(k)} = a_{ij}{}^{\sharp(k+1)}$$

◇

PROBLEM 6 *(T-I) Check the validity of Theorem 3 by determining the number of paths of length $k = 4, 5, \ldots, 8$ on the influence graph in Fig. 3.2(b).*

The influence graph assumes a particular relevance in the case of positive systems. In such systems, in fact, all the influences are positive so that compensations are not possible. In contrast, when dealing with nonpositive systems, it might happen that a variable j is insensitive to a variable i even if there exist paths from i to j in the influence graph (for this to be true, at least two paths between i and j must exist, with a zero sum effect on x_j). We will show next that a number of properties of positive systems are exclusively related to the topology of the influence graph regardless of the numerical values of the nonzero elements of A, b, and c^T. It is therefore particularly appropriate to term these properties "structural" properties.

22 INFLUENCE GRAPHS

Whenever two positive systems, identified by their influence graphs G_1 and G_2, are connected in series, the influence graph of the resulting system can be obtained quite easily. Indeed, the graph G_x is the union of the graphs G_{x1} and G_{x2} with the addition of the arcs linking the nodes of G_{x1} influencing the output y_1 with the nodes of G_{x2} influenced by the input u_2. Similar rules hold for parallel and feedback connections.

A quite simple procedure allows one to determine the influence graph of a sampled-data system. In principle, in order to obtain the graph $\tilde{G}$ of a system obtained by sampling a continuous-time positive system (A, b, c^T), one should first determine the sampled-data system

$$\tilde{A} = e^{AT} \quad \tilde{b} = \left(\int_0^T e^{A\xi} d\xi\right) b \quad \tilde{c}^T = c^T$$

(which is, obviously, positive), and then determine the influence graph $\tilde{G}$ of $(\tilde{A}, \tilde{b}, \tilde{c}^T)$. As a matter of fact, $\tilde{G}$ can be directly obtained from the influence graph G of (A, b, c^T) by adding to it all the missing self-loops to the nodes $i = 1, 2, \ldots, n$ and one arc between i and j; $i = 0, 1, \ldots, n$; $j = 1, \ldots, n$ if a path of length $k \geq 2$ exists between i and j. As an example, Fig. 3.3 shows the influence graph G of a double integrator [the Newton law: $\ddot{y}(t) = (1/m)\, u(t)$ is equivalent to $\dot{x}_2(t) = x_1(t)$ and $\dot{x}_1(t) = (1/m)\, u(t)$] and the influence graph $\tilde{G}$ of the corresponding sampled-data system.

Figure 3.3 The influence graph G of a double integrator (a) and the influence graph $\tilde{G}$ of the corresponding sampled-data system (b).

PROBLEM 7 *(T-II) Prove the above mentioned rule for the determination of the influence graph $\tilde{G}$ of the discrete-time system $(\tilde{A}, \tilde{b}, \tilde{c}^T)$ obtained by sampling a continuous-time positive system (A, b, c^T).*

4
Irreducibility, Excitability, and Transparency

In this chapter, we define irreducible, excitable, and transparent systems which are, as will become clear in the following, of relevant interest in the study of positive systems. The definitions require specific properties of the system (A, b, c^T). As a matter of fact, these properties depend only on the system structure, that is, they can be expressed in terms of the influence graph or, equivalently, of the triple $(A^\sharp, b^\sharp, c^{\sharp T})$.

DEFINITION 3 *(reducible and irreducible system)*

> A positive system (A, b, c^T) of dimension $n \geq 2$ is said to be *reducible* if and only if the free evolution of a set of n_1 state variables is independent of the free evolution of the remaining $n_2 = n - n_1$ state variables. A system that is not reducible is said to be *irreducible*.

Systems composed of two positive systems connected in series [*Fig. 4.1(a)*] or in parallel [*Fig. 4.1(b)*] are reducible. In both cases, in fact, the free motion of the first subsystem does not depend on the free motion of the second one.

From the above definition, it follows that the influence graph G_x of a reducible system can be split into two subgraphs G_{x1} and G_{x2} (containing n_1 and n_2 nodes, respectively), with no arcs connecting nodes of G_{x2} to nodes of G_{x1}. The free motion of a reducible system can be determined by calculating first the free motion of the subsystem with n_1 variables and, then, the evolution of the subsystem with n_2 variables, by taking into account the influence (if any) of the first subsystem on the second one. This means, for example, that a reducible continuous-time positive system

24 IRREDUCIBILITY, EXCITABILITY, AND TRANSPARENCY

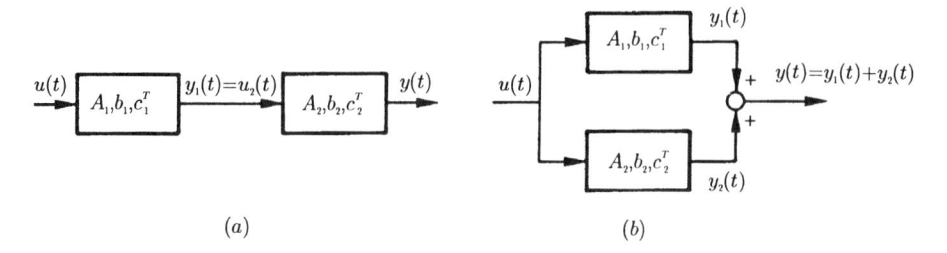

Figure 4.1 Two subsystems connected in series (a) and in parallel (b).

$$\dot{x}(t) = Ax(t)$$

can be viewed, by properly renumbering the state variables, as the acyclical interconnection of two subsystems

$$\dot{z}_1(t) = A^*{}_{11} z_1(t)$$
$$\dot{z}_2(t) = A^*{}_{21} z_1(t) + A^*{}_{22} z_2(t)$$

with $A^*{}_{11}$ and $A^*{}_{22}$ of dimension $(n_1 \times n_1)$ and $(n_2 \times n_2)$, respectively. It is also possible that the first and/or the second subsystems themselves are reducible, so that a further decomposition can be performed.

In order to obtain the ultimate decomposition of a reducible system, the concept of *communicating states* can be used: two states x_i and x_j are said to communicate if there exists at least one path from i to j and at least one path form j to i in the influence graph G_x. This means that the two state variables x_i and x_j influence each other (directly or indirectly). Obviously, if x_i communicates with x_j and x_j communicates with x_l, then x_i communicates with x_l. Thus, a *partition* over the set of state variables can be induced in the following manner. After choosing a first state variable, say x_1, one can determine the set C_1 of all the states communicating with x_i. Since, by assumption, the system is reducible, C_1 does not contain all the state variables, so that it is possible to select a state variable, say x_j, not belonging to C_1 and determine the set C_2 of all the states communicating with x_j. The process can be repeated until all the state variables are clustered in disjoint sets $C_1, C_2, \ldots, C_k$ with cardinality $n_1, n_2, \ldots, n_k$ (with $\sum_i n_i = n$). Each set C_i defines an irreducible subsystem of dimension n_i and the system is the interconnection (necessarily not cyclic) of these subsystems. In conclusion, it can be said that by properly reordering the sets C_i and, within each set, the state variables (*i.e.* using an appropriate permutation matrix P) the system

$$\dot{x}(t) = Ax(t)$$

can be written in the equivalent form (recall that $P^{-1} = P^T$, if P is a permutation matrix)

$$\dot{z}(t) = PAP^T z(t)$$

where PAP^T is block triangular, that is,

$$PAP^T = \begin{pmatrix} A^*_{11} & 0 & \cdots & 0 \\ A^*_{21} & A^*_{22} & \cdots & 0 \\ \vdots & \vdots & & \vdots \\ A^*_{k1} & A^*_{k2} & \cdots & A^*_{kk} \end{pmatrix}$$

EXAMPLE 4 (working processes with scraps and reworks)
Consider the positive system described in *Fig. 4.2* through its influence graph G.

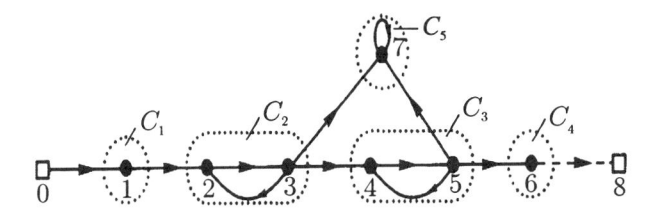

Figure 4.2 The influence graph of a working process with scraps and reworks.

Such a graph describes a working process with scraps and reworks composed of six elementary operations (stages) of the same duration performed successively starting from a raw item (node 0). If $t = 0, 1, 2$, and so on, are the working intervals, the state variable $x_i(t)$ represents the number of items processed in the ith stage during the working interval t. At the end of the third and fifth stage, a quality control with three possible consequences is performed: scrap [arcs (3, 7) and (5, 7)], rework [arcs (3, 2) and (5, 4)], and acceptance [arcs (3, 4) and (5, 6)]. Finally, the nodes 0 and 8 of the graph represent the number $u(t)$ of raw items introduced into the working process and the number $y(t)$ of items produced.

Obviously, the system is reducible, so that we can apply the above described procedure in order to decompose it in irreducible subsystems. To do this, let us select x_1 as the first variable. The set C_1 is clearly given by $C_1 = \{x_1\}$, since node 1 is not reachable from nodes $2, 3, \ldots, 7$ of the graph. By continuing with the variable x_2, it follows that $C_2 = \{x_2, x_3\}$, since there exist paths (arcs in this specific case) linking in both directions nodes 2 and 3. Proceeding like this, by simply inspecting the graph, it is possible to determine all the communicating classes $C_1, C_2, \ldots, C_5$ shown in *Fig. 4.2*. In this particular case, the classes C_i are already well ordered so that it is not necessary to perform any permutation in order to put the matrix A in triangular form. In fact, the system is described by the following triple (A, b, c^T)

$$A = \begin{pmatrix} 0 & 0 & 0 & 0 & 0 & 0 & 0 \\ 1 & 0 & r_3 & 0 & 0 & 0 & 0 \\ 0 & 1 & 0 & 0 & 0 & 0 & 0 \\ 0 & 0 & a_3 & 0 & r_5 & 0 & 0 \\ 0 & 0 & 0 & 1 & 0 & 0 & 0 \\ 0 & 0 & 0 & 0 & a_5 & 0 & 0 \\ 0 & 0 & s_3 & 0 & s_5 & 0 & 1 \end{pmatrix} \quad b = \begin{pmatrix} 1 \\ 0 \\ 0 \\ 0 \\ 0 \\ 0 \\ 0 \end{pmatrix}$$

$$c^T = \begin{pmatrix} 0 & 0 & 0 & 0 & 0 & 1 & 0 \end{pmatrix}$$

where s_i, r_i, and $a_i (= 1 - s_i - r_i)$, $i = 3, 5$ are the fractions of items being *scrapped*, *reworked*, and *accepted* at the end of the third and fifth stage.

♣

What has been said until now can be further specified by stating the irreducibility condition of a system (A, b, c^T) in terms of its influence graph G_x or, equivalently, in terms of its matrix $A^\sharp$.

THEOREM 4 *(condition for irreducibility)*

A positive system (A, b, c^T) is irreducible if and only if its influence graph G_x is connected, or, equivalently, if and only if

$$I + A^\sharp + A^{\sharp^2} + \cdots + A^{\sharp^{n-1}} \gg 0$$

Proof. A positive system is irreducible if each state variable influences and is influenced by any other variable. This is equivalent to the connection of the influence graph G_x. On the other hand, a graph with n nodes is connected if and only if there exists at least one path between each pair of distinct nodes. Recalling that $(A^{\sharp^h})_{ij}$ represents the number of paths of length h from i to j and noticing that among these paths at least one has a length $\leq n - 1$, we can conclude that G_x is connected if and only if all the nondiagonal entries of the matrix $(A^\sharp + A^{\sharp^2} + \cdots + A^{\sharp^{n-1}})$ are positive. This condition is equivalent to the strict positivity of $I + A^\sharp + A^{\sharp^2} + \cdots + A^{\sharp^{n-1}}$.

◊

DEFINITIONS   27

Irreducible systems can be further divided into cyclic and primitive systems as follows:

DEFINITION 4 *(cyclic and primitive system)*

> An irreducible system (A, b, c^T) is said to be *cyclic* if and only if there exists a partition $C_1, C_2, \ldots, C_r$ with $r > 1$ of the set of the state variables for which the following properties hold: Every variable of the class C_i has a direct influence only on variables of the following class C_{i+1}, $i = 1, 2, \ldots, r-1$ and every variable of the class C_r has a direct influence only on variables of the class C_1. The integer r is called the *cyclicity index* (or *imprimitivity index*). A noncyclic irreducible system is said to be *primitive*.

Figure 4.3 shows the influence graph G_x of a cyclic system with $r = 3$. Among the nodes of the same class there are no arcs (or self-loops), and at least one arc connects each node to a node of the following class.

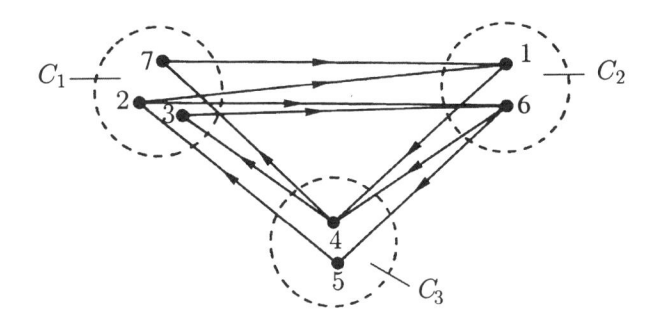

Figure 4.3   The influence graph G_x of a cyclic system with $r = 3$.

The influence graph G_x of a cyclic system with cyclicity index r is characterized by the fact that the lengths of all its cycles (simple or not) are equal to r or to multiples of r. The cyclicity index r of the system is therefore equal to the greatest common divisor of the lengths of all the cycles in the graph G_x. On the other hand, if the greatest common divisor M of the length of the simple cycles in a connected graph G_x is > 1, then the cycles necessarily have a length equal to M or to a multiple of M, and this implies (the proof is left to the reader as an exercise) that the graph G_x has the properties required by *Definition 4*. In conclusion, *Theorem 5* holds.

THEOREM 5 *(condition for cyclicity and primitivity)*

> An irreducible system (A, b, c^T) is primitive if and only if the greatest common divisor M of the lengths of all the simple cycles in its influence graph G_x is 1. Otherwise, the system is cyclic and its cyclicity index r is equal to M.

28   IRREDUCIBILITY, EXCITABILITY, AND TRANSPARENCY

Although a number of algorithms for the determination of the lengths of the simple cycles in a graph are available (note that the presence of a cycle of length l in a graph G_x is revealed by the existence of a nonzero element on the diagonal of $A^{\#l}$), the following rules are worth keeping in mind.

THEOREM 6 *(sufficient conditions for primitivity)*

> If the graph G_x of an irreducible system (A, b, c^T) contains a self-loop or a triangle of arcs $(i, j), (i, h), (h, j)$, then the system is primitive.

Proof. The proof is by contradiction. Suppose the system is not primitive. Thus, the system is cyclic with a cyclicity index $r > 1$. This means that cycles of length kr with $k = 1, 2$, and so on, pass through node i of G_x. But the existence of a self-loop (i, i) or of a triplet of arcs $(i, j), (i, h), (h, j)$ implies that there are also cycles of length $r + 1$ or $kr + 1$ passing through node i. This is not possible since all the cycles in the graph G_x of a cyclic system must have a length equal to r or to multiples of r and $kr + 1$ is not a multiple of r since $r > 1$.

PROBLEM 8 *(A-I) Consider all positive systems discussed in Chapters 1–3 and determine if they are reducible or irreducible (primitive or cyclic).*

PROBLEM 9 *(T-II) Prove that a sampled irreducible continuous-time system is a primitive discrete-time system.*

What has been said about cyclic systems leads naturally to the following interpretation in terms of state equations: By renumbering the state variables of a cyclic system so that the first n_1 variables are in class C_1, the next n_2 variables in class C_2, and so forth, the free motion of the new state vector $z(t) = Px(t)$ can be described by the equation $\dot{z}(t) = A^* z(t)$ with $A^* = PAP^T$, where

$$A^* = \begin{pmatrix} 0 & 0 & \cdots & 0 & A_{1r}^* \\ A_{21}^* & 0 & \cdots & 0 & 0 \\ 0 & A_{32}^* & & 0 & 0 \\ \vdots & \vdots & & \vdots & \vdots \\ 0 & 0 & \cdots & A_{r,r-1}^* & 0 \end{pmatrix}$$

A cyclic system in the above form is said to be in the *Frobenius canonical form*. Primitive systems are characterized by the following very simple property:

THEOREM 7 *(conditions for primitivity)*

> A system (A, b, c^T) is primitive if and only if there exists a positive integer $m > 0$ such that $A^{\#m} \gg 0$ or, equivalently, if there exist paths of the same length m between each pair of nodes of G_x.

Proof. Suppose $A^{\#m} \gg 0$ for some positive integer m. Then, since $(A^{\#m})_{ij}$ is the number of paths of length m between i and j, the graph G_x is connected.

Therefore, the system is irreducible and, consequently, it must be either primitive or cyclic. Suppose the system is cyclic. Then, the length of a path between two nodes in C_1 is equal to r or to multiples of r, while the length of a path between a node in C_1 and a node in $C_2 = 1$ or to $1 + r$ or to 1 plus a multiple of r. Since $r > 1$, the paths of the first kind have different lengths from those of the second kind. Thus, there are no paths of the same length (m) between each pair of nodes in G_x and this contradicts the assumption $A^{\sharp m} \gg 0$. From this it follows that the system must be primitive.

The proof that primitivity implies $A^{\sharp m} \gg 0$ for some $m > 0$ is left to the reader as a useful exercise.

$\diamondsuit$

PROBLEM 10 *(T-I) Prove that a primitive system cannot have finite memory.*

PROBLEM 11 *(T-III) Prove that the minimal integer m satisfying Theorem 7 can be $> n$. Then, prove that $m_{\min} \leq n^2 - 2n + 2$ (Wielandt formula) by constructing an example in which the relationship holds with the equality sign.*

We end this part dealing with reducibility, irreducibility, primitivity, and cyclicity with an example.

EXAMPLE 5 (working processes with total reworks)
Consider the class of working processes in which an initially raw item is transformed sequentially through a certain number of stages. At the end of some of these stages, a test is performed with the aim of detecting the items that have been improperly processed. Assume that for technical reasons the item that has not passed the test is transformed into a raw item and routed back to the first stage. *Figure 4.4* shows the influence graphs G of four possible working processes with five stages (nodes $1, \ldots, 5$).

In all cases, the graph G_x can be obtained by eliminating the input and output nodes 0 and 7 and the arcs $(0, 1)$ and $(5, 7)$. In the first two cases G_x is connected so that the system is irreducible. Process a is cyclic with an index of cyclicity $r = 6$, and process b is cyclic with an index of cyclicity $r = 2$ since the simple cycles in G_x have length 4 or 6 (so that the greatest common divisor is 2). The classes C_1 and C_2, which determine the Frobenius canonical form of process b, are $C_1 = \{x_1, x_3, x_5\}$ and $C_2 = \{x_2, x_4, x_6\}$. Processes c and d are reducible and classes C_i are easily found (see *Fig. 4.4*). The subsystems associated with the classes C_i are, by construction, irreducible, and they can be either cyclic or primitive. Obviously, the subsystems determined by classes C_2 and C_3 are primitive since they are composed of a unique state variable. The subsystem determined by class C_1 is primitive in case c [a triangle of arcs $(i, j), (i, h), (h, j)$ is present] and cyclic with $r = 4$ in case d.

♣

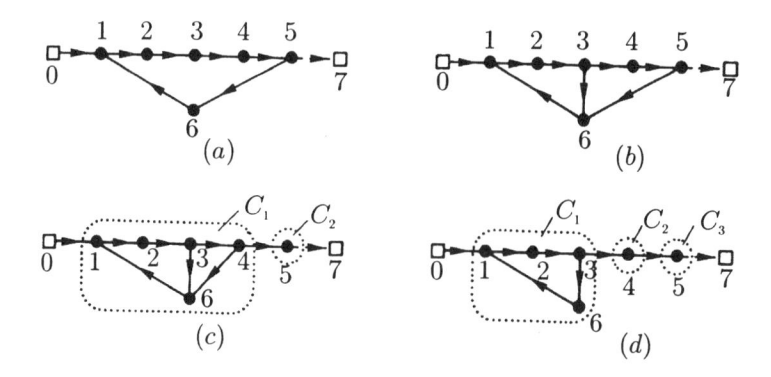

Figure 4.4 Working processes with total reworks.

We consider now a new class of positive systems, namely, *excitable systems*. These systems have a peculiar structural property that is a topological characteristic of their graph G_{ux}.

DEFINITION 5 *(excitable system)*

A positive system (A, b, c^T) is said to be excitable if and only if each state variable can be made positive by applying an appropriate nonnegative input to the system initially at rest $[x(0) = 0]$.

The condition for excitability, when formulated on the influence graph G_{ux} or on the pair $(A^\sharp, b^\sharp)$, is the following:

THEOREM 8 *(condition for excitability)*

A positive system (A, b, c^T) is excitable if and only if there exists at least one path from the input node 0 to each node $i = 1, \ldots, n$, of its influence graph G_{ux} or, equivalently, if and only if

$$b^\sharp + A^{\sharp T} b^\sharp + A^{\sharp T^2} b^\sharp + A^{\sharp T^{n-1}} b^\sharp \gg 0$$

Proof. Excitability of the system means that each state variable x_i can be influenced directly or indirectly from the input u of the system. This implies that there exists at least one path from node 0 to node i in the graph G_{ux}. Recalling that the ith component of the vector $A^{\sharp T^{h-1}} b^\sharp$ represents the number of paths of length h from node 0 to node i and noting that among these paths there exists at least one simple path (of length not $> n$) we can conclude that if the system is excitable, the sum of the vectors $b^\sharp, A^{\sharp T} b^\sharp, A^{\sharp T^2} b^\sharp, A^{\sharp T^{n-1}} b^\sharp$ must be a strictly positive vector.

On the other hand, if G_{ux} contains at least one path from node 0 to any other node i, then the input influences directly or indirectly each state variable. It is then possible to determine, for every variable x_i, a time interval t_i and an input function $u^i(\cdot)$ that, applied to the system at rest, yields $x_i(t_i) > 0$.

$\diamondsuit$

The definition of excitability does not say anything about the possibility of jointly exciting all state variables, that is, about the possibility of "invading" the positive orthant $\mathbb{R}^n_+$ of the state space starting from the origin. Furthermore, nothing is said about the way this invasion can be performed. *Theorem 9* shows that excitability and the possibility of invading the positive orthant are, actually, the same thing and that the invasion in guaranteed in a short time by any strictly positive input.

THEOREM 9 *(excitability and invasion)*

> If a strictly positive input $u(\cdot)$ is applied to an excitable system (A, b, c^T) initially at rest, then the state of the system immediately invades the positive orthant in the case of continuous-time systems and at most after n transitions in the case of discrete-time systems.

Proof. Suppose the system is continuous-time and $x(0) = 0$. The application of a strictly positive input $u(\cdot)$, yields $\dot{x}_i(0) = b_i u(0) > 0$ for each i such that $b_i > 0$ (note that b cannot be the null vector since the system is excitable). Denote by I_1 the set of these indices, i, so that $x_i(t) > 0$ for $t > 0$ and $i \in I_1$. Then, determine the set I_2 of indices j for which $\dot{x}_j$ does not depend on u but depends on $x_i, i \in I_1$. The set $I_1 \bigcup I_2$ is the set of nodes in G_{ux} reachable in one or two steps from the input node 0. Since $x_i(t) > 0$ for $t > 0$, then $\dot{x}_j(t) > 0$ for $t > 0$. All state variables belonging to $I_1 \bigcup I_2$ are, therefore, positive for $t > 0$. By iterating this procedure, one obtains the sets $I_3, I_4, \ldots, I_h$ and $x_i(t) > 0$ for $t > 0$ and $i \in I_1 \bigcup I_2 \ldots \bigcup I_h$. The set $I_1 \bigcup I_2 \ldots \bigcup I_h$ is the set of nodes of G_{ux} reachable from node 0 in at most h steps. Thus, the excitability of the system implies that for $h = n$ such a set coincides with the set $\{1, 2, \ldots, n\}$. Consequently, $x(t) \gg 0$ for $t > 0$.

A similar proof can be given for discrete-time systems. In this case, $x_i(h) > 0$ for $t \geq h$ and $i \in I_1, \bigcup I_2 \ldots \bigcup I_h$. Therefore, $x(n) \gg 0$.

$\diamondsuit$

The last property we discuss in this chapter is *transparency*, which is, in a way, the dual property of excitability. In fact, transparent systems are characterized by a peculiar topology of the graph G_{xy}.

DEFINITION 6 *(transparent system)*

> A positive system (A, b, c^T) is said to be *transparent* if and only if its free output response $y(\cdot)$ [*i.e.* its output for $u(\cdot) = 0$] is positive for every $x(0) > 0$.

In particular, the free output response of a transparent system is positive for $x(0) = e_i$, $i = 1, 2, \ldots, n$, where e_i is the unit vector of the ith axis of the state space. This implies that every state variable x_i directly or indirectly influences the output y of the system. Thus, in the influence graph G_{xy} of a transparent system there necessarily exists at least one path between each node $i, i = 1, 2, \ldots, n$ and the output node $n+1$. On the other hand, this topologic property of the graph G_{xy} enables us to conclude that the system is transparent. In fact, if the n free output responses corresponding to the vectors $x(0) = e_i$, $i = 1, 2, \ldots, n$ are positive, by linearity, the free output response corresponding to any $x(0) > 0$ is positive. Taking into account that the ith component of the vector $A^{\sharp^{h-1}} c^{\sharp}$ represents the number of paths of length h from node i to the output node $(n+1)$ in the graph G_{xy}, the previous discussion can be summarized by the following *Theorem 10* (dual of *Theorem 8*).

THEOREM 10 *(condition for transparency)*

A positive system (A, b, c^T) is transparent if and only if there exists at least one path from each node $i = 1, 2, \ldots, n$ to the output node $n + 1$ in its influence graph G_{xy} or, equivalently, if and only if

$$c^{\sharp} + A^{\sharp} c^{\sharp} + A^{\sharp^2} c^{\sharp} + \cdots + A^{\sharp^{n-1}} c^{\sharp} \gg 0$$

It might be interesting to know how long one has to wait until the free output response of a transparent system becomes positive, thus revealing the positivity of the initial state. By dualizing *Theorem 9* we can conclude that no more than n transitions are needed in discrete-time systems, while the free output response of a continuous-time system is positive for any $t > 0$.

It is worth noting that excitability and transparency are quite weak properties that are usually met in applications. Furthermore, although excitability and transparency recall, even formally, reachability and observability of linear systems, such properties are not equivalent. In other words, there exist unreachable and unobservable (in the usual sense) positive systems that are excitable and transparent. A simple example is the system composed of a pair of identical reservoirs connected in parallel [*Fig. 2.4(a)*], which is excitable and transparent, although it is neither reachable nor observable.

Part II

Properties

5
Stability

This chapter and Chapter 6 are devoted to the stability of positive systems. The results we discuss in this chapter are quite general and do not refer to the structure of the system. In contrast, Chapter 6 presents some remarkable properties (known since 1912) of the spectrum of the A matrix of irreducible systems (cyclic and primitive).

Before we begin, we recall that it is quite usual to consider two kinds of stability, called, respectively, "internal" and "external". The first refers to asymptotic properties of the free motion of the state vector $x(t) = \Phi(t)x(0)$, where $\Phi(t)$ is the transition matrix (e^{At} for continuous-time systems and A^t for discrete-time systems). Precisely, a linear system is said to be *asymptotically stable*, when, for each initial state $x(0)$, $\Phi(t)x(0) \to 0$ as $t \to \infty$, while it is *unstable* if there exists an initial state $x(0)$ such that $\Phi(t)x(0)$ is unbounded as $t \to \infty$ and *marginally stable* otherwise. On the contrary, external stability refers to the forced output response. More precisely, a linear system is said to be *externally stable* when its forced output [its output for $x(0) = 0$] is bounded for every bounded input $u(\cdot)$. The simplest necessary and sufficient conditions for asymptotic (eigenvalues criterion and Liapunov equation) and for external stability (poles criterion) are reported in Appendix B.

It would be natural to define asymptotic stability of positive systems by requiring that $\Phi(t)x(0)$ tends to zero for any nonnegative $x(0)$, instead of for any $x(0)$. But any state vector $x(0)$ can be split into the difference of two vectors $x'(0) \geq 0$ and $x''(0) \geq 0$. Thus, from linearity, it follows that $\Phi(t)x(0)$ tends to zero for every $x(0)$ whenever it tends to zero for every $x(0) \geq 0$. Analogously, each input function $u(\cdot)$ can be written as $u(\cdot) = u'(\cdot) - u''(\cdot)$, where $u'(\cdot) \geq 0$ and $u'' \geq 0$ so that, from linearity, the output of an externally positive system with $x(0) = 0$ is bounded for every bounded $u(\cdot)$, whenever it is bounded for every bounded nonnegative $u(\cdot)$. From these remarks, it follows that all the conditions for internal and external stability of linear systems, hold for positive systems. Nevertheless, positive systems have some peculiar properties, which are quite useful for stability analysis. The most famous of these properties concerns the dominant eigenvalue and eigenvector.

THEOREM 11 *(dominant (Frobenius) eigenvalue and eigenvector)*

> The dominant eigenvalue of a continuous-time positive system is real (and, hence, unique). Discrete-time positive systems can have more than one dominant eigenvalue. One of them, however, is real and nonnegative. This dominant eigenvalue is called the *Frobenius eigenvalue* and is denoted by λ_F. Among the eigenvectors associated with λ_F, some are positive. They are called *Frobenius eigenvectors* and are denoted by x_F. Furthermore, the Frobenius eigenvector can be strictly positive but no other eigenvector can be such.

Proof. Whenever a continuous-time system $\dot{x}(t) = Ax(t)$ has, among its dominant eigenvalues, one (or more) pair of complex eigenvalues, the state of the system, for a generic initial condition, has an oscillatory behavior in the long run. This means that some state variables must, at some instant of time, change sign, and this implies that the system cannot be positive. Therefore, a continuous-time positive system must have a real (and, hence, unique) dominant eigenvalue (even if its algebraic multiplicity is > 1). Such an eigenvalue λ_F is called the Frobenius eigenvalue.

In discrete-time systems $x(t+1) = Ax(t)$, the existence of negative (or complex conjugate) dominant eigenvalues implies that the components of the state vector $x(t)$ along the dominant eigenvector change sign over time. This fact does not imply, however, that the state vector cannot remain positive. This is what happens, for example, in the positive system $x(t+1) = Ax(t)$ with

$$A = \begin{pmatrix} 0 & \rho \\ \rho & 0 \end{pmatrix} \quad \rho > 0$$

Such a system has, in fact, $\lambda_1 = \rho$, $\lambda_2 = -\rho$ and the corresponding eigenvectors are $x^{(1)} = (1 \ \ 1)^T, x^{(2)} = (1 \ \ -1)^T$ so that the component of the state vector $x(t)$ along the eigenvector $x^{(2)}$ changes sign at each transition, even though $x(t)$ remains positive for every t provided $x(0)$ is such.

One of the dominant eigenvalues, however, must be nonnegative. In fact, if the dominant eigenvalue were unique and negative, the state of the system would tend to align for $t \to \infty$, with the dominant eigenvector, and the components of the state vector along this eigenvector would change sign at each transition, which is in contrast with the positivity of the system. This allows us to define the Frobenius eigenvalue λ_F as the nonnegative dominant eigenvalue.

The eigenvector associated with λ_F (or, at least, one of the eigenvectors associated with λ_F) must be positive, since, otherwise, the trajectory starting from a state $x(0) > 0$ (which tends to align for $t \to \infty$ with the eigenvector) would leave, in the long run, the positive orthant $\mathbb{R}^n_+$.

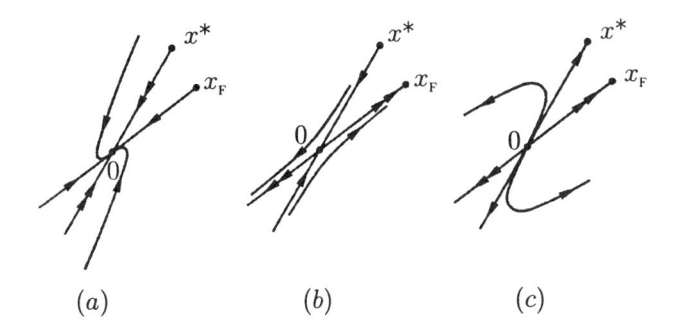

Figure 5.1 Trajectories of a continuous-time system in the plane generated by x^* and x_F in the cases: (a)- $\lambda^* < \lambda_F < 0$; (b)- $\lambda^* < 0 < \lambda_F$; (c)- $0 < \lambda^* < \lambda_F$ (the double arrow indicates fast motion along the trajectory).

Finally, we prove that the eigenvectors associated with eigenvalues different from λ_F, cannot be strictly positive. For this, we consider continuous-time systems and leave to the reader, as an exercise, the proof of the discrete-time case. In view of the previous discussion we must assume that the system has a positive Frobenius eigenvector $x_F > 0$. Suppose, by contradiction, that there exists an eigenvector $x^* \gg 0$ corresponding to an eigenvalue $\lambda^* \neq \lambda_F$. From the previous arguments, it follows that $\lambda^* < \lambda_F$. Thus, in the plane determined by the vectors x^* and x_F, which form an acute angle since $x^* \gg 0$ and $x_F > 0$, the trajectories tend to align for $t \to \infty$ with x_F as $t \to \infty$ as shown in *Fig. 5.1*. In all cases some trajectories leave the positive orthant, so that the system cannot be positive. This contradicts the assumption $x^* \gg 0$.

$\diamond$

Although the previous proof clearly explains why the dominant eigenvalue is unique for continuous-time systems and not necessarily unique for discrete-time systems, it is also worth investigating this difference from a purely algebraic point of view. To this end note, first, that if $\Lambda_A = \{\lambda_1, \lambda_2, \ldots, \lambda_n\}$ is the spectrum of a matrix A, then the spectrum of the matrix $A + \alpha I$ is $\Lambda_{A+\alpha I} = \{\lambda_1 + \alpha, \lambda_2 + \alpha, \ldots, \lambda_n + \alpha\}$. In other words, by adding [subtracting] to each element of the diagonal of A a positive number, the spectrum shifts to the right (left) of the same quantity. This property is a consequence of

$$\Delta_{A+\alpha I}(\lambda) = \det(\lambda I - A - \alpha I) = \det((\lambda - \alpha)I - A) = \Delta_A(\lambda - \alpha)$$

so that, if λ_i is a root of $\Delta_A(\lambda)$, then $\lambda_i + \alpha$ is a root of $\Delta_{A+\alpha I}(\lambda)$. Let us then apply this property to a Metzler matrix A noting that for $\alpha \geq 0$ sufficiently large, the matrix $A + \alpha I$ is nonnegative. Thus, the spectrum Λ_A of the Metzler matrix is obtained by shifting the spectrum $\Lambda_{A+\alpha I}$ of $A^* = A + \alpha I \geq 0$ to the left of a quantity equal to $\alpha \geq 0$. The situation is depicted in *Fig. 5.2*, where the discrete-time system $x(t+1) = A^* x(t)$ has three dominant eigenvalues $(\lambda_F^*, \lambda_1^*, \lambda_2^*)$ and

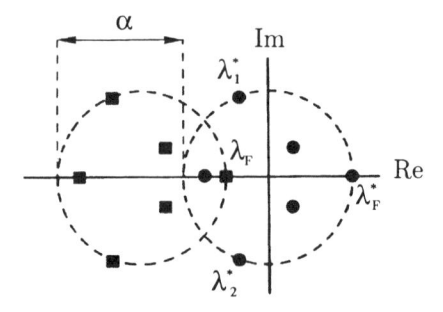

Figure 5.2 The spectrum $\Lambda_{A^*}(\bullet)$ of matrix $A^* \geq 0$ with a Frobenius eigenvalue λ_F^* and two other dominant eigenvalues (λ_1^* and λ_2^*) and the spectrum $\Lambda_A(\blacksquare)$ of the Metzler matrix $A = A^* - \alpha I$ with one dominant eigenvalue λ_F.

the continuous-time system $\dot{x}(t) = Ax(t) = (A^* - \alpha I)x(t)$ has only one dominant eigenvalue ($\lambda_F = \lambda_1^* - \alpha$).

In conclusion, the dominant eigenvalue of a continuous-time positive system is unique, since the Frobenius eigenvalue of a discrete-time positive system (besides having maximum modulus) is the one with a maximal real part.

A simple rule establishes a lower and an upper bound for the Frobenius eigenvalue of a positive linear system (continuous or discrete -time).

THEOREM 12 *(lower and upper bounds for the Frobenius eigenvalue)*

Let r_i^+ and c_i^+ denote the sum of the elements of the *i*th row and column of the matrix A of a positive system. Then, the dominant eigenvalue λ_F of the system belongs to the following interval

$$\max\{\min_i r_i^+, \min_i c_i^+\} \leq \lambda_F \leq \min\{\max_i r_i^+, \max_i c_i^+\}$$

Proof. Denote by λ_F the Frobenius eigenvalue of a positive system $\dot{x}(t) = Ax(t)$ or $x(t+1) = Ax(t)$. In view of *Theorem 11*, there exists one positive eigenvector x_F associated to λ_F. Clearly, this eigenvector can be normalized so that the sum of its components is equal to 1. Thus

$$\sum_{i=1}^n x_{F_i} = 1 \quad 0 \leq x_{F_i} \leq 1 \quad i = 1, 2, \ldots, n$$

By definition, the eigenvector x_F satisfies the equation $Ax_F = \lambda_F x_F$, so that

$$
\begin{aligned}
a_{11}x_{F_1} + a_{12}x_{F_2} + \cdots + a_{1n}x_{F_n} &= \lambda_F x_{F_1} \\
a_{21}x_{F_1} + a_{22}x_{F_2} + \cdots + a_{2n}x_{F_n} &= \lambda_F x_{F_2} \\
\vdots \qquad \vdots \qquad \qquad \vdots \qquad & \quad \vdots \\
a_{n1}x_{F_1} + a_{n2}x_{F_2} + \cdots + a_{nn}x_{F_n} &= \lambda_F x_{F_n}
\end{aligned}
$$

By summing up these equations, one obtains

$$c_1^+ x_{F_1} + c_2^+ x_{F_2} + \cdots + c_n^+ x_{F_n} = \lambda_F \sum_{i=1}^{n} x_{F_i} = \lambda_F$$

The dominant eigenvalue λ_F is then a convex combination of the $c_i^+, i = 1, 2, \ldots, n$ since

$$0 \le x_{F_i} \le 1 \quad \sum_{i=1}^{n} x_{F_i} = 1$$

so that

$$\min_i c_i^+ \le \lambda_F \le \max_i c_i^+$$

The same result holds for the system $\dot{x}(t) = A^T x(t)$ or $x(t+1) = A^T x(t)$ so that, recalling that A and A^T have the same eigenvalues, one obtains

$$\min_i r_i^+ \le \lambda_F \le \max_i r_i^+$$

By writing the inequalities found for λ_F in a compact form, the theorem is proved.
◊

Theorems 11 and *12* allow us to establish a simple procedure for the evaluation of the dominant eigenvalue of a positive system. In fact, after calculating the coefficients of the characteristic polynomial $\Delta_A(\lambda)$, one can study the function $\Delta_A(\lambda)$ in the interval defined in *Theorem 12*. The largest root of the function in such an interval is λ_F, and the continuous-time [discrete-time] system is asymptotically stable if and only if $\lambda_F < 0$ [$\lambda_F < 1$].

PROBLEM 12 *(T-III) Let a_{ij} be the elements of the matrix A. Prove that the dominant eigenvalue λ_F belongs to the interval*

$$\min_i \left(\frac{1}{r_i^+} \sum_{j=1}^{n} a_{ij} r_j^+ \right) \le \lambda_F \le \max_i \left(\frac{1}{r_i^+} \sum_{j=1}^{n} a_{ij} r_j^+ \right)$$

Then, determine the bounds as a function of c_i^+. Finally, verify by means of some numerical example, that such intervals of uncertainty are often contained in the interval defined by Theorem 12.

Many necessary or sufficient conditions for the stability of linear systems become simpler in the case of positive systems. For example, the *trace condition*

$$\operatorname{tr} A > 0 \Rightarrow \dot{x}(t) = Ax(t) \text{ is unstable}$$
$$|\operatorname{tr} A| > n \Rightarrow x(t+1) = Ax(t) \text{ is unstable}$$

in the case of positive systems becomes

$$\exists i : a_{ii} > 0 \Rightarrow \dot{x}(t) = Ax(t) \text{ unstable}$$
$$\exists i : a_{ii} > 1 \Rightarrow x(t+1) = Ax(t) \text{ unstable}$$

(the proof is straightforward: Consider the solution of the ith state equation with $x_i(0) > 0$ and $x_j(0) = 0$, for $j \neq i$). Also, the Hurwitz and the Routh criteria for testing asymptotic stability become simpler. We recall (see Appendix B) that such criteria consist of verifying the positivity of n coefficients obtained from an appropriate table obtained from the coefficients α_i of the characteristic polynomial $\Delta_A(\lambda)$. In the case of positive systems, the answer can be obtained without computing the coefficients of the characteristic polynomial since the above mentioned criteria can be directly applied to the elements of the matrix A (see *Theorem 13*). Moreover, if the coefficients α_i of the characteristic polynomial are known, then it suffices to check their positivity (*Theorem 14*).

THEOREM 13 *(conditions for asymptotic stability)*

> A continuous-time $\dot{x}(t) = Ax(t)$ [discrete-time $x(t+1) = Ax(t)$] positive system is asymptotically stable if and only if the first n leading minors of the matrix $(-A)$ $[(I - A)]$ are positive.

We recall that the leading minors m_{ih} of a matrix M of dimension $n \times n$ are the determinants of the matrices of dimensions $i \times i$ obtained from M by deleting $n - i$ rows and the corresponding columns, and that the "first" leading minor m_{i1} is the determinant of the matrix obtained by deleting from M the last $(n - i)$ rows and the corresponding columns.

THEOREM 14 *(asymptotic stability and characteristic polynomial)*

> A continuous-time [discrete-time] positive system is asymptotically stable if and only if the coefficients of the characteristic polynomial $\Delta_A(\lambda)$ $[\Delta_{A-I}(\lambda)]$ are positive.

Proof. We prove the theorem for the continuous-time case, leaving the extension to discrete-time systems to the reader (*Hint*: Consider the shifted spectrum of A).

Asymptotic stability of any linear system $\dot{x}(t) = Ax(t)$, implies $\alpha_i > 0$, since $\Delta_A(\lambda) = \Pi_i(\lambda - \lambda_i)$ with $\text{Re}(\lambda_i) < 0$. On the other hand, $\alpha_i > 0, i = 1, 2, \ldots, n$ implies $\lambda^n + \alpha_1 \lambda^{n-1} + \cdots + \alpha_n > 0$, for $\lambda \geq 0$ so that the Frobenius eigenvalue of a positive system, which is real, is necessarily negative. From this, it follows that the system is asymptotically stable.

$\diamondsuit$

Even the Liapunov theorem takes on a particular form in the case of positive systems (see *Theorem 15*). We recall (see Section 13 of Appendix B) that the Liapunov theorem states that a linear system $\dot{x}(t) = Ax(t)$ or $x(t+1) = Ax(t)$ is asymptotically stable if and only if there exists a positive definite quadratic function $V(x) = x^T P x$, which is strictly decreasing ($\dot{V} < 0$ or $\Delta V < 0$) along each perturbed trajectory of the system. In the case of positive systems, the corresponding theorem states that we can restrict our attention to the class of "pure" quadratic functions, that is, quadratic functions not containing the terms $x_i x_j$ with $i \neq j$. This means that the matrix P can be assumed to be diagonal with positive diagonal elements. Such a matrix is called a strictly positive diagonal matrix.

THEOREM 15 *(Liapunov theorem for positive systems)*

> A continuous-time [discrete-time] positive system is asymptotically stable if and only if there exists a strictly positive diagonal matrix P such that the matrix $A^T P + PA$ [$A^T PA - P$] is negative definite.

Proof. We report the proof for continuous-time systems $\dot{x}(t) = Ax(t)$, asking the reader to extend the proof to discrete-time systems.
Sufficiency.
Since $A^T P + PA$ is negative definite and P is positive definite, in view of the Liapunov theorem for linear systems (see Appendix B.13), the system is asymptotically stable.
Necessity. By assumption, A is asymptotically stable. Thus, the state of the positive system $\dot{x}(t) = Ax(t) + bu(t)$, with $b = (1 \; 1 \; \ldots \; 1)^T$ and $u(t) = \bar{u} > 0$, tends asymptotically toward the same equilibrium state $\bar{u}$ whatever the initial state $x(0)$ is. If $x(0) \geq 0$, the positivity of the system implies, $x(t) \geq 0$, for $t \geq 0$. Hence, also $\bar{x} \geq 0$. Actually $\bar{x} \gg 0$ since, otherwise, denoting by i a zero component of $\bar{x}$, we would have

$$\dot{\bar{x}}_i = \sum_j a_{ij} \bar{x}_j + \bar{u} = \sum_{j \neq i} a_{ij} \bar{x}_j + \bar{u} > 0$$

which contradicts the assumption that $\bar{x}$ is an equilibrium for the system. Thus, the system of equations $A\bar{x} = -b\bar{u}$ admits a strictly positive solution $\bar{x} = (\bar{x}_1 \; \bar{x}_2 \; \ldots \bar{x}_n)^T$. Analogously, the equation $A^T \bar{y} = -b\bar{u}$ admits a strictly positive solution $\bar{y} = (\bar{y}_1 \; \bar{y}_2 \; \ldots \bar{y}_n)^T$.
Consider, then, the matrix B with entries

$$b_{ij} = -\bar{y}_i a_{ij} \bar{x}_j$$

and note that $b_{ii} \geq 0$ and $b_{ij} \leq 0, i \neq j$, since $a_{ii} \leq 0$ and $a_{ij} \geq 0$. Hence,

$$b_{ii} - \sum_{j \neq i} |b_{ij}| = \sum_j b_{ij} = -\bar{y}_i \sum_j a_{ij} \bar{x}_j = -\bar{y}_i (A\bar{x})_i = -\bar{y}_i(-\bar{u}) = \bar{u}\bar{y}_i > 0$$

and, similarly,

$$b_{ii} - \sum_{j \neq i} |b_{ji}| > 0$$

As a consequence, both B and B^T are matrices with strictly dominant diagonal elements and, therefore, also $B + B^T$ is such. Hence, $B + B^T$ is symmetric and positive definite, so that there exists $\alpha > 0$ such that $z^T B z \geq \alpha \|z\|^2 \ \forall z$. Consider now a diagonal matrix P with diagonal entries

$$p_{ii} = \frac{\bar{y}_i}{\bar{x}_i} > 0$$

and write

$$\begin{aligned}-z^T P A z &= \sum_{i,j} p_{ii}(-a_{ij}) z_i z_j = \sum_{i,j} (\bar{y}_i)(\bar{x}_i)^{-1} b_{ij} (\bar{y}_i)^{-1} (\bar{x}_j)^{-1} z_i z_j \\ &= \sum_{i,j} b_{ij}(\bar{x}_i^{-1} z_i)(\bar{x}_j^{-1} z_j) \geq \alpha \sum_i |\bar{x}_i^{-1} z_i|^2 \geq \beta \sum_i |z_i|^2\end{aligned}$$

This means that $z^T P A z$ is negative for every $z \neq 0$ as well as $z^T A^T P z$. Hence, $z^T(A^T P + PA)z$ is negative for every $z \neq 0$ and the matrix $A^T P + PA$ is negative definite.

◇

Another important property of positive systems concerns their robustness, namely, the fact that their stability is maintained while perturbing the matrix A of the system. Obviously, the perturbation of A cannot be generic: It must satisfy certain rules that should be "reasonable" with regard to applications. In particular, the perturbation must preserve positivity.

One such rule consists of reducing the interactions among the components of the state vector by decreasing the value of the elements a_{ij} with $i \neq j$. In order to preserve the positivity of the system, the perturbation must be such that matrix A becomes $\hat{A} = A - \delta A$ with

$$0 \leq \delta a_{ij} \leq a_{ij}, i \neq j, \quad \delta a_{ii} = 0, \quad i = 1, \ldots, n$$

If all the δa_{ij} are $< a_{ij}$, the influence graph remains unchanged, that is, $\hat{G}_x = G_x$. On the contrary, if for some (i, j) $\delta a_{ij} = a_{ij}$, the influence graph $\hat{G}_x$ is obtained by deleting from G_x the arc (j, i). It is therefore possible to consider perturbations that transform a given system into a collection of noninteracting subsystems. By contrast, it is impossible to add new interconnections since the conditions $a_{ij} = 0$ and $0 \leq \delta a_{ij} \leq a_{ij}$ imply $\delta a_{ij} = 0$. This class of perturbations describes pretty well what happens in a real system with components subject to ageing, malfunctioning, or breaking down. Whenever, for example, the pipe connecting reactor j with reactor i of a chemical process becomes narrower because of the deposition of some substance along the pipe, one must decrease a_{ij}, while the breakdown of the pipe can be modeled by means of the perturbation $\delta a_{ij} = a_{ij}$, namely, by annihilating the influence coefficient $\hat{a}_{ij}$. Clearly, it is relevant to know whether an asymptotically stable system remains such after any perturbation of this kind or, equivalently, whether an asymptotically stable system is composed of asymptotically stable subsystems. Systems (not necessarily positive) possessing such a property are said to be *connectively stable*. What makes this discussion interesting is that all asymptotically stable positive systems are connectively stable. The proof of this property is a direct consequence of the law of comparative dynamics (see *Problem 2*). In fact, denoting by $\hat{x}(t)$ and $x(t)$ the free motions of the perturbed system $(\hat{A})$ and of the unperturbed system (A) with $\hat{x}(0) = x(0)$, it follows that

$$x(t) \geq \hat{x}(t)$$

since A can be obtained from $\hat{A}$ by positively perturbing the nonnegative elements of $\hat{A}$. Since the unperturbed system (A) is asymptotically stable, $\hat{x}(t) \to 0$ as $t \to \infty$. In conclusion, we can say that all the asymptotically stable positive systems are particularly robust since they remain asymptotically stable even if the interactions among the state variables are decreased (or annihilated).

Another class of reasonable perturbations of a positive system is related to the concept of D-stability. Under this perturbation, the continuous-time system $\dot{x}(t) = Ax(t)$ becomes $\dot{x}(t) = DAx(t)$, where D is a strictly positive diagonal matrix. This perturbation preserves the positivity of the system and is particularly interesting when the derivative $\dot{x}_i(t)$ of each state variable is known up to a multiplicative term $d_{ii} > 0$. In the case of discrete-time systems, from the state equation $x(t + 1) = Ax(t)$ it follows that the increment of the state $x(t + 1) - x(t)$ is given by $(A - I)x(t)$. It might therefore be of interest to assume that such an increment can be modified, by multiplying each component by a term $d_{ii} > 0$, thus obtaining $x(t + 1) - x(t) = D(A - I)x(t)$, which can be rewritten as

$$x(t + 1) = (I - D + DA)x(t)$$

Clearly, the diagonal matrix D must be strictly positive and such that $I - D + DA \geq 0$ so that the perturbed system remains positive. Asymptotically stable systems that remain such after any perturbation of the above kind are said to be *D-stable*.

THEOREM 16 *(D-stability of positive systems)*
> Asymptotically stable positive linear systems are D-stable. Hence, if a continuous-time positive system $\dot{x}(t) = Ax(t)$ is asymptotically stable, all the perturbed systems $\dot{x}(t) = DAx(t)$, where D is a strictly positive diagonal matrix, are asymptotically stable. Similarly, the asymptotic stability of a discrete-time positive system $x(t+1) = Ax(t)$ implies the asymptotic stability of every system of the form $x(t+1) = (I - D + DA)x(t)$ with $I - D + DA \geq 0$, where D is a strictly positive diagonal matrix.

Proof. First, we prove the theorem for the continuous-time case. Since the positive system $\dot{x}(t) = Ax(t)$ is asymptotically stable, from *Theorem 15* it follows that there exists a strictly positive diagonal matrix P such that $A^T P + PA$ is negative definite. Thus, consider the perturbed system $\dot{x}(t) = DAx(t)$ and the function $V(x) = x^T P D^{-1} x$. The time derivative of this function is

$$\dot{V}(x) = \dot{x}^T P D^{-1} x + x^T P D^{-1} \dot{x} = x^T A^T D P D^{-1} x + x^T P D^{-1} D A x$$

But, $PD = DP$, since P and D are diagonal. Hence,

$$\dot{V}(x) = \dot{x}^T (A^T P + PA) x$$

In conclusion, $\dot{V}(x)$ is negative for every $x \neq 0$ and the state $x(t)$ of the system $\dot{x}(t) = DAx(t)$ tends to zero for every $x(0) \neq 0$, namely, the system $\dot{x}(t) = DAx(t)$ is asymptotically stable.

The proof of the theorem for the discrete-time case can be obtained using the property of the shifted spectrum. By hypothesis, the system $x(t+1) = Ax(t)$ is asymptotically stable so that its spectrum Λ_A lies inside the unit circle. The spectrum Λ_{A-I} of the matrix $A - I$ is, then, inside the unit circle centered at point $(-1, 0)$ of the complex plane. The positive system $\dot{x}(t) = (A - I)x(t)$ is therefore asymptotically stable. From the previous results on D-stability of continuous-time positive systems, we can say that the system $\dot{x}(t) = D(A - I)x(t)$ is asymptotically stable for every strictly positive diagonal matrix D and then, in particular, for every D satisfying the relationship $I - D + DA \geq 0$. The spectrum $\Lambda_{D(A-I)}$ of the matrix $D(A - I)$ is, therefore, inside the left half-plane of the complex plane. If we note that $D(A - I) + I = I - D + DA$, we can conclude that the spectrum Λ_{I-D+DA} of the perturbed system $x(t+1) = (I - D + DA)x(t)$, which is positive, is characterized by a Frobenius eigenvalue < 1. The perturbed system is, therefore, asymptotically stable.

$\diamond$

PROBLEM 13 *(T-II)* *Prove that if a continuous-time positive system $\dot{x}(t) = Ax(t)$ is unstable, the perturbed system $\dot{x}(t) = DAx(t)$, where D is a strictly positive diagonal matrix, cannot be asymptotically stable. Then, reformulate the statement for the discrete-time case and prove it using the property of the shifted spectrum.*

The importance of *Theorem 16* is highlighted by *Examples 6* and *7*.

EXAMPLE 6 (price dynamics)
Consider an economic system composed of n sectors and denote by $x_i(t)$ the price of the good, which characterizes the ith sector at time t. Suppose that the demand D_i and the supply S_i of good i depend on all prices and that the derivative $\dot{x}_i(t)$ of the price is proportional to the surplus of demand, that is,

$$\dot{x}_i(t) = d_{ii}[D_i(x(t)) - S_i(x(t))]$$

where d_{ii} is a positive proportionality factor. When analyzing complex economic systems it is unrealistic to imagine that one knows the proportionality factors d_{ii} between the surplus of demand and the growth rate of the prices. One is therefore interested in knowing whether there exist cases in which the price dynamics is, at least qualitatively, insensitive to the unknown proportionality factors. In particular, it is important to know whether there exist demand and supply functions that guarantee that the prices tend toward constant values $\bar{x}_i$, where demand and supply are balanced, for all values of the proportionality factors. A simple answer to this question can be given by assuming that demand and supply functions are linear and that the influence coefficients of x_i on $D_i[S_i]$ are negative [positive], while the others (influence of x_i on $D_j[S_j], i \neq j$) are nonnegative [nonpositive]. Thus, by means of an appropriate transformation, price dynamics can be modeled by a linear system of the form

$$\dot{x}(t) = DA(x(t) - \bar{x})$$

where A is a Metzler matrix and D is a strictly positive diagonal matrix with diagonal elements equal to the proportionality factors d_{ii}. By letting

$$\delta x(t) = x(t) - \bar{x}$$

the system is described by the state equations

$$\delta \dot{x}(t) = DA\delta x(t)$$

where A is a Metzler matrix. From *Theorem 16* one can, therefore, conclude that whenever the prices are stable for a particular setting of the proportionality factors, then they are such for any other setting. Clearly, the system $\delta \dot{x}(t) = A\delta x(t)$ is asymptotically stable if the demand and supply functions satisfy suitable conditions. Consider, for example, the case in which the price x_i influences much more the demand D_i and the supply S_i than any other demand or supply. Then, $-\alpha_{ii} \gg \alpha_{ij}, j \neq i$, so that $A^T + A$ is negative definitive. This means that there exists a strictly positive diagonal matrix P (the matrix $P = I$) such that $A^T P + PA$ is negative definite. In view of the previous discussion and from *Theorem 15*, we are then allowed to conclude that the prices always converge toward an equilibrium.

EXAMPLE 7 (stage structured populations)
Consider a population of animals characterized by a vital cycle with n distinct stages. Denote by $x_i(t)$ the number of individuals in stage i in year t and suppose that the transition from one stage to the next occurs only in a particular season of the year. Moreover, suppose that only the individuals in the last stage are fertile [as, e.g., in the case of many insect (may bugs, cicadas, etc.)] and that each individual produces an average number of f individuals of the first stage and then dies. A population of this kind can be described by a discrete-time linear model $x(t+1) = Ax(t)$ with

$$A = \begin{pmatrix} q_1 & 0 & 0 & \cdots & 0 & f \\ p_1 & q_2 & 0 & \cdots & 0 & 0 \\ 0 & p_2 & q_3 & \cdots & 0 & 0 \\ \vdots & \vdots & \vdots & & \vdots & \vdots \\ 0 & 0 & 0 & \cdots & q_{n-1} & 0 \\ 0 & 0 & 0 & \cdots & p_{n-1} & 0 \end{pmatrix}$$

where the parameters p_i and q_i, $i = 1, 2 \ldots, n-1$, are, respectively, the probability that an individual in stage i steps into the next stage before the end of the year and the probability of remaining in the same stage i. Such parameters p_i and q_i are, therefore, between 0 and 1, but their sum is not 1 since each individual in stage i can possibly die during the year (with probability m_i) so that

$$p_i + q_i + m_i = 1 \quad i = 1, 2, \ldots, n-1$$

The matrix A is, therefore, positive and the system is asymptotically stable whenever the population is damned to extinction and unstable whenever it is exponentially growing. It might be of interest to know whether a perturbation of the environment, like the defoliation of the area where the population lives or the spreading of a chemical compound killing prey or predators, might have a serious impact on the population by changing its stability into instability or vice versa. Such information might be of great importance whenever one wishes to protect a valuable population or to eliminate a pest population.

We show, now, how *Theorem 16* can be effectively used in order to give an answer to the previous questions. To do this, suppose that the action performed on the environment has the consequence of slowing down the physiologic processes involved in the transition from one stage to the next and to decrease, with the same proportion $d < 1$, both fertility f and mortality m_i. Thus, if the parameters of the population, after the perturbation, are given by

$$f^* = df \quad p_i^* = dp_i \quad m_i^* = dm_i$$

from the equality $p_i^* + q_i^* + m_i^* = 1$, we obtain

$$q_i^* = 1 - d + dq_i \quad i = 1, 2, \ldots, n-1$$

Hence, the perturbed matrix A^* is

$$A^* = \begin{pmatrix} 1-d+dq_1 & 0 & 0 & \ldots & 0 & df \\ dp_1 & 1-d+dq_2 & 0 & \ldots & 0 & 0 \\ 0 & dp_2 & 1-d+dq_3 & \ldots & 0 & 0 \\ \vdots & \vdots & \vdots & & \vdots & \vdots \\ 0 & 0 & 0 & \ldots & 1-d+dq_{n-1} & 0 \\ 0 & 0 & 0 & \ldots & dp_{n-1} & 0 \end{pmatrix}$$

and the perturbation is of the kind considered when dealing with D-stability, since

$$A^* = I - D + DA$$

with $D = dI$. We can therefore conclude, in view of *Theorem 16*, that this perturbation cannot destabilize a stable population. In other words, the action taken on the environment is useless if the target is to save a population that is going toward extinction but, on the other hand, the same action does not imply the risk of facilitating the demographic explosion of a pest population that, in the past, was unable to survive in that environment.

Similarly (see *Problem 13*), the same kind of perturbation cannot stabilize an unstable population. Consequently, in order to eliminate a florishing pest population, we must conceive and undertake a different action on the environment.

♣

6
Spectral Characterization of Irreducible Systems

The results presented in Chapter 5 are not linked to the structure of the system. If the system is irreducible, more can be said about its Frobenius eigenvalue and eigenvector. First, recall that reducible systems can be decomposed (by determining the classes of communicating states) in k irreducible acyclically connected systems. This implies that a continuous [discrete]-time system $\dot{x}(t) = Ax(t)$ [$x(t+1) = Ax(t)$] can be transformed, through a permutation $z(t) = Px(t)$ of the state variables, into a system $\dot{z}(t) = PAP^T z(t)$ [$z(t+1) = PAP^T z(t)$] with the matrix PAP^T in triangular form

$$PAP^T = \begin{pmatrix} A^*_{11} & 0 & \cdots & 0 \\ A^*_{21} & A^*_{22} & \cdots & 0 \\ \vdots & \vdots & & \vdots \\ A^*_{k1} & A^*_{k2} & \cdots & A^*_{kk} \end{pmatrix}$$

where $A^*_{ij} \geq 0$, for $i \neq j$ and A^*_{ii} is a Metzler [nonnegative] matrix. Thus, the spectrum of A is the union of the spectra of the matrices A^*_{ii}, and each one of these spectra is characterized by a Frobenius eigenvalue λ_{F_i}. The Frobenius eigenvalue λ_F of A is, therefore, given by

$$\lambda_F = \max_i \lambda_{F_i}$$

We now concentrate on irreducible systems, which enjoy a number of peculiar properties. The first one refers to the Frobenius eigenvalue and eigenvector.

THEOREM 17 *(Frobenius eigenvalue and eigenvector of irreducible systems)*

> The Frobenius eigenvector x_F of an irreducible system is unique and strictly positive ($x_F \gg 0$) and the Frobenius eigenvalue λ_F has an algebraic multiplicity equal to 1 (and is positive in discrete-time systems). Moreover, there are no positive eigenvectors other than the Frobenius eigenvector.

Proof. We prove the theorem in five steps.

i. *The Frobenius eigenvector is strictly positive ($x_F \gg 0$).*

Suppose the Frobenius eigenvector x_F is not strictly positive. Then $n_1 < n$ components $x_{Fi} \in I$ must be zero. Since $Ax_F = \lambda_F x_F$, then $a_{ij} = 0$ for $i \in I$ and $j \notin I$. Therefore, the free evolution of the n_1 state variables is not influenced by the remaining $n_2 = n - n_1$ state variables, which contradicts the irreducibility hypothesis.

ii. *The Frobenius eigenvector is unique.*

Suppose there exist two independent Frobenius eigenvectors x'_F and x''_F. Then, all the vectors belonging to the two-dimensional subspace determined by x'_F and x''_F are eigenvectors associated to λ_F. Among them there is certainly a not strictly positive eigenvector, which contradicts what has been proved in step i.

iii. *The Frobenius eigenvalue has a unitary algebraic multiplicity.*

We show that the algebraic multiplicity of λ_F cannot be equal to 2 asking the reader to complete the proof. Assume the Frobenius eigenvalue λ_F has an algebraic multiplicity equal to 2. Then, there exists an invariant two-dimensional subspace in which the trajectories are, in the case of a continuous-time system, as shown in *Fig. 6.1*.

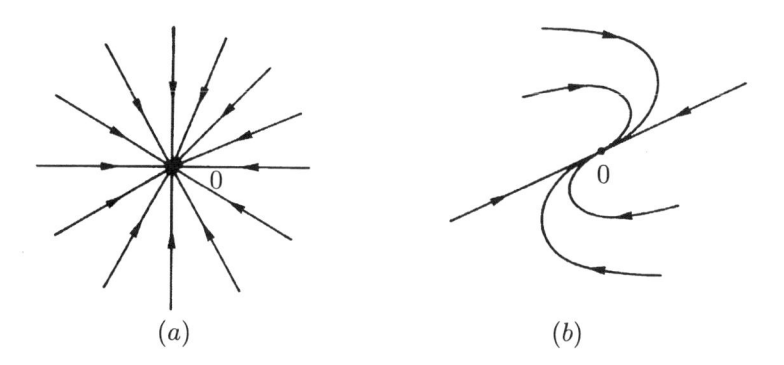

(a) (b)

Figure 6.1 Trajectories of a second-order continuous-time system with a double eigenvalue $\lambda < 0$.

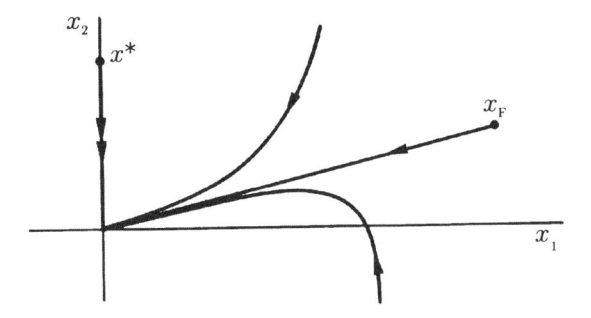

Figure 6.2 Trajectories of a second-order continuous-time system with two positive eigenvectors.

The state portrait of *Fig. 6.1(a)* contradicts the uniqueness of the Frobenius eigenvector. But the state portrait of *Fig. 6.1(b)* is also contradictory, since from the positivity of the system it follows that the straight trajectory (associated to the eigenvector x_F) must coincide with one axis of $\mathbb{R}_+^2$. In other words, x_F cannot strictly positive, which contradicts point i.

iv. *In discrete-time-systems $\lambda_F > 0$.*

Assume λ_F is zero. Then, $Ax_F = 0$ with $x_F \gg 0$. Thus, the matrix A is the null matrix. But this contradicts the irreducibility hypothesis.

v. *There are no positive eigenvectors other than the Frobenius eigenvector.*

We report the proof for a second-order continuous-time system, asking the reader to extend the proof to the general case. Suppose that, besides the Frobenius eigenvector x_F, there exists another positive eigenvector x^*. Such an eigenvector cannot be dominant since we have already proved that the Frobenius eigenvalue is the unique dominant eigenvalue and that the Frobenius eigenvector is unique. On the other hand, from *Theorem 11*, x^* cannot be strictly positive. The remaining possibility is that depicted in *Fig. 6.2*, where the trajectory with a single arrow is determined by the dominant eigenvector x_F while that with a double arrow is determined by the subdominant eigenvector x^*. From the previous discussion, we deduce that $x^* = (0\ 1)^T$ and from $Ax^* = \lambda x^*$ it follows that $a_{12} = 0$. But a system with $a_{12} = 0$ is reducible, which contradicts the hypothesis.

◇

EXAMPLE 8 (Peter's principle)
We now formally prove the famous *Peter's principle*, which states that in hierarchical structures, the proportion of incompetent individuals grows with the hierarchical

52 SPECTRAL CHARACTERIZATION OF IRREDUCIBLE SYSTEMS

level. In order to simplify the discussion, we consider the case of individuals who can be partitioned into *competent* and *incompetent*. Moreover, the *levels* of increasing responsibility (and salary) are labeled as $i = 1, 2, 3$, and so on. The promotion from level i to the next $(i+1)$ might occur even if the employee is incompetent at level i but the possibility that an incompetent employee at level i becomes competent after being promoted is excluded. The transition from incompetence to competence is possible only by acquiring experience within the same level.

The system is described by the influence graph G_{ux} of *Fig. 6.3*, where $u(t)$ is the number of new employees [most of them are competent at level 1, i.e. $\gamma \gg (1-\gamma)$] and $x_1^i(t)$ and $x_2^i(t)$ represent the number of competent and incompetent employees at level i during year t.

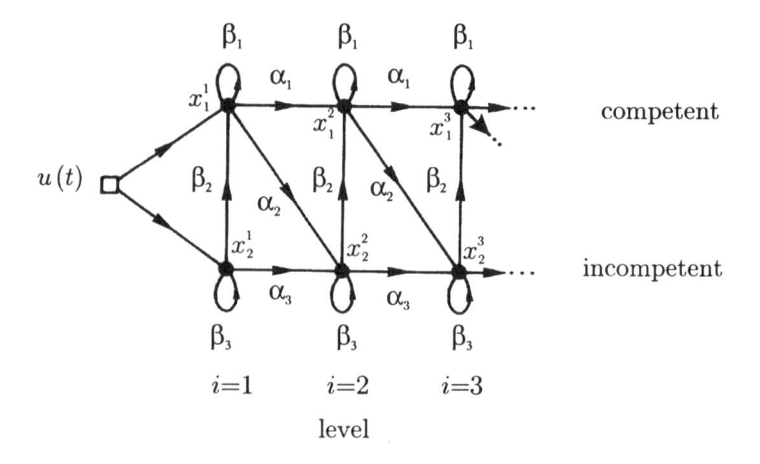

Figure 6.3 Influence graph of a hierarchical promotion structure.

The state equations of the system $x(t+1) = Ax(t) + bu(t)$ are the following:

$$\begin{pmatrix} x^1(t+1) \\ x^2(t+1) \\ x^3(t+1) \\ \vdots \end{pmatrix} = \begin{pmatrix} R_1 & 0 & 0 & \cdots \\ P_1 & R_2 & 0 & \cdots \\ 0 & P_2 & R_3 & \cdots \\ & \ddots & & \ddots \end{pmatrix} \begin{pmatrix} x^1(t) \\ x^2(t) \\ x^3(t) \\ \vdots \end{pmatrix} + \begin{pmatrix} b \\ 0 \\ 0 \\ \vdots \end{pmatrix} u(t)$$

where

$$x^i(t) = (x_1^i(t) \ x_2^i(t))^T \quad b = (\gamma \ 1-\gamma)^T$$

and P_i, R_i are 2×2 matrices called, respectively, the *promotion* and *recycling* matrix. For the sake of simplicity, consider the case of matrices with equal entries for each i, that is

$$P = \begin{pmatrix} \alpha_1 & 0 \\ \alpha_2 & \alpha_3 \end{pmatrix} \quad R = \begin{pmatrix} \beta_1 & \beta_2 \\ 0 & \beta_3 \end{pmatrix}$$

where the elements α_i and β_i represent, respectively, the probabilities of a change of level and of a change of competence of an individual during 1 year.

The reducibility of the system, pointed out by the triangular form of matrix A, enables us to formally conclude that the system is asymptotically stable, since β_1 and β_3, which are the eigenvalues of R, are both < 1. This implies that if the number of new employees is fixed $[u(t) = \bar{u}]$, the system tends toward an equilibrium $\bar{x}$, characterized by a particular two-dimensional vector $\bar{x}^i$ for each level i. Since the system is reducible, such vectors can be obtained sequentially by solving the equations

$$\bar{x}^1 = R\bar{x}^1 + b\bar{u}$$
$$\bar{x}^{i+1} = R\bar{x}^{i+1} + P\bar{x}^i \quad i = 1, 2, \text{and so on}$$

The matrix $(I - R)$ is clearly invertible

$$(I - R)^{-1} = \begin{pmatrix} \dfrac{1}{1-\beta_1} & \dfrac{\beta_2}{(1-\beta_i)(1-\beta_3)} \\ 0 & \dfrac{1}{1-\beta_3} \end{pmatrix}$$

so that we can write

$$\bar{x}^{i+1} = (I - R)^{-1} P \bar{x}^i$$

In other words, the vectors, which specify the numbers of competent and incompetent employees belonging to each level at equilibrium, are linked by the linear relationship

$$\bar{x}^{i+1} = A^* \bar{x}^i$$

which is an irreducible positive linear system, since A^* is strictly positive. Therefore, the Frobenius eigenvalue λ_F of A^* is positive and the corresponding eigenvector is strictly positive. Moreover, the eigenvector associated with the subdominant eigenvalue λ^* cannot be positive, and

$$\det A^* = \lambda_F \lambda^* = \det(I - R)^{-1} \det P = \frac{\alpha_1 \alpha_3}{(1-\beta_1)(1-\beta_3)} > 0$$

so that we can conclude that even the subdominant eigenvalue λ^* is positive. From this, it follows that the vector $\bar{x}^i$ varies with i, as shown in *Fig. 6.4*.

Thus, we have proved that, whenever at the low levels the number of competent individuals is greater than the number of incompetent individuals, the ratio

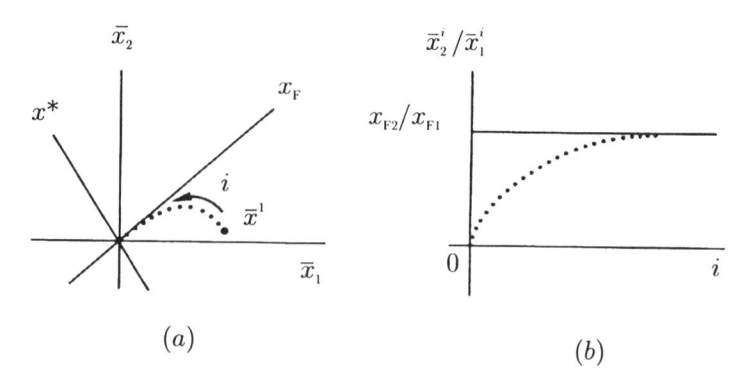

Figure 6.4 The vector $\bar{x}^i$ (a) and the ratio of incompetent/competent individuals (b) versus the responsibility level i.

incompetent/competent, increases with the level of responsibility toward the ratio x_{F_2}/x_{F_1} of the two components of the Frobenius eigenvector of the matrix A^*.

♣

Further details on the eigenvalues of irreducible systems are given in *Theorem 18* (due to Frobenius).

THEOREM 18 *(dominant eigenvalues of irreducible systems)*

> The number of eigenvalues of cyclic discrete-time systems is a multiple of the cyclicity index r, while the number of dominant eigenvalues is exactly r. The eigenvalues λ_i are regularly distributed on circles of radius $|\lambda_i|$. If the system is primitive, then the dominant eigenvalue is unique.

PROBLEM 14 *(T-III)* Prove Theorem 18 by verifying that the rth root of λ_F^r is an eigenvalue of A. *(Hint: Use the Frobenius canonical form.)*

EXAMPLE 9 (populations in periodic environments)
Consider a population living on a territory characterized by two seasons (summer and winter) and assume that the survival and fertility rates of the individuals are a function of the date (season) of birth. By measuring the time t and the age i in terms of seasons, we are then allowed to describe such a population by means of two variables $S_i(t)$ and $W_i(t)$, where

$S_i(t)$ = number of individuals born in summer and of age i at the end of season t
$W_i(t)$ = number of individuals born in winter and of age i at the end of season t.

In the case where all individuals are fertile after one season from their birth and survive at most for n seasons, the system can be described by linear equations

$$x(t+1) = Ax(t)$$

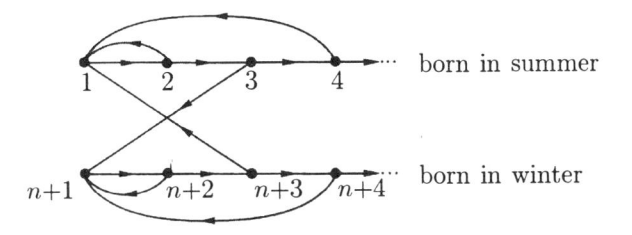

Figure 6.5 Influence graph G_x of a population in a periodic environment.

where $x(t) = (S_1(t) \ldots S_n(t) W_1(t) \ldots W_n(t))^T$. The corresponding influence graph G_x is depicted in *Fig. 6.5*.

Such a graph G_x is connected and contains cycles of length $2, 4, 6$, and so on. Since the greatest common divisor of the length of the cycles is 2, the system is cyclic with an index of cyclicity $r = 2$. From *Theorem 18* we can infer that such a system has two dominant eigenvalues. One of them, the Frobenius eigenvalue λ_F, is positive and the other, $-\lambda_F$, is negative.

♣

We suggest that the reader compare the general results of *Theorem 11* with the more specific ones contained in *Theorems 17* and *18* concerning irreducible systems (cyclic and primitive). In order to facilitate this comparison, we summarize the results in a table. It is worth noting that the properties of cyclic and primitive systems are the same, except the uniqueness of the dominant eigenvector in the case of discrete-time systems (*Theorem 18*).

SPECTRAL PROPERTIES OF POSITIVE SYSTEMS

	Generic disc / cont	Irreducible disc / cont	Cyclic disc / cont	Primitive disc / cont
x_{dom} unique	?	? / yes	no / yes	yes
λ_F real	yes	yes	yes	yes
$\lambda_F \geq 0$	yes / ?	yes / ?	yes / ?	yes / ?
$\lambda_F > 0$	?	yes / ?	yes / ?	yes / ?
multiplicity of λ_F	?	1	1	1
x_F unique	?	yes	yes	yes
$x_F > 0$	yes	yes	yes	yes
$x_F \gg 0$	?	yes	yes	yes
$\exists$ eigenvectors $x > 0 \ x \neq x_F$	?	no	no	no
$\exists$ eigenvectors $x \gg 0$ and $x \neq x_F$	no	no	no	no

7
Positivity of Equilibria

In many applications, it is interesting to know whether the state of the system tends toward a positive or a strictly positive equilibrium state $\bar{x}$, when a constant input $\bar{u} > 0$ is applied. If no information is available on the structure of the system, the best answer to this question is the following.

THEOREM 19 *(asymptotic stability and positivity of equilibria)*

> The equilibrium state $\bar{x}$ of an asymptotically stable positive system (A, b, c^T) with $b > 0$ and $\bar{u} > 0$ is positive, that is, $\bar{x} > 0$.

Proof. Since the system is asymptotically stable, for any input $\bar{u}$ there exists a unique equilibrium state $\bar{x}$. Moreover, the state vector $x(t)$ tends, as $t \to \infty$, toward $\bar{x}$, for every initial state $x(0)$. For $x(0) \geq 0$ also $x(t) \geq 0$ so that $\bar{x} \geq 0$. On the other hand, it is not possible that $\bar{x} = 0$, since $b\bar{u} = 0$ would contradict the hypothesis, $b > 0$ and $\bar{u} > 0$. Therefore, $\bar{x} > 0$.
$\diamondsuit$

This result can be strengthened if the system is excitable. In fact, *Theorem 20* holds.

THEOREM 20 *(asymptotic stability and strict positivity of equilibria)*

> The equilibrium state $\bar{x}$ of an asymptotically stable and excitable positive system with input $\bar{u} > 0$ is strictly positive, that is, $\bar{x} \gg 0$.

Proof. We present two different proofs: one for discrete-time systems and one for continuous-time systems.

Discrete-time systems
If the system $x(t+1) = Ax(t) + b\bar{u}$ is asymptotically stable and $\bar{u} > 0$, then $x(t)$ tends as $t \to \infty$ to $\bar{x}$ for every $x(0)$. Suppose $x(0) = 0$. This implies

$$\bar{x} = (b + Ab + \cdots + A^{n-1}b + A^n b + \cdots)\bar{u}$$

In view of *Theorem 8*, we can write

$$b^\sharp + A^{\sharp T} b^\sharp + A^{\sharp T^2} b^\sharp + \cdots + A^{\sharp T^{n-1}} b^\sharp \gg 0$$

so that

$$b + Ab + \cdots + A^{n-1}b \gg 0$$

Hence, $\bar{x} \gg 0$.

Continuous-time systems
If the system $\dot{x}(t) = Ax(t) + b\bar{u}$ is asymptotically stable and $\bar{u} > 0$, from *Theorem 19*, $x(t)$ tends, as $t \to \infty$, toward an equilibrium state $\bar{x} > 0$. Suppose $\bar{x}$ is not strictly positive, that is, $\bar{x}_i = 0$ for $i \in I$ and $\bar{x}_j > 0$ for $j \notin I$. Thus, it follows that

$$\sum_{j \notin I} a_{ij} \bar{x}_j + b_i \bar{u} = 0 \quad i \in I$$

which implies that $b_i = 0$ and $a_{ij} = 0$ for $i \in I$ and $j \notin I$. This means that the nodes $i \in I$ in the influence graph G_{ux} are not reachable in one step from node 0 and from nodes $j \notin I$. In other words, there does not exist in G_{ux} a path from the input node to the nodes $i \in I$ so that the system is not excitable. This is a contradiction.

◇

PROBLEM 15 *(A-II) Consider a network of reservoirs, possibly with pumping stations recycling part of the water. Suppose that the inflow reaches each reservoir and that a small fraction of water is lost in each reservoir due to evaporation. Prove that no reservoir can be empty at equilibrium.*

When dealing with excitable systems, the asymptotic stability, which is a sufficient condition for strict positivity of the equilibria, as stated in *Theorem 20*, is also a necessary condition. The following theorem establishes a link between asymptotic stability, excitability, and strict positivity of equilibria.

THEOREM 21 *(properties of standard systems)*

> In a positive system, two of the following properties: (*a*) excitability, (*b*) asymptotic stability, and (*c*) strict positivity of the equilibria $\bar{x}$ corresponding to inputs $\bar{u} > 0$ imply the third. Systems satisfying properties $a, b,$ and c are called standard.

Proof. We give hereafter the proofs for the discrete-time case. The reader can easily extend these proofs to the continuous-time case.

a and b imply c.
This has already been proved (see *Theorem 20*).

b and c imply a.
The asymptotic stability of $x(t+1) = Ax(t) + b\bar{u}$ guarantees that $x(t)$ tends, as $t \to \infty$, toward $\bar{x}$ for every $x(0)$. If $x(0) = 0$, we can write

$$x(t) = \sum_{i=1}^{t} A^{t-i} b\bar{u}$$

so that

$$\bar{x} = (b + Ab + \cdots + A^{n-1}b + A^n b + \cdots)\bar{u}$$

If the vector $b + Ab + \cdots + A^{n-1}b$ would have a component, say the ith one, equal to zero, then the ith component of all the reachability vectors $b, Ab, \ldots, A^{n-1}b$ would be zero. Thus, from the Cayley–Hamilton theorem $(A^n + \alpha_1 A^{n-1} + \alpha_2 A^{n-2} + \cdots + \alpha_n I) = 0$ the ith component of the vector

$$A^n b = -\alpha_1 A^{n-1} b - \alpha_2 A^{n-2} b - \cdots - \alpha_n b$$

would be zero, as well as that of all the vectors $A^t b$ with $t \geq n+1$. Then, taking into account that $\bar{x} = (b + Ab + \cdots + A^{n-1}b + A^n b + \cdots)\bar{u}$, the ith component of $\bar{x}$ would be zero, which contradicts condition (*c*). Thus, the vector $b + Ab + \cdots + A^{n-1}b$ cannot have a zero component, so that

$$b^{\sharp} + A^{\sharp T} b^{\sharp} + A^{\sharp T^2} b^{\sharp} + \cdots + A^{\sharp T^{n-1}} b^{\sharp} \gg 0$$

which is the excitability condition (property *a*).

a and c imply b.
Let $x(0) = \bar{x}$, and $x(t+1) = Ax(t) + b\bar{u}$. We can write

$$\bar{x} = A^n \bar{x} + (b + Ab + \cdots + A^{n-1}b)\bar{u}$$

and by transposition we obtain

$$\bar{x}^T (I - (A^T)^n) = (b + Ab + \cdots + A^{n-1}b)^T \bar{u}$$

By multiplying both sides of this equation by the Frobenius eigenvector x_F^* of A^T (satisfying the relationship $A^T x_F^* = \lambda_F x_F^*$), one obtains

$$(1 - \lambda_F^n)\bar{x}^T x_F^* = (b + Ab + \cdots + A^{n-1}b)^T x_F^* \bar{u}$$

But $\bar{x}^T x_F^* > 0$ since $\bar{x} \gg 0$ (property (c)) and $x_F^* > 0$. Also $(b + Ab + \cdots + A^{n-1}b)x_F^* > 0$ since $x_F^* > 0$ and, from excitability, $b + Ab + \cdots + A^{n-1}b \gg 0$. Therefore $(1 - \lambda_F^n) > 0$ and, consequently, $\lambda_F < 1$, which implies the asymptotic stability of the system (property b).

◊

The value of *Theorem 21* is pointed out by the following examples:

EXAMPLE 10 (coexistence of symbiotic species)
Consider an ecological system composed of n species, some of which are living in a symbiotic relationship. This means that, denoting by $x_i(t)$ the number of individuals of the species i, the growth rate per capita $\dot{x}_i(t)/x_i(t)$ is positively correlated with the biomass $x_j(t)$ of all the species j, which are in symbiosis with species i. By assuming that such a relationship is linear, we can write

$$\dot{x}_i(t) = x_i(t)\left(a_{ii}x_i(t) + \sum_{j \neq i} a_{ij}x_j(t) + b_i\right) \quad (7.1)$$

where the terms a_{ii} are negative and represent a measure of the mortality due to intraspecific competition (struggle among individuals of the same species for food or space), while the terms b_i represent the per capita growth due to external (inexhaustible) sources of energy. From the previous discussion, we can assume

$$a_{ii} < 0 \quad i = 1, 2, \ldots, n$$
$$a_{ij} \geq 0 \quad j \neq i$$
$$b_i \geq 0$$

Matrix A is, therefore, a Metzler matrix. Moreover, suppose that the energy coming from the environment is provided, directly or indirectly, through the symbiotic relationships to all species. This is equivalent to saying that the positive system

$$\dot{z}(t) = Az(t) + bu(t) \quad (7.2)$$

is excitable.

We are now ready to prove that, if n species can coexist at equilibrium, then such equilibrium is asymptotically stable. For this, note that (7.1) can be rewritten in the form

$$\dot{x}(t) = D(x(t))(Ax(t) + bu(t)) \quad (7.3)$$

where $u(t) = 1$ and $D(x(t)) = \text{diag}(x_i(t))$ is a diagonal matrix with the diagonal element $d_{ii} = x_i(t)$. Since, by hypothesis, the n species can coexist at equilibrium, system (7.1) has an equilibrium $\bar{x} \gg 0$, and then $A\bar{x} + b = 0$. This means that system (7.2) with $u(t) = \bar{u} = 1$ admits a strictly positive equilibrium. But system (7.2) is also excitable, so that, in view of *Theorem 21*, it is also asymptotically stable. Thus, the matrix A is an asymptotically stable Metzler matrix, and from *Theorem 16* on D-stability of positive systems, it follows that all the matrices DA with $D \gg 0$ diagonal, are asymptotically stable. In particular, this is true for the matrix $D(\bar{x})A$. On the other hand, by letting

$$\delta x_i(t) = x_i(t) - \bar{x}_i \quad i = 1, 2, \ldots, n$$

and linearizing system (7.3) around the equilibrium $\bar{x}$, one obtains

$$\delta \dot{x}(t) = D(\bar{x})A\delta x(t) \qquad (7.4)$$

so that the linearized system (7.4) is asymptotically stable. From a well-known theorem on the stability of the linearized system, it follows that the equilibrium state $\bar{x}$ of system (7.1) is asymptotically stable.

♣

EXAMPLE 11 (stock exchange)
Consider a set of firms, $i = 1, 2, \ldots, n$, quoted in the stock market, and denote by $b_i > 0$ and $a_{ij} \geq 0$ the shares of firm i held by an outside shareholder and by firm $j, j = 1, 2, \ldots, n$. Clearly,

$$b_i + \sum_{j=1}^{n} a_{ij} = 1 \qquad (7.5)$$

so that the positive matrix A has a Frobenius eigenvalue < 1, because the sum of the elements of every row of A is $1 - b_i$. Thus, matrix A is asymptotically stable.

If we denote the influence level of the outside shareholder on each firm by $\bar{x}_i$, the following must hold:

$$\bar{x}_i = b_i + a_{i1}\bar{x}_1 + a_{i2}\bar{x}_2 + \cdots + a_{in}\bar{x}_n$$

This is equivalent to saying that the vector $\bar{x}$ is the equilibrium state of the dynamic system

$$x(t+1) = Ax(t) + b \qquad (7.6)$$

Clearly, $\bar{x} \gg 0$, since system (7.6) is asymptotically stable and excitable ($b \gg 0$). In the case a firm, say the i-th one, is not quoted in the stock exchange, the term b_i is zero, so that the sum of the shares a_{ij} held by all firms is unitary [see (7.5)]. Thus, we cannot immediately state that the Frobenius eigenvalue of A is < 1 and that the vector $\bar{x}$ is strictly positive. If, however, one assumes that there are no

62 POSITIVITY OF EQUILIBRIA

subsets of firms holding themselves, then the outside shareholder has an influence level $\bar{x}_i > 0$ on each firm. If we take Eq. (7.5) into account, the above assumption is equivalent to the excitability of system (7.6). Therefore, system (7.6) is excitable and has a strictly positive equilibrium. Consequently, from *Theorem 21* it follows that the Frobenius eigenvalue of matrix A is < 1.

♣

It is worth noting that a standard system remains such for sufficiently small perturbations of the element of the matrix A and of the vector b. In fact, small perturbations of A imply small perturbations of the eigenvalues, so that if a system is asymptotically stable, after the perturbation, it is still asymptotically stable. The same holds for strict positivity of the equilibrium, so that the perturbed system is standard. A strong property of standard systems is the following theorem:

THEOREM 22 (*component of greatest relative change*)

> Whenever one (or more) elements of the rth row of the matrix A and of vector b of a standard system are slightly increased, the rth state variable is, at equilibrium, maximally affected in relative terms by such perturbation, that is
>
> $$\frac{\delta \bar{x}_r}{\bar{x}_r} \geq \frac{\delta \bar{x}_i}{\bar{x}_i} \quad i = 1, 2, \ldots, n$$

Proof. We prove the theorem for the case of discrete-time systems, leaving to the reader, as an exercise, the proof for the continuous-time case.

Since the system is standard, then $\bar{x} \gg 0$. If we denote the perturbations by δ, we can write

$$(\bar{x} + \delta\bar{x}) = (A + \delta A)(\bar{x} + \delta\bar{x}) + (b + \delta b)\bar{u}$$

from which, noting that $\bar{x} = A\bar{x} + b\bar{u}$, we obtain

$$\delta\bar{x} = A\delta\bar{x} + b^{(r)} \tag{7.7}$$

where the positive vector

$$b^{(r)} = \delta A(\bar{x} + \delta\bar{x}) + \delta b \bar{u}$$

has only the rth component different from zero since, by hypothesis, the perturbation affects only the elements of the rth row of A and b. Consider the influence graph G_x corresponding to the matrix A and denote by I_k the set of nodes of G_x reachable from node r in k steps. We can write [see system (7.7)]

$$\delta\bar{x}_i > 0 \quad \text{for } i \in I = \{r\} \cup I_1 \cup I_2 \cup \ldots \cup I_{n-1}$$
$$\delta\bar{x}_i = 0 \quad \text{for } i \notin I = \{r\} \cup I_1 \cup I_2 \cup \ldots \cup I_{n-1}$$

By letting
$$\rho_i = \frac{\delta \bar{x}_i}{\bar{x}_i} \quad i = 1, 2, \ldots, n$$

we have $\rho_i = 0$ for $i \notin I$ and $\rho_i > 0$ for $i \in I$. We must now prove that

$$\rho_r \geq \rho_i \quad \forall i \in I$$

Assume, by contradiction, that the maximum value of ρ is obtained for some index $i \neq r$. From (7.7) it follows that

$$\delta \bar{x}_i = \sum_j a_{ij} \delta \bar{x}_j$$

and, from $\delta \bar{x}_j = \rho_j \bar{x}_j$,

$$\bar{x}_i = \sum_j \frac{\rho_j}{\rho_i} a_{ij} \bar{x}_j$$

Since $\rho_j \leq \rho_i$ and $\rho_r < \rho_i$ it follows that

$$\frac{\rho_r}{\rho_i} a_{ir} \bar{x}_r < a_{ir} \bar{x}_r$$

so that

$$\bar{x}_i < \sum_j a_{ij} \bar{x}_j$$

which contradicts the hypothesis

$$\bar{x}_i = \sum_j a_{ij} \bar{x}_j + b_i \bar{u}$$

$\diamond$

It is worth noting that *Theorem 22* does not provide any information about the absolute change of the state variables. In particular, it is possible that the state variable that is maximally affected in relative terms is not maximally affected in absolute terms.

PROBLEM 16 (A-II) *Each year, $\bar{u}$ males and females of a game population are stocked into a reserve, where they are able to breed but cannot survive > 1 year. In this reserve, hunters shoot only adult individuals and, without stocking, the game population would go extinct. Show that if the hunting licenses are decreased, the individuals that take most advantage in relative terms are the adult individuals. Show also that the young individuals can increase more than the adults.*

8
Reachability and Observability

This chapter is devoted, to a large extent, to the problem of controlling a positive system. We are therefore interested in characterizing the possibility of driving a system from a nonnegative initial state to a nonnegative final state. This means that we are dealing with reachability and controllability.

In a linear system with unconstrained input, the reachability set $X_r(t)$, that is the set of states reachable from the origin at time t, is a linear subspace and the set X_r of the states reachable in finite time is the subspace spanned by the first n reachability vectors $b, Ab, \ldots, A^{n-1}b$. The reachability set of a system with constrained inputs is, in general, a subset of such a subspace which, in the case of positive systems, is contained in the positive orthant $\mathbb{R}_+^n$.

The reachability sets $X_r(t)$ and X_r of a positive system have peculiar geometrical characteristics. First of all, such sets are *cones*, that is sets X with the property

$$x \in X \Rightarrow \alpha x \in X \quad \forall \alpha \geq 0$$

This property is a direct consequence of linearity. In fact, if a state $x \geq 0$ is reachable from the origin, then there exists a nonnegative input function $u(\cdot)$ that, applied to the system initially at rest, steers the state toward x. If we apply the nonnegative input function $\alpha u(\cdot)(\alpha \geq 0)$ to the same system, by linearity the state αx is reached in the same time interval.

In *Fig. 8.1*, three reachability cones in a three-dimensional space are shown. Such cones belong to the positive orthant and they are *solid* (*i.e.*, contain an open ball of $\mathbb{R}^n$) and *convex* (*i.e.*, the segment connecting two points of the cone belongs to the cone itself). Moreover, the cone (a) is polyhedral, that is, it has a finite number of edges.

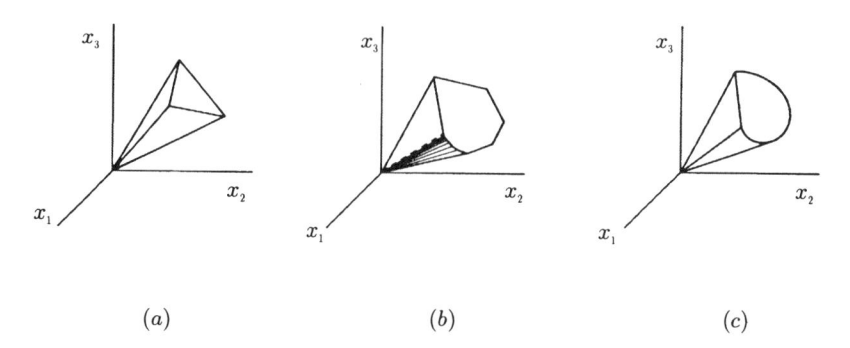

Figure 8.1 Three cones in $\mathbb{R}^3$.

The second and third cone are not polyhedral. In the sequel, we will show that the cones depicted in *Fig. 8.1* are common reachability cones of positive systems. More precisely, (a) and (b) are typical in discrete-time systems and (c) in continuous-time systems. We will also find conditions under which the reachability cone coincides or almost coincides with $\mathbb{R}^n_+$ (complete or almost complete reachability).

We now state the first result on reachability cones.

THEOREM 23 *(convexity and solidness of reachability cones)*

> The reachability set $X_r(t)$ of a positive system (A, b, c^T) is a convex cone of $\mathbb{R}^n_+$. Moreover, it is solid (for every $t \geq 0$ in the case of continuous-time systems and for $t \geq n$ in the case of discrete-time systems) if and only if the first n reachability vectors $b, Ab, \ldots, A^{n-1}b$ are linearly independent [*i.e.*, if and only if system (A, b, c^T) is reachable in the usual sense].

Proof. It has been already shown that $X_r(t)$ is a cone of $\mathbb{R}^n_+$. On the other hand, from the Lagrange formula (see Appendix B) $x(t) = \Phi(t)x(0) + \Psi(t)u_{[0,t)}(\cdot)$, it follows that the elements of $X_r(t)$ are vectors of the kind $\Psi(t)u_{[0,t)}(\cdot)$ with nonnegative $u_{[0,t)}(\cdot)$. In order to prove that the cone $X_r(t)$ is convex, we have to show that if $x' = \Psi(t)u'_{[0,t)}(\cdot)$ and $x'' = \Psi(t)u''_{[0,t)}(\cdot)$ are two points of the cone, then the segment connecting x' with x'' belongs to $X_r(t)$. For this, letting $\tilde{x}$ be a point of such a segment

$$\tilde{x} = \alpha x' + (1-\alpha)x'' \quad 0 \leq \alpha \leq 1$$

from linearity it follows that $\tilde{x} = \Psi(t)\tilde{u}_{[0,t)}(\cdot)$ with

$$\tilde{u}_{[0,t)}(\cdot) = \alpha u'_{[0,t)}(\cdot) + (1-\alpha)u''_{[0,t)}(\cdot)$$

We prove now that $X_r(t)$ is solid, for t sufficiently large, if and only if the vectors $b, Ab, \ldots, A^{n-1}b$ are linearly independent. To do this, suppose first that such vectors are linearly independent and consider a generic strictly positive function $u_{[0,t)}(\cdot)$ on the interval $[0,t)$ with $t \geq n$, in the case the system is discrete-time. Denote by x the state reached at t when applying the function $u_{[0,t)}(\cdot)$ to the system initially at rest, that is, $x = \Psi(t)u_{[0,t)}(\cdot)$. By definition $x \in X_r(t)$. Consider, then, a ball $S(0,\varepsilon)$ of $\mathbb{R}^n$ with radius ε centered in the origin of the state space. Because the vectors $b, Ab, \ldots, A^{n-1}b$ are linearly independent, the unconstrained system (A,b) is reachable. This means that each point of the ball $S(0,\varepsilon)$ can be reached from the origin (in at most n steps for discrete-time systems) by applying an appropriate input $\delta u_{[0,t)}(\cdot)$. From the superposition principle, by applying to the system initially at rest the input $u_{[0,t)}(\cdot) + \delta u_{[0,t)}(\cdot)$, a point in the ball $S(x,\varepsilon)$ with radius ε centered in x is reached. The norm of the function $\delta u_{[0,t)}(\cdot)$ can be reduced at will by decreasing ε, so that it is always possible to find an ε such that the function $u_{[0,t)}(\cdot) + \delta u_{[0,t)}(\cdot)$ is nonnegative. This means that $S(x,\varepsilon) \subset X_r(t)$ and, therefore, $X_r(t)$ is solid.

On the order hand, in the case $X_r(t)$ contains a ball $S(x,\varepsilon)$, there exist functions $\delta u_{[0,t)}(\cdot)$ able to drive the system from 0 to any point of $S(0,\varepsilon)$. This means that the unconstrained system (A,b) is completely reachable, so that the vectors $b, Ab, \ldots, A^{n-1}b$ are linearly independent.

$\diamond$

In order to determine the set X_r of the states reachable in finite time,

$$X_r = \cup_t X_r(t)$$

we need to know how the cone $X_r(t)$ changes in time. *Theorem 24* is an extension of a well-known property of the reachability subspaces of linear systems with unconstrained inputs.

THEOREM 24 *(expansion of the reachability cone $X_r(t)$)*

> The reachability cone $X_r(t)$ grows or remains the same as time goes on, that is, $X_r(t_2) \supset X_r(t_1)$ $\forall t_2 \geq t_1$. Moreover, if $X_r(t)$ stops growing for a certain time interval, it does it forever. Finally, for discrete-time systems $X_r(t)$ becomes solid and, at the same time, stops growing (at time $t = n$) if and only if the coefficients of the characteristic polynomial of A are nonpositive (*i.e.*, $\alpha_i \leq 0, i = 1, 2, \ldots, n$) and the vectors $b, Ab, \ldots, A^{n-1}b$ are linearly independent.

Proof. The proof of the first statement is obvious: If x is reachable at time t_1 with an input function $u(\cdot)$, it suffices to remain at the origin (with a zero input) for an interval of time $t_2 - t_1$, and then apply the input $u(\cdot)$ to reach x at time t_2.

We shall see now, in the case of discrete-time systems, that if there exists a time $t_2 > t_1$ for which $X_r(t_2) = X_r(t_1)$, then $X_r(t) = X_r(t_1), \forall t \geq t_1$. For this, first we write $x(t) = \Psi(t)u_{[0,t)}(\cdot)$ in the form

$$x(t) = \begin{pmatrix} b & Ab & \cdots & A^{t-1}b \end{pmatrix} u_0^{t-1}$$

where

$$u_0^{t-1} = (u(t-1) \quad u(t-2) \quad \ldots \quad u(0))^T$$

Since $u(t-1) \geq 0, u(t-2) \geq 0, \ldots, u(0) \geq 0$, the reachability cone $X_r(t)$ is the set of all nonnegative linear combinations of the first t reachability vectors $b, Ab, \ldots, A^{t-1}b$. From this it follows that, if $X_r(t_2) = X_r(t_1)$ with $t_2 > t_1$, the vectors $A^{t_1}b, A^{t_1+1}b, \ldots, A^{t_2-1}b$ are nonnegative linear combinations of $b, Ab, \ldots, A^{t_1-1}b$. Therefore, we can write

$$A^{t_1}b = \alpha_1' A^{t_1-1}b + \alpha_2' A^{t_1-2}b + \cdots + \alpha_{t_1}' b$$

where $\alpha_i' \geq 0, i = 1, 2, \ldots, t_1$. By multiplying both sides of the above expression by A, one obtains

$$\begin{aligned}A^{t_1+1}b &= \alpha_1' A^{t_1}b + \alpha_2' A^{t_1-1}b + \cdots + \alpha_{t_1}' Ab \\ &= \alpha_1'(\alpha_1' A^{t_1-1}b + \alpha_2' A^{t_1-2}b + \cdots + \alpha_{t_1}' b) + \alpha_2' A^{t_1-1}b + \cdots + \alpha_{t_1}' Ab \\ &= (\alpha_1'^2 + \alpha_2') A^{t_1-1}b + (\alpha_1'\alpha_2' + \alpha_3') A^{t_1-2}b + \cdots + \alpha_1'\alpha_{t_1}' b \\ &= \alpha_1'' A^{t_1-1}b + \alpha_2'' A^{t_1-2}b + \cdots + \alpha_{t_1}'' b\end{aligned}$$

where $\alpha_i'' \geq 0, i = 1, 2, \ldots, t_1$. The iteration of this procedure yields a sequence of coefficients $\alpha_i', \alpha_i'', \alpha_i''', \ldots$ representing the weights of the first t_1 reachability vectors that can be used to construct the reachability vectors $A^{t_1}b, A^{t_1+1}b, A^{t_1+2}b$, and so on. Since all these coefficients $\alpha_i', \alpha_i'', \alpha_i'''$, and so on, are nonnegative, the reachability cone remains unchanged, that is $X_r(t) = X_r(t_1), \forall t \geq t_1$.

In the case of continuous-time systems, we show now, using an example, that the cone $X_r(t)$ may actually grow for a certain time interval $[0, t^*]$, and then remain unchanged forever. For this, consider the expression

$$x(t) = \int_0^t e^{A(t-\xi)} bu(\xi) d\xi$$

which shows that $x(t)$ can be viewed as the weighted integral of the vectors $e^{A(t-\xi)}b$ with nonnegative weights $u(\xi)$. This implies that the reachability cone $X_r(t)$ is generated by the trajectory corresponding to the impulse response. *Figure 8.2* shows how the cone $X_r(t)$ may be constructed. In particular, *Fig. 8.2(a)* shows the trajectory corresponding to the impulse response $(e^{At}b)$ of a third-order system with a complex conjugate pair of eigenvalues, which must be subdominant, while *Fig. 8.2(b)* shows the projection (from the origin) on a plane (the *simplex* $x_1+x_2+x_3 = 1$) of such a trajectory. The line thus obtained is parameterized by t and the convex hull of the line on the simplex increases until time t^*. *Figure 8.2(c)* shows that the cone $X_r(t)$ has a flat face for $t \geq t^*$ [see *Fig. 8.1(c)*].

Consider, now, a discrete-time system whose coefficients of the characteristic polynomial $\alpha_i, i = 1, 2, \ldots, n$ are nonpositive and whose first n reachability vectors $b, Ab, \ldots, A^{n-1}b$ are linearly independent. From *Theorem 23*, the cone $X_r(t)$ gets

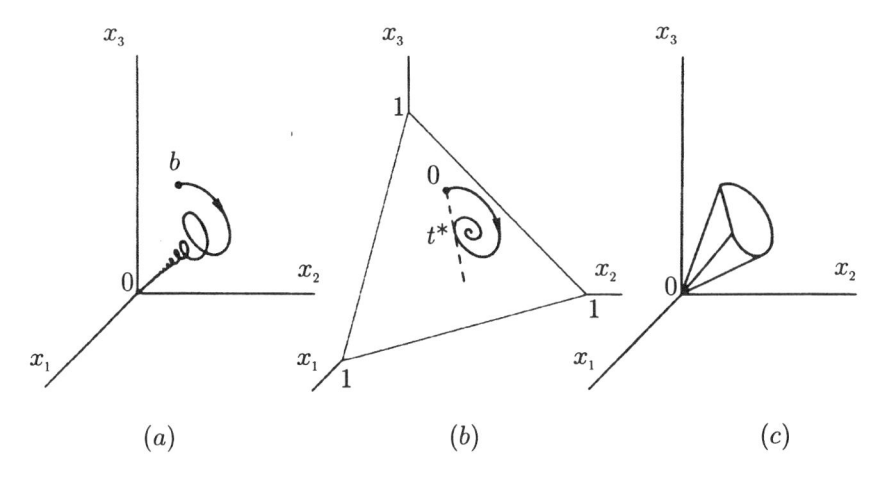

Figure 8.2 Construction of the cone $X_r(t)$ for a continuous-time positive system with a complex conjugate pair of eigenvalues: (a) trajectory corresponding to the impulse response; (b) projection on the simplex and determination of the instant t^*; (c) cone $X_r(t)$ for $t \geq t^*$.

solid at time $t = n$. On the other hand, from the Cayley–Hamilton theorem it follows that

$$A^n = -\alpha_1 A^{n-1} - \alpha_2 A^{n-2} - \cdots - \alpha_n I$$

is a nonnegative linear combination of the first n powers of A, so that also $A^n b$ is a nonnegative linear combination of the first n reachability vectors $b, Ab, \ldots, A^{n-1}b$. From the above considerations, we see that $X_r(n+1) = X_r(n)$ and consequently, $X_r(t) = X_r(n), \forall t \geq n$. If, on the other hand, the cone $X_r(t)$ becomes solid and stops growing at time t, this means that (see *Theorem 23*) $t = n$. If $X_r(n) = X_r(n+1)$ then

$$A^n b = a_1 A^{n-1} b + a_2 A^{n-2} + \ldots + a_n b$$

where $a_i \geq 0, i = 1, 2, \ldots, n$. From the Cayley-Hamilton theorem and from the linear independence of the vectors $b, Ab, \ldots, A^{n-1}b$, it follows that $a_i = -\alpha_i$, so that $\alpha_i \leq 0, i = 1, 2, \ldots, n$.

$\diamond$

It is worth noting that condition $\alpha_i \leq 0, i = 1, 2, \ldots, n$ considered in *Theorem 24*, may be, sometimes, checked by a simple inspection of the influence graph G_x. In fact, it can be shown that the non existence of disjoint cycles in the graph G_x implies $\alpha_i \leq 0, i = 1, 2, \ldots, n$ [to prove this, one may think to the denominator of Mason's formula (see Appendix B): $1 - \sum_i L_i + \sum_i \sum_j L_i L_j + \cdots$ that, in this case, contains only the sum of the transfer functions (L_i) of all the cycles]. For example, the influence graphs of *Fig. 4.4* do not posses disjoint cycles so that,

EXAMPLE 12 (medical center)

Suppose $u(t)$ individuals go to a medical center during the week t to ascertain whether they are affected by a certain disease or not. The test consists of blood drawn followed by an analysis, and has three outcomes: positive, negative, or ambiguous, in which case it must be repeated after 1 week of rest.

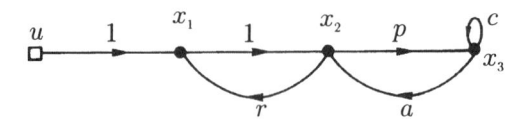

Figure 8.3 Influence graph G_{ux} of a medical center.

The system is described with the influence graph G_{ux}, of *Fig. 8.3*, where $x_1(t), x_2(t)$, and $x_3(t)$ represent the number of individuals waiting for the test, the number of individuals performing the test and the number of individuals under treatment (for whom the disease has been detected). The graph shows that a certain percentage a of the individuals under treatment need to perform the test (arc 3, 2) once more, while a percentage c of the individuals remain under treatment (obviously, $a + c \leq 1$).

The state equations of the system are

$$\begin{pmatrix} x_1(t+1) \\ x_2(t+1) \\ x_3(t+1) \end{pmatrix} = \begin{pmatrix} 0 & r & 0 \\ 1 & 0 & a \\ 0 & p & c \end{pmatrix} \begin{pmatrix} x_1(t) \\ x_2(t) \\ x_3(t) \end{pmatrix} + \begin{pmatrix} 1 \\ 0 \\ 0 \end{pmatrix} u(t)$$

where p is the probability that the test is positive and r is the probability that the test has to be repeated because ambiguous (obviously $p + r < 1$).

The reachability cone $X_r(t)$ represents, in this case, all possible populations of patients (x_1, x_2, x_3), which can be obtained after t weeks from the opening of the medical center. To determine $X_r(t)$ we evaluate the characteristic polynomial

$$\Delta_A(\lambda) = \det(\lambda I - A) = \lambda^3 - c\lambda^2 - (r + ap)\lambda + rc$$

Thus, $\alpha_1 \leq 0, \alpha_2 < 0$ and $\alpha_3 \geq 0$. The three reachability vectors

$$b = \begin{pmatrix} 1 \\ 0 \\ 0 \end{pmatrix} \quad Ab = \begin{pmatrix} 0 \\ 1 \\ 0 \end{pmatrix} \quad A^2 b = \begin{pmatrix} r \\ 0 \\ p \end{pmatrix}$$

are linearly independent for every value of a and c.

On the basis of *Theorems 23* and *24*, we can assert that the reachability cone $X_r(t)$ is solid from the third week on. Moreover, if $a = 1$ and $c = 0$, that is if all the individuals quit the treatment after 1 week and perform the test, we have $\alpha_i \leq 0, i = 1, 2, 3$, and the cone $X_r(t)$ remains unchanged from the third week on. Its three edges coincide with the first three reachability vectors, as highlighted in *Fig. 8.4 (a)* where a section of the cone $X_r(t), t \geq 3$ is depicted on the simplex. If, on the contrary, $c > 0$, that is if some people remain under treatment for more than one week, we have $\alpha_3 > 0$ and the reachability cone continues to grow after 3 weeks, as shown in *Fig. 8.4(b)* for $a = 0.5$ and $c = 0.4$. In this case, the cone $X_r(t)$ has t edges that accumulate as $t \to \infty$ toward the Frobenius eigenvector x_F of A.

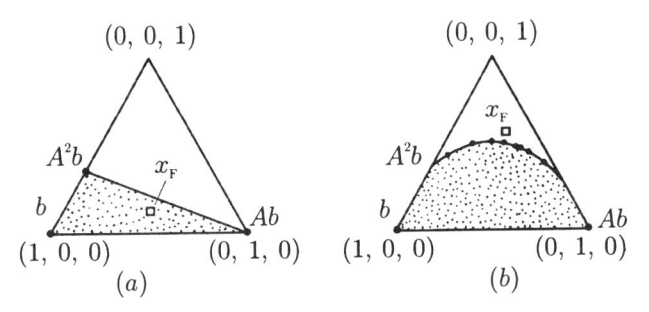

Figure 8.4 Sections of the cone $X_r(t)$ for $p = 0.4$ and $r = 0.5$: (a) $X_r(t)$ for $t \geq 3$ in the case $a = 1, c = 0$; (b) $X_r(12)$ for $a = 0.5, c = 0.4$.

Now, let us study the particular case in which the test is not repeated at the end of the treatment. Therefore, a is zero and c is positive (the bigger c is, the longer, on the average, the treatment lasts). As we shall see in the sequel (*Theorem 28*), the structure of the influence graph [see *Fig. 8.5(a)*] is particularly interesting. In this case, the cone $X_r(t)$ for $t \geq 4$ has only four edges determined by the first two reachability vectors b and Ab and by the last two $A^{t-2}b$ and $A^{t-1}b$. This is due to the fact that all the reachability vectors $A^t b$ with t even [odd] lie on the left [right] edge of the simplex and "climb" with t (easy to check). The system [see *Fig. 8.5(a)*] is composed of a second-order cyclic system (variables x_1 and x_2) and of a first-order primitive system (variable x_3). In fact, the matrix

$$A = \begin{pmatrix} 0 & r & 0 \\ 1 & 0 & 0 \\ 0 & p & c \end{pmatrix}$$

is in (block) triangular form and its eigenvalues are $\lambda_{1,2} = \pm\sqrt{r}, \lambda_3 = c$. Therefore, the Frobenius eigenvalue of the system is

$$\lambda_F = \max\{\sqrt{r}, c\}$$

In the case, $r > c^2$ the Frobenius eigenvalue is $\lambda_F = \sqrt{r}$ and the corresponding eigenvector

REACHABILITY AND OBSERVABILITY

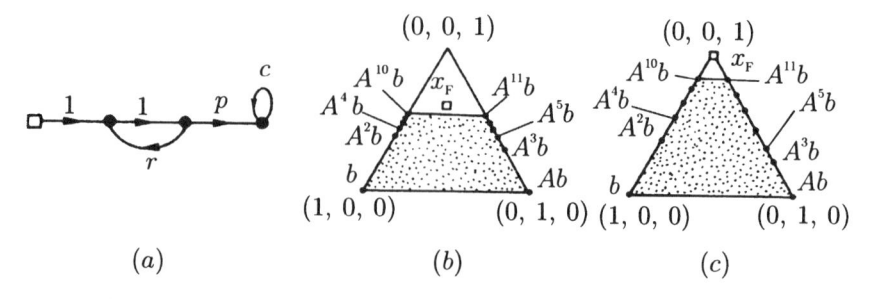

Figure 8.5 Influence graph G_{ux} for $a = 0$ (a) and sections of the cone $X_r(t)$ on the simplex for $p = 0.4$ and $r = 0.5$: (b) $X_r(12)$ for $a = 0, c = 0.6$, (c) $X_r(12)$ for $a = 0, c = 0.8$.

$$x_F = ((\sqrt{r} - c)\sqrt{r} \quad \sqrt{r} - c \quad p)^T$$

is represented by a point in the interior of the simplex. In the case $r \leq c^2$, we have $\lambda_F = c$ and the Frobenius eigenvector

$$x_F = (0 \quad 0 \quad 1)^T$$

is a vertex of the simplex. The reachability cones $X_r(t)$ are shown for $r > c^2$ and $r < c^2$ in Fig. 8.5 (b) and (c). In the first case, the reachability cone tends as $t \to \infty$ toward a cone with a quadrangular section, while in the second case $X_r(t)$ tends asymptotically toward the positive orthant (axis x_3 not included). It is, therefore, only in this latter case that each combination (x_1, x_2, x_3) with $x_i > 0, i = 1, 2, 3$ can be generated, by appropriately "feeding" the medical center.

♣

PROBLEM 17 *(T-II) Show that the reachability cone $X_r(t)$ of discrete-time systems may stop to grow at some instant $t^* > n$. In case one is not able to find a general justification, compute the reachability cone $X_r(t)$ of the system*

$$A = \begin{pmatrix} 0.2 & 0.1 & 0.2 \\ 1 & 0.2 & 0 \\ 0 & 1 & 0.6 \end{pmatrix} \quad b = \begin{pmatrix} 1 \\ 0 \\ 0 \end{pmatrix}$$

for $t = 1, 2, \ldots, 5$.

PROBLEM 18 *(T-III) Prove that a discrete-time positive system with linearly independent reachability vectors $b, Ab, \ldots, A^{n-1}b$ and $\alpha_i > 0, i = 2, 3, \ldots, n$, has a reachability cone $X_r(t)$, which continues to grow in time but always has n edges.*

What has been said until now enables us to better outline the nature of the reachability set X_r (the set of states reachable from the origin in finite time). Such a set is, by definition, given by $X_r = \cup_t X_r(t)$, so that, from *Theorem 24* one has that the closure $\bar{X}_r$ of X_r is given by

$$\bar{X}_r = \lim_{t \to \infty} X_r(t)$$

Therefore, X_r is a positive convex cone which, unlike $X_r(t)$, which is always closed, may be open. This happens when $X_r(t)$ continues to grow in time, as shown, for instance, in *Figs. 8.4* and *8.5*. The reachability cone X_r possesses several properties, some of which recall properties holding for linear systems with unconstrained inputs. *Theorem 25* is an example.

THEOREM 25 *(reachability of the equilibria)*

> The equilibria $\bar{x}$ of an asymptotically stable positive system are reachable from the origin in finite time (*i.e*, $\bar{x} \in X_r$).

Proof. For an asymptotically stable system with $x(0) = 0$ and constant input $\bar{u} > 0$, the state $x^*(t)$, which is an element of X_r, tends toward $\bar{x}$. By continuity, $\bar{x}$ must be either a point in the interior of X_r or a point on its boundary. The second hypothesis is false. In fact, perturbing from time t_1 onward the input $\bar{u}$ with an input $\delta u(t)$ (possibly negative) with $|\delta u(t)| \leq \bar{u}$, it is possible to obtain a perturbation $\delta x(t)$ on $x^*(t)$ in any desired direction in the subspace generated by the reachability vectors $b, Ab, \ldots, A^{n-1}b$ (where the cone X_r lies) and of a certain amplitude, say Δ. By considering that $\|x^*(t_1) - \bar{x}\|$ tends to zero as $t_1 \to \infty$, one can choose t_1 large enough such that $\|x^*(t_1) - \bar{x}\| < \Delta$. This fact would lead to the conclusion that, by applying a nonnegative input $\bar{u} + \delta u(t)$ to the system at rest, one could reach states not contained in the reachability set, which is a contradiction.
◇

Another property of the reachability cone X_r is described in *Problem 19*.

PROBLEM 19 *(T-III) Prove that positive systems with linearly independent reachability vectors $b, Ab, \ldots, A^{n-1}b$, have Frobenius eigenvectors contained in the closure of X_r (Hint: Treat the case of cyclic and primitive systems separately).*

Not all the properties holding for reachability subspaces of linear systems with unconstrained inputs have, however, a counterpart in the case of positive systems. For example, when no constraints are imposed on inputs, it is always possible to drive a linear system from one point to another of its reachability subspace in finite time. An analogous property does not hold for positive systems. In fact, if a continuous-time [discrete-time] system is positive and has, for example, all its eigenvalues with positive real part [modulus > 1] the free evolution $\Phi(t)x(0)$ is nonnegative but cannot tend toward zero for any initial condition $x(0) \geq 0$. Since the forced evolution $\Psi(t)u_{[0,t)}(\cdot)$ is nonnegative, the state $x(t) = \Phi(t)x(0) + \Psi(t)u_{[0,t)}(\cdot)$ of the system cannot be driven to zero. In other words, any displacement within the reachability cone may be only away from the origin. If, on the contrary, the system is asymptotically stable, any state of the reachability cone can be reached in finite time from any nonnegative initial state, as specified in *Theorem 26*, the proof of which is left to the reader.

THEOREM 26 *(reachability cone)*

> Any state of the reachability cone X_r of an asymptotically stable positive system with linearly independent reachability vectors $b, Ab, \ldots, A^{n-1}b$ can be reached in finite time from any state $x \geq 0$.

The reachability cone X_r (or its closure) may coincide with the positive orthant $\mathbb{R}_+^n$. In this case, a positive system will be said to be *completely reachable* or *almost completely reachable*, accordingly to *Definition 7*.

DEFINITION 7 *(complete and almost complete reachability)*

> A positive system is said to be *completely reachable* if and only if all states $x \geq 0$ are reachable in finite time from the origin, that is, if and only if $X_r = \mathbb{R}_+^n$. If all states $x \gg 0$ are reachable in finite time but some states $x > 0$ are not, the positive system is said to be *almost completely reachable*.

Completely reachable and almost completely reachable discrete-time systems are very particular and easily characterized in terms of their structure, as shown by the *Theorems 27* and *28*.

THEOREM 27 *(conditions for complete reachability)*

> A discrete-time positive system is completely reachable if and only if it is possible to reorder its state variables in such a way that the input u directly influences only x_1 and x_i directly influences only x_{i+1} for $i = 1, 2, \ldots, n-1$ (note that the influences of x_n do not play any role).

Before we report the proof of this theorem, note that a completely reachable discrete-time system has a very particular structure represented by the influence graph G_{ux} in *Fig. 8.6* or by the corresponding pair $(A^\sharp, b^\sharp)$

$$A^{\sharp^T} = \begin{pmatrix} 0 & 0 & \cdots & 0 & * \\ 1 & 0 & \cdots & 0 & * \\ 0 & 1 & \cdots & 0 & * \\ \vdots & \vdots & & \vdots & \vdots \\ 0 & 0 & \cdots & 1 & * \end{pmatrix} \quad b^\sharp = \begin{pmatrix} 1 \\ 0 \\ 0 \\ \vdots \\ 0 \end{pmatrix}$$

where $*$ in the last column of $A^{\sharp^T}$ may be either 0 or 1.

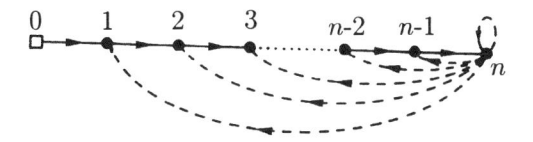

Figure 8.6 The influence graph of a completely reachable system (the dashed arcs may not exist).

Proof.
Sufficiency. It is readily seen that the vectors $b^\sharp, A^{\sharp T} b^\sharp, \ldots, A^{\sharp T^{n-1}} b^\sharp$ are the unit vectors $e_i, i = 1, 2, \ldots, n$. Therefore, the first n reachability vectors $b, Ab, \ldots, A^{n-1}$ lie on the axis $x_1, x_2, \ldots, x_n$ of the state space, so that $X_r(n) = \mathbb{R}^n_+$.
Necessity. If $X_r = \mathbb{R}^n_+$, there exist n reachability vectors $A^{k_1-1}b, A^{k_2-1}b, \ldots, A^{k_n-1}b$ with $1 \le k_1 < k_2 < \cdots < k_n$ lying on the axis of the state space. By reordering the state variables, we can suppose that $A^{k_i-1}b$ lies on the x_i axis, so that the vector $\left(A^{\sharp T}\right)^{k_i-1} b^\sharp$ is equal to $n_i e_i$, where n_i is a positive integer. Consequently, in the influence graph G_{ux} there exist n_i paths of length k_i from the input node 0 to node i, while paths of the same length from node 0 to nodes $j \ne i$ do not exist. This is clearly possible only if $n_i = 1$ and $k_i = i$, that is, if $\left(A^{\sharp T}\right)^{i-1} b^\sharp = e_i$, which implies that $A^\sharp$ and $b^\sharp$ are as in (8).
$\diamond$

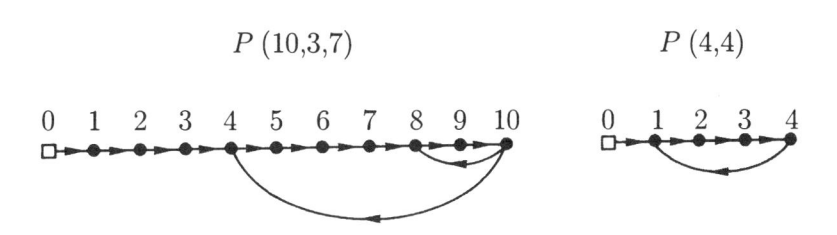

$P(10,3,7)$ $\qquad\qquad\qquad\qquad$ $P(4,4)$

Figure 8.7 Influence graphs of $P(10, 3, 7)$ and $P(4, 4)$.

Theorem 27 allows one to verify by inspection of the influence graph G_{ux}, whether a system is completely reachable or not. For example, only the first system of *Fig. 4.4* is completely reachable. On the contrary, it is not so simple to check whether a system is almost completely reachable or not since, as shown by the following theorem, this property does not depend solely on the influence graph. Before we state the theorem on almost complete reachability, we note that a completely reachable system (see *Fig. 8.6*) may be "encoded" by an integer n, which represents its dimension and by a set (possibly empty) of positive integers $\tau_1 < \tau_2 <$ and so on representing the lengths of the simple cycles in the influence graph. As a

76 REACHABILITY AND OBSERVABILITY

$$P(n - \tau,\ h_1\tau,\ h_2\tau,\ldots)$$

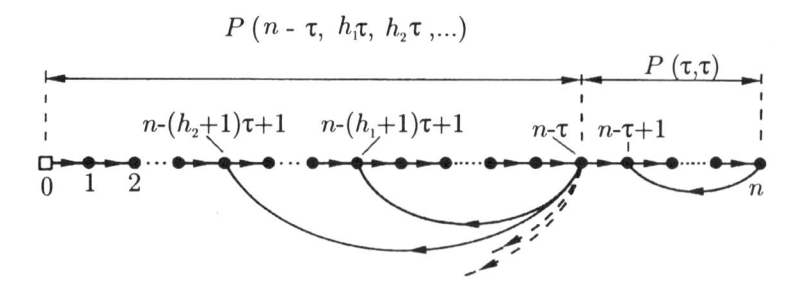

$$P(\tau,\tau)$$

$n-(h_2+1)\tau+1 \quad n-(h_1+1)\tau+1 \quad n-\tau \ \vert\ n-\tau+1$

0 1 2 .... n

Figure 8.8 Influence graph G_{ux} of a system $\sum$ composed of the series connection of $P(n - \tau, h_1\tau, h_2\tau,$ etc.) with $P(\tau, \tau)$.

consequence, any completely reachable system can be denoted by $P(n, \tau_1, \tau_2,$ etc.). For example, $P(n)$ is a system composed of a cascade of n delays. Two examples, the influence graph of $P(10, 3, 7)$ and that of $P(4, 4)$, are depicted in *Fig. 8.7*.

We now show that an almost completely reachable system is composed of the series connection of two particular completely reachable systems.

THEOREM 28 *(conditions for almost complete reachability)*

> A discrete-time positive system $\sum$ of order n is almost completely reachable if and only if the following conditions hold:
>
> (a) $\sum$ is composed of the series connection of a system $P(n - \tau, h_1\tau, h_2\tau, \ldots)$ (h_i is a positive integer and $h_1 < h_2 < \ldots$) and a system $P(\tau, \tau)$
>
> (b) the Frobenius eigenvalue of $P(\tau, \tau)$ is greater than or equal to the Frobenius eigenvalue of $P(n - \tau, h_1\tau, h_2\tau, \ldots)$.

Before proving this theorem, note that only condition (*a*) is a structural one. It states that the influence graph G_{ux} of $\sum$ is as in *Fig. 8.8*.

Proof.

Sufficiency. Consider *Fig. 8.8* where the system composed of the series connection of $P(n - \tau, h_1\tau, h_2\tau, \ldots)$ with $P(\tau, \tau)$ is shown. To facilitate the understanding of the proof, imagine that the influence graph is an electrical circuit and that each node corresponds to a bulb switched on or off during each elementary time interval t. Thus, the state vector $x(t)$ corresponds to an electrical configuration: $x_i(t) > 0$ means that bulb i is on during the interval t and $x_i(t) = 0$ means that bulb i is off during the same interval. A state transition $x(t) \to x(t+1)$ corresponds to a change of configuration in the circuit. The transitions $x(t) \to x(t+1) = Ax(t)$ are particularly simple to interpret on the influence graph since the following rule can be applied: if bulb i is on during interval t then in the subsequent interval bulb j is on if the arc (i, j) is present in the graph. If, for example, at time t bulbs $1, n - \tau$

and n, are on, at time $(t+1)$ bulbs $2, n-(h_1+1)\tau+1, n-(h_2+1)\tau+1, \ldots$ and $(n-\tau+1)$ are on.

The reachability vectors $b, Ab, A^2b, \ldots$ which identify the reachability cone can be determined by switching on bulb 1 alone (this configuration corresponds to the vector $x(1) = b$) and by allowing the circuit to evolve freely. It is clear, then, that in the first $(n-\tau)$ time intervals only one bulb will be on: at each transition one bulb turns off and the subsequent one is switched on. This means that the reachability vectors $b, Ab, \ldots, A^{n-\tau-1}$ have only one nonzero component and that they lie on the axis $x_1, x_2, \ldots, x_{n-\tau}$ of the state space (note that the i-th unit vector $x \equiv e_i$ corresponds to "only the bulb i is on"). Let us now continue our study of the circuit starting from the time at which only the bulb $n\tau$ is on. At the subsequent instant of time we find that more than one bulb is on, precisely the $n-\tau+1$-th, the $n-(h_1+1)\tau+1$-th, the $n-(h_2+1)\tau+1$-th, ... Thus, the corresponding reachability vector $A^{n-\tau}$ is not proportional to a unit vector. Moreover, it is clear from the graph in *Fig. 8.8*, that from this time onward more than one bulb will always be on.

To prove that the system is almost completely reachable we have to show that the vectors $A^t b$ tend, as $t \to \infty$, to align with the remaining unit vectors $e_{n-\tau+1}, e_{n-\tau+2}, \ldots, e_n$. We use, now, hypothesis (b) and, for simplicity, we assume that the Frobenius eigenvalue of $P(\tau, \tau)$ is greater than the Frobenius eigenvalue of $P(n-\tau, h_1\tau, h_2\tau, \ldots)$ (the analysis of the case in which the two Frobenius eigenvalues are equal is left to the reader). Under this assumption, the dynamics of the first subsystem $[P(n-\tau, h_1\tau, h_2\tau, \ldots)]$ is dominated, in the long run, by the one of the second subsystem $[P(\tau, \tau)]$. This can be, again, interpreted in electrical terms by saying that, as time goes on, the bulbs of the first subsystem shine less and less brightly with respect to those of the second one. In other words, the light intensity of the bulbs of the first part of the circuit gradually vanishes so that, in the long run, only the bulbs of the second circuit are, practically, on. To complete the proof it remains to verify that two (or more) bulbs of the second subsystem cannot be on at the same time. This is guaranteed by the fact that the cycles of the first subgraph have a length $h_i\tau$ which is a multiple of the length of the cycle of the second subsystem. In other words, the bulb $n-\tau$ is on only when the n-th bulb is on, and this implies that only the bulb $n-\tau+1$ is on in the second circuit during the next time interval.

Necessity. Instead of giving a complete proof, we will only show, using the previously used "enlightening" technique, that complete reachability is lost when the structure of the graph in *Fig. 8.8* is modified.

In fact, the addition of an arc from $n-\tau$ to $j < n\tau$ creating a cycle of a length which is not a multiple of τ, implies that, soon or late, bulb $n-\tau$ will switch on at time instants different from $n+k\tau, k=1,2,\ldots$: Therefore, soon or late, two bulbs of the second circuit will be on at the same time so that the vectors $A^t b$ will not tend to align with the unit vectors $e_i, i \geq n-\tau+1$. On the other hand, if we delete arc $(n-\tau, n-\tau+1)$ and lump together nodes $(n-\tau)$ and $(n-\tau+1)$, the behavior of the first subsystem will be influenced by that of the second one, and the brightness of the bulbs of the first circuit will not asymptotically vanish.

If we add an arc between the graphs of the two subsystems, the added bulb i^* will switch on at the same time of the last one, so that none of the reachability vectors $A^t b$ will tend to align with the unit vectors e_i. Finally, if the Frobenius eigenvalue of $P(\tau, \tau)$ would be smaller than that of $P(n - \tau, h_1 \tau, h_2 \tau, \ldots)$, the brightness of the bulbs of the second circuit would asymptotically vanish and the vectors $A^t b$ would not tend to align with the unit vectors.

◇

On the basis of *Theorem 28* it is immediate to recognize whether a given system has or not the structure required by almost complete reachability. For example, the third-order system depicted in *Fig. 8.3* can be almost completely reachable only if $a = 0$ since, in that case, it is composed of $P(2, 2)$ in cascade with $P(1, 1)$. For the system to be almost completely reachable, it is also necessary that the Frobenius eigenvalue of $P(1, 1)$ be greater than or equal to the Frobenius eigenvalue of $P(2, 2)$. This is consistent with *Fig. 8.5*, which shows two reachability cones, where only the second one corresponds to an almost completely reachable system.

Theorems 27 and *28* on complete and almost complete reachability, concern discrete-time systems. The reachability cone of continuous-time systems is not, in general, polyhedral, and hence cannot coincide with the positive orthant $\mathbb{R}_+^n$. This is not true for second-order systems, which can be completely reachable as shown in *Example 13*.

EXAMPLE 13 (river pollution)
The most popular model for describing river pollution is, still now, the model proposed by Streeter and Phelps in 1925. The model is the following

$$\dot{x}_1(t) = -k_1 x_1(t) + u(t)$$
$$\dot{x}_2(t) = k_1 x_1(t) - k_2 x_2(t)$$

where x_1 and x_2 are, respectively, the concentration of biodegradable matter contained in the water and the so-called *dissolved oxygen deficit* (DOD), that is, the difference between the highest concentration of dissolved oxygen in water and the actual one (measured in [mg/l]). In the Streeter and Phelps model, the concentration of biodegradable matter is, for simplicity and by convention, expressed in terms of *biochemical oxygen demand* (BOD). By definition, a unit of BOD is the concentration of biodegradable matter requiring one unit of oxygen to be degraded by bacteria. This is the reason why the same term appears in both state equations with an opposite sign. The term $-k_2 x_2$ in the second equation represents reaeration, namely, the diffusion of oxygen from the air into the water body, which takes place through the surface of the river at a rate proportional to the oxygen deficit. The "time" t is the so-called flow time, that is, the time needed for each drop of water to reach the generic section starting from the initial one. In other words, t can be interpreted as the "distance" of the generic section of the river from a reference upstream section (see *Fig. 8.9*).

PROPERTIES   79

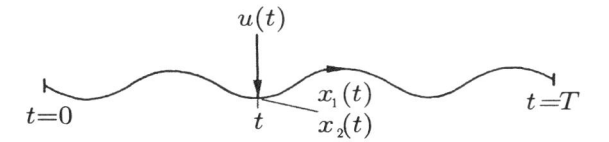

Figure 8.9 A river stretch between an upstream section $t = 0$ and a downstream section $t = T$.

Thus $x_1(t)$ and $x_2(t)$ are BOD and DOD concentrations in section t and $u(t)dt$ is the BOD discharged into the river from section t to section $t + dt$ [$u(t)$ is an impulse function if a "lumped" discharge is present in section t].

The Streeter and Phelps model

$$A = \begin{pmatrix} -k_1 & 0 \\ k_1 & -k_2 \end{pmatrix} \quad b = \begin{pmatrix} 1 \\ 0 \end{pmatrix}$$

is a positive system (A is a Metzler matrix). The reachability cone $X_r(t)$ is the set of all pairs of BOD and DOD that can be generated in section t by distributed discharges along the river when the water is pure ($x_1 = 0$) and saturated by oxygen ($x_2 = 0$) in the upstream section. Analogously, the cone X_r represents all the pairs (x_1, x_2) that can be obtained in the sections very far from the upstream one. The reachability cones $X_r(t)$ and X_r can be easily obtained since it is easy to determine the trajectory corresponding to the impulse response of the system. In fact, in case $k_2 > k_1$ (reaeration constant greater than biodegradation constant) $\lambda_F = -k_1$, and $x_F = (k_2 - k_1 \quad k_1)^T$, while in the opposite case, $k_1 > k_2$, one has $\lambda_F = -k_2$ and $x_F = (0 \quad 1)^T$. The trajectories corresponding to the impulse responses are, therefore, the ones shown in *Fig. 8.10*.

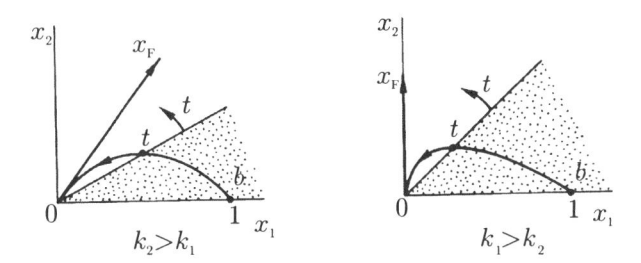

Figure 8.10 Reachability cone $X_r(t)$ as t varies. In case $k_1 > k_2$ the system is almost completely reachable.

In both cases, the trajectories tend toward the origin and are tangent to the Frobenius eigenvectors x_F. Therefore, the reachability cone X_r is strictly contained in $\mathbb{R}_+^2$ in case $k_2 > k_1$, while in the opposite case, the system is almost completely reachable. This means that water quality is less "variable" when substances are less easily biodegradable.

♣

What has been said about reachability of positive systems can be expressed in terms of observability using the duality principle, which states that a system $\Sigma = (A, b, c^T)$ is reachable if and only if its dual system $\Sigma^* = (A^T, c, b^T)$ is observable. The observability set $X_o^\Sigma(t)$ of a system Σ is, therefore, equal to the reachability set $X_r^{\Sigma^*}(t)$ of the dual system Σ^*. From this it follows (see *Theorem 23*) that $X_o^\Sigma(t)$ is a positive convex cone that is solid if and only if the first n observability vectors $c, A^T c, \ldots, (A^T)^{n-1} c$ are linearly independent. A system Σ is completely observable if and only if (see *Theorem 27*) its influence graph G_{xy} is like that shown in *Fig. 8.6* with reversed arrows and the input node replaced by the output one. A result analogous to *Theorem 28* holds for almost complete observability.

It is worth noting that it is possible to prove that the set X_y^+ of all initial states $x(0)$ giving rise to nonnegative free output evolutions, that is, $X_y^+ = \{x(0) : c^T \Phi(t) x(0) \geq 0, \forall t\}$, coincides with the observability cone X_o. This implies that a positive system is completely observable if and only if the initial states that generate nonnegative free output evolutions are the states of $\mathbb{R}_+^n$.

9
Realization

In Chapter 2, we defined external positivity (*Definition 1*), and then we showed (*Theorem 1*) that a system is externally positive if and only if its impulse response $g(t)$ is nonnegative. The external positivity of a system is, therefore, a property of its input–output relationship for zero initial state (impulse response, transfer function, reduced ARMA model). This means that, the knowledge of the coefficients α_i and β_i, $i = 1, \ldots, n$ of the reduced ARMA model of a discrete-time system

$$y(t) = \sum_{i=1}^{n}(-\alpha_i)y(t-i) + \sum_{i=1}^{n}\beta_i u(t-i)$$

or of a continuous-time system

$$y^{(n)}(t) = \sum_{i=1}^{n}(-\alpha_i)y^{(n-i)}(t) + \sum_{i=1}^{n}\beta_i u^{(n-i)}(t)$$

is, in principle, sufficient to ascertain whether the system is externally positive or not. The same holds, obviously, if one knows the transfer function

$$G(p) = \frac{\beta_1 p^{n-1} + \beta_2 p^{n-2} + \cdots + \beta_n}{p^n + \alpha_1 p^{n-1} + \cdots + \alpha_n}$$

of the system, which is determined by the same coefficients α_i and β_i of the ARMA model or by the so-called Markov coefficients g_i, $i = 1, 2$, and so on, which uniquely identify the transfer function

$$G(p) = g_1 p^{-1} + g_2 p^{-2} + \ldots = \sum_{i=1}^{\infty} g_i p^{-i}$$

82 REALIZATION

of the system. The Markov coefficients g_i can be obtained from the coefficients α_i and β_i by dividing the polynomial $(\beta_1 p^{n-1} + \cdots + \beta_n)$ by the polynomial $(p^n + \alpha_1 p^{n-1} + \cdots + \alpha_n)$. For the first n Markov coefficients $g_1, \ldots, g_n$, the following holds:

$$\begin{aligned}
\beta_1 &= g_1 \\
\beta_2 &= g_2 + \alpha_1 g_1 \\
\beta_3 &= g_3 + \alpha_1 g_2 + \alpha_2 g_1 \\
&\vdots \\
\beta_n &= g_n + \alpha_1 g_{n-1} + \cdots + \alpha_{n-1} g_1
\end{aligned} \tag{9.1}$$

which can be written as

$$\beta = Mg$$

with

$$M = \begin{pmatrix} 1 & 0 & 0 & \cdots & 0 & 0 \\ \alpha_1 & 1 & 0 & \cdots & 0 & 0 \\ \alpha_2 & \alpha_1 & 1 & \cdots & 0 & 0 \\ \vdots & \vdots & \vdots & & \vdots & \vdots \\ \vdots & \vdots & \vdots & & \vdots & \vdots \\ \alpha_{n-1} & \alpha_{n-2} & \alpha_{n-3} & \cdots & \alpha_1 & 0 \end{pmatrix}$$

In short, we can assert that an externally positive system has its $2n$ coefficients (α_i, β_i), $i = 1, \ldots, n$ or, equivalently, its $2n$ coefficients $(\alpha_i, g_i), i = 1, \ldots, n$, which satisfy suitable conditions that guarantee the nonnegativity of the impulse response.

In the course of Chapter 2, we also defined the (internal) positivity of a system (A, b, c^T) and we noted that a positive system is always externally positive. Moreover, we noted that there are externally positive systems that cannot be internally positive no matter how the basis of the state space is chosen. This means that the *realization* problem [*i.e.*, the determination of a triple (A, b, c^T) of a given transfer function $G(p)$] is far from being trivial in the case where the triple (A, b, c^T) is constrained to be that of an internally positive system. If this constraint would not exist, then the realization problem would be trivial, since many "canonical" realizations of a transfer function are known as, for example, the control canonical form

$$A_c = \begin{pmatrix} 0 & 1 & 0 & \cdots & 0 \\ 0 & 0 & 1 & \cdots & 0 \\ \vdots & \vdots & \vdots & & \vdots \\ 0 & 0 & 0 & \cdots & 1 \\ -\alpha_n & -\alpha_{n-1} & -\alpha_{n-2} & \cdots & -\alpha_1 \end{pmatrix} \quad b_c = \begin{pmatrix} 0 \\ 0 \\ \vdots \\ 0 \\ 1 \end{pmatrix}$$

$$c_c^T = \begin{pmatrix} \beta_n & \beta_{n-1} & \beta_{n-2} & \cdots & \beta_1 \end{pmatrix}$$

the reconstruction canonical form

$$A_r = \begin{pmatrix} 0 & 0 & \cdots & 0 & -\alpha_n \\ 1 & 0 & \cdots & 0 & -\alpha_{n-1} \\ \vdots & \vdots & & \vdots & \vdots \\ 0 & 0 & \cdots & 1 & -\alpha_1 \\ 0 & 1 & \cdots & 0 & -\alpha_1 \end{pmatrix} \qquad b_r = \begin{pmatrix} \beta_n \\ \beta_{n-1} \\ \beta_{n-2} \\ \vdots \\ \beta_1 \end{pmatrix}$$

$$c_r^T = \begin{pmatrix} 0 & 0 & \cdots & 0 & 1 \end{pmatrix}$$

or the Markov form

$$A_M = \begin{pmatrix} 0 & 0 & \cdots & 0 & -\alpha_n \\ 1 & 0 & \cdots & 0 & -\alpha_{n-1} \\ 0 & 1 & \cdots & 0 & -\alpha_{n-2} \\ \vdots & \vdots & & \vdots & \vdots \\ 0 & 0 & \cdots & 1 & -\alpha_1 \end{pmatrix} \qquad b_M = \begin{pmatrix} 0 \\ 0 \\ 0 \\ \vdots \\ 0 \\ 1 \end{pmatrix}$$

$$c_M^T = \begin{pmatrix} g_1 & g_2 & \cdots & g_{n-1} & g_n \end{pmatrix}$$

and its dual

$$A_{M^*} = \begin{pmatrix} 0 & 1 & 0 & \cdots & 0 \\ 0 & 0 & 1 & \cdots & 0 \\ \vdots & \vdots & \vdots & & \vdots \\ 0 & 0 & 0 & \cdots & 1 \\ -\alpha_n & -\alpha_{n-1} & -\alpha_{n-2} & \cdots & -\alpha_1 \end{pmatrix} \qquad b_{M^*} = \begin{pmatrix} g_1 \\ g_2 \\ \vdots \\ g_{n-1} \\ g_n \end{pmatrix}$$

$$c_{M^*}^T = \begin{pmatrix} 1 & 0 & 0 & \cdots & 0 \end{pmatrix}$$

A transfer function $G(p)$ is *positively realizable* if it admits a realization (A, b, c^T) with b and c^T nonnegative and A nonnegative, in the discrete-time case, or Metzler in the continuous-time case. The external positivity condition, that is, the nonnegativity of the impulse response $g(t)$, is a necessary condition for the existence of a positive realization since, as just stated, a positive system is always externally positive.

It is worth noting that the realization problem does not necessarily ask for the triplet (A, b, c^T) to be of order n. Once specific requirements on the properties of the triplet are made (such as positivity) it is natural to ask whether considering realizations of dimension $N > n$ allows us to fulfill such requirements. We will see in the sequel that the answer is often positive, even if there exist cases in which increasing the order of the realization does not give any advantage, as shown in *Problem 20*.

84 REALIZATION

PROBLEM 20 *(T-III) Prove, by appropriately choosing the parameters, that the system with transfer function*

$$G(s) = \frac{\beta_1 s^2 + \beta_2 s + \beta_3}{(s-\alpha)(s-\alpha-i\omega)(s-\alpha+i\omega)}$$

may be externally positive while no positive realization exists for such a transfer function, no matter how large its dimension N is.

It is straightforward to verify that all the first-order ($n = 1$) externally positive systems admit a positive realization of the same dimension. The same is true for second-order systems, that is, all transfer functions

$$G(p) = \frac{\beta_1 p + \beta_2}{p^2 + \alpha_1 p + \alpha_2}$$

of externally positive systems are realizable with a second-order positive system (A, b, c^T). We show how this can be proven in the discrete-time case. Consider the following two realizations of $G(z)$. The first is in dual Markov form

$$A_{M^*} = \begin{pmatrix} 0 & 1 \\ -\alpha_2 & -\alpha_1 \end{pmatrix} \quad b_{M^*} = \begin{pmatrix} g_1 \\ g_2 \end{pmatrix}$$
$$c_{M^*}^T = \begin{pmatrix} 1 & 0 \end{pmatrix}$$

where, from (9.1) $g_1 = \beta_1$ and $g_2 = \beta_2 - \alpha_1\beta_1$. The second is the following triplet [the reader can easily verify that $c^T(zI - A)^{-1}b = G(z)$]

$$A = \begin{pmatrix} p_2 & 0 \\ \beta_2 + \beta_1 p_1 & p_1 \end{pmatrix} \quad b = \begin{pmatrix} 1 \\ 0 \end{pmatrix}$$
$$c^T = \begin{pmatrix} \beta_1 & 1 \end{pmatrix}$$

where p_1 and p_2 are the two poles of the system.

External positivity implies that g_1 and g_2 are nonnegative and that the dominant pole, say p_1, is positive and greater than or equal to p_2 in modulus (*i.e.*, $p_1 \geq |p_2| > 0$). Since $\alpha_1 = -p_1 - p_2$, we conclude that $\alpha_1 \leq 0$. Moreover, external positivity also implies

$$\beta_2 + \beta_1 p_1 > 0$$

[the reader can prove this inequality taking into account that the impulse response of a second-order positive system is of the form $g(t) = r_1 p_1^{t-1} + r_2 p_2^{t-1}$ so that for large values of t the sign of $g(t)$ is determined by the sign of r_1]. Thus in conclusion, the first realization is positive if $\alpha_2 \leq 0$ while the second one is positive if $p_2 \geq 0$. But $\alpha_2 = p_1 p_2$ with $p_1 > 0$ so that one of the two realizations is always positive.

Problem 20 shows, without ambiguity, that what is true for $n = 1$ and 2 does not hold, in general, for $n = 3$. In the case $n \geq 3$, external positivity is not a sufficient condition for the existence of a positive realization. A first result, which allows us to construct an interesting class of positive realizations of dimension $N = n$, is *Theorem 29*.

THEOREM 29 (*positive realization of $G(z)$ in Markov canonical form*)

> The transfer function $G(z)$ of an externally positive discrete-time system with $\alpha_i \leq 0, i = 1, 2, \ldots, n$ is realizable with a positive system of dimension n in Markov canonical form (A_M, b_M, c_M^T) or in its dual form $(A_{M^*}, b_{M^*}, c_{M^*}^T)$. The first one of these two realizations is completely reachable and the second one is completely observable.

Proof. For an externally positive discrete-time system, we have $g_i \geq 0, i = 1, 2, \ldots$ and the conditions $\alpha_i \leq 0, i = 1, 2, \ldots, n$ guarantees that the Markov canonical realization and its dual are positive. On the other hand, the system (A_M, b_M, c_M^T) is completely reachable since its influence graph G_{ux} satisfies the conditions of *Theorem 27*. By duality, the system $(A_{M^*}, b_{M^*}, c_{M^*}^T)$ is completely observable.
◇

It is worth noting that a transfer function $G(z)$ with $\alpha_i \leq 0, \beta_i \geq 0, i = 1, 2, \ldots, n$, can be positively realized in control [reconstruction] canonical form. However, there exist transfer functions with $\alpha_i \leq 0, i = 1, 2, \ldots, n$ and $\beta_i < 0$ for some i, which have $g_i \geq 0, i = 1, 2, \ldots, n$ so that they are positively realizable in Markov canonical form but not in control [reconstruction] canonical form.

As for continuous-time systems, we must keep in mind that once a positive realization of $G(s)$ is known, a positive realization of $g(s - \sigma)$ with σ real can be easily found. In fact, if (A, b, c^T) is a positive realization of $G(s)$, one has $G(s) = c^T(sI - A)^{-1}b$ and $G(s - \sigma) = c^T[(s - \sigma)I - A]^{-1}b = c^T[sI - (A + \sigma I)]^{-1}b$ so that $[(A + \sigma I), b, c^T]$ is a positive realization of $G(s - \sigma)$. By using this property, *Theorem 30* can be easily proved:

THEOREM 30 (*positive realization of $G(s)$*)

> The transfer function $G(s)$ of an externally positive continuous-time system is realizable with a positive system of dimension n if, for some σ, the Markov parameters $g_i(\sigma)$ of $G(s - \sigma)$ are nonnegative and the coefficients $\alpha_i(\sigma)$ are nonpositive.

Obviously, the realization is not completely reachable [observable] because, in general, the reachability [observability] cone of a continuous-time system is not polyhedral. Let us now state a rather general result relative to the case of transfer functions with a unique (possibly multiple) dominant pole.

THEOREM 31 (*positive realization of $G(z)$ with unique dominant pole*)

> The transfer function $G(z)$ of an externally positive discrete-time system with a unique (possibly multiple) dominant pole is realizable with a positive system (A, b, c^T) of dimension $N \geq n$.

For simplicity, we omit the proof of this theorem (the interested reader will find useful information in the *Annotated Bibliography* in this book). However, a

86   REALIZATION

characteristic of the proof of *Theorem 31* is that the dimension N of the positive realization is not derived in a straightforward manner. For this reason, we present a sufficient condition for positive realizability of a transfer function in which the dimension N of the positive realization can be obtained very easily. For this we first need *Definition 8*.

DEFINITION 8 *(Markov cone $\mathcal{C}_M$ of $G(p)$)*

> The cone $\mathcal{C}_M$ generated by the vectors
>
> $$m_i = \begin{pmatrix} g_i \\ g_{i+1} \\ \vdots \\ g_{i+n-1} \end{pmatrix} \quad i = 1, 2, \ldots$$
>
> is called the *Markov cone* of a system described by the transfer function $G(p) = \sum_{i=1}^{\infty} g_i p^{-i}$.

It is important to note that the reachability cone X_r of a discrete-time system in dual Markov form $(A_{M^*}, b_{M^*}, c_{M^*}^T)$ coincides with its Markov cone and is invariant w.r.t. A_{M^*}, that is, $A_{M^*} X_r \subseteq X_r$. Moreover, if the system is externally positive, the Markov cone is contained in $\mathbb{R}_+^n$. We are now ready to state *Theorem 32*.

THEOREM 32 *(positive realization of $G(z)$ when $\mathcal{C}_M$ has N edges)*

> The transfer function $G(z)$ of an externally positive discrete-time system with a Markov cone $\mathcal{C}_M$ with N edges is realizable with a completely (or almost completely) reachable positive system (A, b, c^T) of dimension N. The realization (A, b, c^T) can be computed by solving the following linear equations:
>
> $$A_{M^*} P = PA \quad Pb = b_{M^*} \quad c^T = c_{M^*}^T P \qquad (9.2)$$
>
> where the columns of the $n \times N$ matrix P are the N edges of $\mathcal{C}_M$ and $(A_{M^*}, b_{M^*}, c_{M^*}^T)$ is the realization of $G(z)$ in dual Markov form.

Proof. The triple (A, b, c^T) satisfying Eqs. (9.2) is a realization of $G(z)$ since its Markov parameters are the same as those of $(A_{M^*}, b_{M^*}, c_{M^*}^T)$. In fact,

$$c_{M^*}^T A_{M^*}^{i-1} b_{M^*} = c_{M^*}^T A_{M^*}^{i-2} A_{M^*} b_{M^*} = c_{M^*}^T A_{M^*}^{i-2} A_{M^*} Pb$$
$$= c_{M^*}^T A_{M^*}^{i-2} PAb = c_{M^*}^T A_{M^*}^{i-3} PA^2 b = \cdots = c_{M^*}^T PA^{i-1} b = c^T A^{i-1} b$$

The triplet (A, b, c^T) has the following properties:

  i. $a \geq 0$ since the Markov cone $\mathcal{C}_M$ is invariant w.r.t. A_{M^*}.

  ii. $b \geq 0$ since $b = (1 \ 0 \ \ldots \ 0)^T$ (b_{M^*} being proportional to the first vertex m_1 of $\mathcal{C}_M$, that is, the first column of P).

iii. $c \geq 0$ since c^T coincides with the first row of P and $P \geq 0$.

Hence, (A, b, c^T) is a positive realization of dimension N. Moreover, its reachability cone is the nonnegative orthant. Consequently, (A, b, c^T) is completely (or almost completely) reachable.

$\Diamond$

EXAMPLE 14 (charge routing networks)
A *charge routing network* (CRN) is a MOS-based technology for digital signal filtering and consists of a collection of *storage cells* (locations where a packet of charge can be stored and maintained isolated from the others) and of a specific periodical *routing procedure* (operations involving the packets of charge stored in the cells). The basic operations consist in applying a charge packet to a storage cell such that the packet's size is proportional to a given positive voltage amplitude (*injection*), in splitting the charge packet of a cell into positive components and transferring these parts into distinct empty cells (*transfer and splitting*), in combining charge packets from different cells and transferring them simultaneously into the same cell (*addition*), and in emptying a cell by removing its charge packet from the network (*extraction*). Every CRN can be completely described by an oriented graph with nodes representing cells and arcs indicating routes of charge transfer between cells. A positive weight is associated with each arc and indicates the fraction of the charge of the starting node transferred along that route. Obviously, a CRN is a discrete-time positive system.

Suppose now one wants to design a CRN that "approximates" the electrical network depicted in *Fig. 9.1* (also considered in *Problem 1*) and described by

$$A = \begin{pmatrix} -\dfrac{1}{R_1 C_1} & 0 & -\dfrac{1}{C_1} \\ 0 & -\dfrac{1}{R_2 C_2} & 0 \\ \dfrac{1}{L} & 1 & 0 \end{pmatrix} \qquad b = \begin{pmatrix} \dfrac{1}{C_1} \\ \dfrac{1}{C_2} \\ 0 \end{pmatrix}$$

$$c^T = \begin{pmatrix} 1 & 1 & 0 \end{pmatrix}$$

Since the element a_{13} of the matrix A is negative for every value of C_1, the network is not a positive system. However (see *Problem 1*) for suitable parameter values the network is externally positive. For example, this is true for $L = C_1 = C_2 = R = 1$ and $R_2 = 10$ and in such a case the transfer function is

$$G(s) = \frac{2s^2 + 1.1s + 1}{s^3 + 1.1s^2 + 1.1s + 0.1}$$

A simple and straightforward way of approximating the given electrical network with a discrete-time system is to "replace" the derivative operator with a (forward) first-order difference

88 REALIZATION

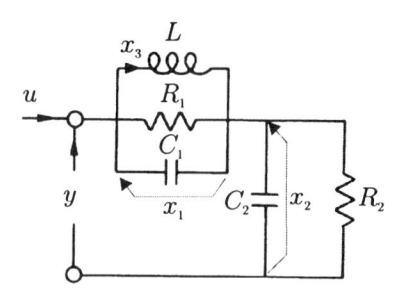

Figure 9.1 Nonpositive electrical network.

$$\dot{x}(t)_{t=kT} \approx \frac{x[(k+1)T] - x(kT)}{T}$$

where T is a "short" time interval. If this approximation is used the impulse response of the resulting discrete-time system is identified (not difficult to check) by the following Markov coefficients

$$g_i = T^i c^T \left(A + \frac{1}{T}I\right)^{i-1} b$$

Thus, in order to find a positive realization, one can use *Theorem 32* and determine the Markov cone $\mathcal{C}_M$, which is depicted in *Fig. 9.2* for $T = 2/3$.

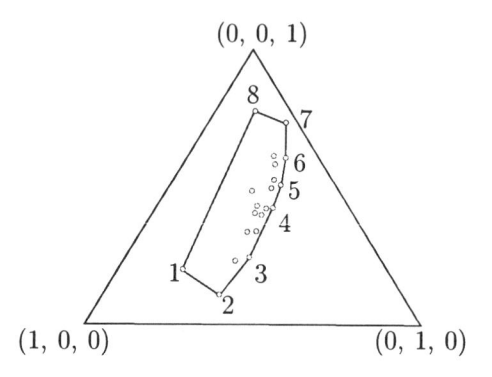

Figure 9.2 Section of the Markov cone $\mathcal{C}_M$ on the simplex. The circles represent the first 20 vectors m_i and those labeled $1, \ldots, 8$ are $m_1, \ldots, m_8$.

This figure clearly shows that the Markov cone has $N = 8$ edges, so that we can apply *Theorem 32*. Then, by solving system (9.2) one obtains a positive realization with $N = 8$, which can be synthesized with a CRN.

Finally, *Theorem 31* can be extended to the continuous-time case provided the impulse response $g(t)$ is positive for $t > 0$. The result is *Theorem 33*.

THEOREM 33 *(positive realization of $G(s)$ with unique dominant pole)*

> The transfer function $G(s)$ of an externally positive continuous-time system with impulse response $g(t) > 0$ for $t > 0$ and a unique (possibly multiple) dominant pole is realizable with a positive system (A, b, c^T) of finite dimension $N \geq n$.

10
Minimum Phase

The *minimum phase* of a linear system Σ (not necessarily a positive one) is a property concerning the zeros of its transfer function. More precisely, consider the transfer function $G(p)$

$$G(p) = \rho \frac{\prod_{i=1}^{n-h}(p - z_i)}{\prod_{i=1}^{n}(p - p_i)}$$

where ρ is the *transfer constant* and $z_i, i = 1, \ldots, n - h$ and $p_i, i = 1, \ldots, n$ are the *zeros* and the *poles*. Then, a system is said to be *minimum phase* if and only if all its zeros are stable, that is, if and only if

$\text{Re}(z_i) < 0, i = 1, \ldots, n - h$ for continuous-time systems

$|z_i| < 1, i = 1, \ldots, n - h$ for discrete-time systems

Moreover, a system with no zeros is minimum phase.

Minimum phase systems are, loosely speaking, more regular than all other systems, from now on called *nonminimum phase systems*, which indeed possess a number of peculiar properties, some of which are recalled hereafter.

a. The step response [*i.e.*, the response $y(t)$ due to the input $u(t) = 1 \; \forall t \geq 0$ and $x(0) = 0$], of an externally stable continuous-time [discrete-time] system with a unique unstable zero, changes its sign after some time.

b. A nonminimum phase system has nonvanishing (bounded or unbounded) *hidden inputs*, that is, inputs that produce an identically zero output when applied to the system in an appropriate initial state.

92 MINIMUM PHASE

c. It is not possible to *asymptotically reconstruct* the input of a nonminimum phase system from a record of its output, that is, it is not possible to obtain an estimate $\hat{u}(t)$ of $u(t)$ with $\hat{u}(t) \to u(t)$ as $t \to \infty$.

Moreover, nonminimum phase systems are particularly hard to control. For example, the error due to an additive disturbance on the output of a nonminimum phase system cannot be compensated through a standard feed-forward control scheme. Also, other regulation schemes as, for example, those suggested by adaptive control theory, are not applicable to systems that are not minimum phase. We shall not study these problems, which are beyond the scope of this book.

Property a can be understood by verifying that the first nonzero Markov coefficient g_h represents the first nonzero value of the step response of a discrete-time system and the first nonzero derivative of the step response of a continuous-time system for $t = 0$. In fact, in a discrete-time system the step response is given by $y(0) = 0$ and $y(t) = c^T(b + Ab + \cdots + A^{t-1}b) = \sum_{i=1}^{t} g_i, t = 1, 2, \ldots$. Analogously, in a continuous-time system with $x(0) = 0$, one has $y(0) = 0, y^{(t)}(0) = c^T A^{t-1} b = g_t$. Moreover, from (9.1) it follows that $g_h = \beta_h$, so that we can conclude that the step response has, initially, the sign of β_h, that is, the sign of the transfer constant ρ, which equals β_h. On the other hand, if the system is externally stable, as assumed in a, its step response tends asymptotically toward

$$\bar{y} = \rho \frac{\prod_{i=1}^{n-h}(-z_i)}{\prod_{i=1}^{n}(-p_i)}$$

for continuous-time systems, and toward

$$\bar{y} = \rho \frac{\prod_{i=1}^{n-h}(1 - z_i)}{\prod_{i=1}^{n}(1 - p_i)}$$

for discrete-time systems. In both cases, the denominator is positive (due to external stability) and the numerator is negative, if there is only one unstable zero that is positive in the continuous-time case and > 1 in the discrete-time case. Therefore, $\bar{y}$ has the sign of $-\rho$ so that the step response tends toward a positive [negative] value if it is initially negative [positive]. Clearly, this cannot happen in a positive system that has, necessarily, a nonnegative step response. Thus, we conclude that an externally positive and stable continuous-time [discrete-time] system cannot have a unique unstable zero that is > 0 [> 1]. Actually, as stated by the following theorem, one can prove more than this, namely, that an externally positive and stable continuous-time [discrete-time] system cannot have real unstable zeros ≥ 0 [≥ 1].

THEOREM 34 *(zeros of a positive system)*

> The real zeros of a positive system are smaller than the Frobenius eigenvalue.

Proof. We report the proof for continuous-time systems and leave to the reader its extension to the discrete-time case.

Consider the transfer function

$$G(s) = \int_0^\infty g(t)e^{-st}dt$$

of the system, and recall that the impulse response $g(t)$ is nonnegative (see *Theorem 1*). Thus, for s real and greater than or equal to the real part of the dominant pole, which coincides with the radius of convergence of the Laplace transform, $G(s)$ is positive because it is the integral of the product of a positive and a nonnegative function. Therefore, real zeros greater than or equal to the real part of the dominant pole do not exist. Since the latter is smaller than or equal to the Frobenius eigenvalue, *Theorem 34* follows.

◊

Let us now make a few remarks for a better understanding of *Theorem 34*.

The output $y(\cdot)$ of a system initially at rest may be evaluated using the convolution formula

$$y(t) = \int_0^t g(t-\tau)u(\tau)d\tau$$

where $u(\cdot)$ is the input function and $g(\cdot)$ is the impulse response of the system [*i.e.*, the inverse Laplace transform of $G(s)$]. If the system is externally positive, its impulse response is nonnegative (see *Theorem 1*). Therefore, $y(t)$ cannot tend to zero if $u(t) \geq \alpha > 0\ \forall t$. On the other hand, the ARMA model of the system is

$$[\rho(s-z_1)(s-z_2)\ldots(s-z_{n-h})]u(t) = [(s-p_1)(s-p_2)\ldots(s-p_n)]y(t) \quad (10.1)$$

where the variable s should be formally interpreted as the derivative operator w.r.t. time. Suppose, then, that there exists a nonnegative real zero, $z_1 \geq 0$, and consider the input $u(t) = e^{z_1 t}$. Since $(s-z_1)e^{z_1 t} = 0$, the left-hand side of (10.1) is identically zero so that the output of the system is the solution of the differential equation

$$[(s-p_1)(s-p_2)\ldots(s-p_n)]y(t) = 0$$

with initial conditions $y(0), y^{(1)}(0)$ and so on corresponding to a zero initial state and $u(t) = e^{z_1 t}$. Such conditions are given by

$$\begin{aligned}
y(0) &= 0 \\
\dot{y}(0) &= c^T b u(0) = g_1 u(0) = g_1 \\
\ddot{y}(0) &= c^T A b u(0) + c^T b \dot{u}(0) = g_2 u(0) + g_1 \dot{u}(0) = g_2 + z_1 g_1 \\
\dddot{y}(0) &= c^T A^2 b u(0) + c^T A b \dot{u}(0) + c^T b \ddot{u}(0) = g_3 u(0) + g_2 \dot{u}(0) + g_1 \ddot{u}(0) = \\
&\quad g_3 + z_1 g_2 + z_1^2 g_1
\end{aligned}$$

⋮

and can therefore be obtained by evaluating the Markov coefficients g_i from α_i and β_i identifying the transfer function. However, the computation of these initial conditions is not even necessary, since in the case of external stability of the system one can conclude that for any initial condition $y(t) \to 0$ as $t \to \infty$. This is in contradiction with the fact already established that $y(t)$ cannot tend to zero with a persistent input. Thus, in conclusion, the hypothesis of existence of a real nonnegative zero z_1 is absurd. In other words, continuous-time externally stable positive systems cannot have real nonnegative zeros, as implied by *Theorem 34*. Nevertheless, *Theorem 34* does not imply that externally stable positive systems are minimum phase. In fact, they can have unstable complex zeros (obviously in conjugate pairs) and, in the discrete-time case, they can also have real zeros smaller than or equal to -1. An important example of this last possibility is given by sampled-data systems. In fact, by sampling a continuous-time system (not necessarily a positive one) with transfer function $G(s)$ characterized by $(n-h)$ zeros $z_1, \ldots, z_{n-h}$ and n poles $p_1, \ldots, p_n$, one obtains a discrete-time system having a transfer function $\tilde{G}(z)$ with $n-1$ zeros $\tilde{z}_1, \ldots, \tilde{z}_{n-h}, \tilde{z}_{n-h+1}, \ldots, \tilde{z}_{n-1}$ and n poles $\tilde{p}_1, \ldots, \tilde{p}_n$. If T is the sampling period, the poles $\tilde{p}_i$ are given by

$$\tilde{p}_i = e^{p_i T} \quad i = 1, \ldots, n$$

and therefore tend to unity for $T \to 0$. As for the zeros, the first $(n-h)$ of them correspond to the $(n-h)$ zeros $z_1, \ldots, z_{n-h}$ of $G(s)$ but are not easily computable. Nevertheless, they also tend to unity for $T \to 0$ since

$$\tilde{z}_i \to e^{z_i T} \quad \text{as} \quad T \to 0 \quad i = 1, \ldots, n-h$$

The above expression shows that positive real zeros of the continuous-time system ($z_i > 0$) are transformed into unstable zeros $\tilde{z}_i > 1$ of the sampled data system if the sampling period is sufficiently small. The remaining $(h-1)$ zeros are a "byproduct" of the sampling operation and one can prove that as $T \to 0$, that is, for high-frequency sampling, these zeros tend toward "universal" zeros, that is, zeros independent of the transfer function, which are hereafter reported for $h = 2, \ldots, 5$

$h = 2$ $\tilde{z}_{n-1} = -1$

$h = 3$ $\tilde{z}_{n-2} = -3.732$ $\tilde{z}_{n-1} = -0.268$

$h = 4$ $\tilde{z}_{n-3} = -1$ $\tilde{z}_{n-2} = -9.899$ $\tilde{z}_{n-1} = -0.101$

$h = 5$ $\tilde{z}_{n-4} = -2.322$ $\tilde{z}_{n-3} = -23.20$ $\tilde{z}_{n-2} = -0.431$ $\tilde{z}_{n-1} = -0.043$

It is readily seen that all the "universal" zeros are real and negative and that for $h \geq 3$ at least one of them is > 1 in modulus. This means that, when high-frequency sampling is performed on a continuous-time system with an excess of poles $h \geq 3$, one obtains a sampled-data system that is a nonminimum phase (in the case $h = 2$, one can prove that, for positive systems, the universal unstable zero located at -1 is approached, for $T \to 0$, from the interior of the stability

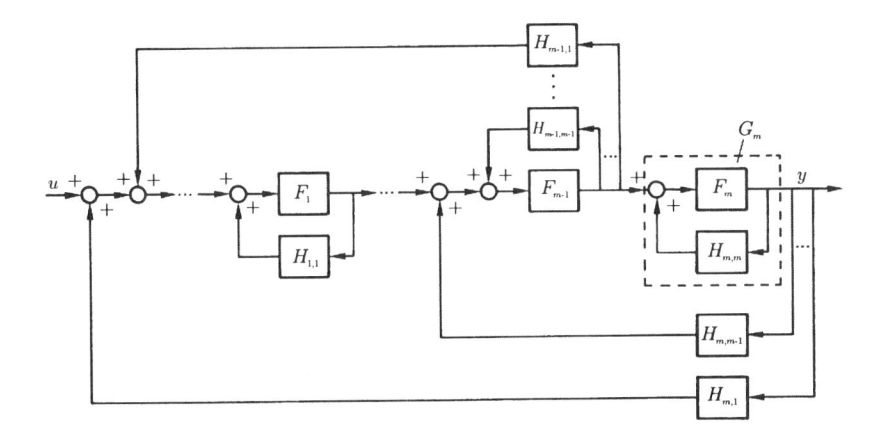

Figure 10.1 Block diagram of a system Σ_m having the same transfer function as a positive system Σ with a unique direct input–output path in its influence graph.

region so that the phase properties of the continuous-time system are preserved in the sampled-data system). This result is consistent with *Theorem 34*, which does not exclude that discrete-time stable positive systems have negative unstable zeros.

Obviously, for some particular classes of systems a number of specific properties dealing with minimum phase may hold and some of them will be reported in the chapters devoted to applications. The only general property known until now is *Theorem 35*.

THEOREM 35 (*sufficient condition for minimum phase*)

> An asymptotically stable positive system with an influence graph containing a unique input–output path is minimum phase.

Proof. For simplicity, given a positive system Σ, denote by $1, 2, \ldots, m$ the nodes of the influence graph G that compose the direct input–output path and consider a pair (i, j) of such nodes with $i < j$. Due to the uniqueness of the path, the subgraph G_{ij} of G, which is the influence graph of the subsystem with input i and output j, contains arcs of the direct path $(1 \to 2 \to \ldots \to m)$. In contrast, there exist subgraphs G_{ji} with $i \leq j \leq m$, which do not contain arcs of the direct input–output path. The transfer function of system Σ is therefore equal to the transfer function of the system Σ_m represented by a block diagram in *Fig. 10.1*.

In such a figure, the transfer function of the subsystem $\Sigma_{ji}, i \leq j \leq m$, corresponding to the influence graph G_{ji}, is indicated by $H_{ji}(p)$, while

$$F_i(p) = \frac{1}{p - a_{ii}}$$

96 MINIMUM PHASE

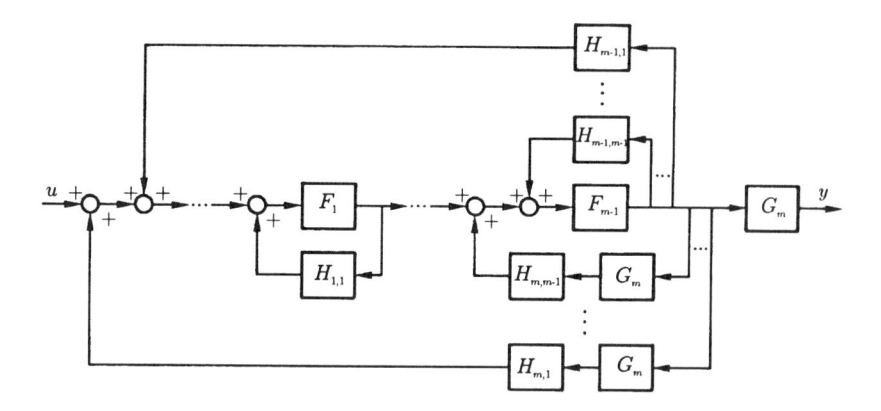

Figure 10.2 Block diagram equivalent to that of *Fig. 10.1*.

denotes the transfer function of the first-order system $\dot{x}_i(t) = a_{ii}x_i(t) + u_i(t)$ or $x_i(t+1) = a_{ii}x_i(t) + u_i(t)$. Although system Σ_m has, by construction, a greater dimension than system Σ, it has the same transfer function. Since system Σ is, by assumption, asymptotically stable, it is also connectively stable. Hence, all its subsystems are stable, so that all transfer functions $H_{ji}(p)$ and $F_i(p)$ have "stable" poles. Moreover, the transfer functions $F_i(p)$ are minimum phase (since they have no zeros). These two properties of the system depicted in *Fig. 10.1*, that is,

a. External stability of all the subsystems.

b. Minimum phase of all the subsystems composing the direct input–output path

are the only properties that will be used in the sequel.

To prove the theorem, we can proceed recursively through a series of successive elaborations of the block diagram of *Fig. 10.1*, which leave the transfer function of the system unchanged. For this, let us first replace the two blocks with transfer functions $F_m(p)$ and $H_{mm}(p)$ with a single block with transfer function

$$G_m(p) = \frac{F_m(p)}{1 - F_m(p)H_{mm}(p)}$$

thus obtaining the block diagram depicted in *Fig. 10.2*.

Since the zeros of $G_m(p)$ are the zeros of $F_m(p)$ and the poles of $H_{mm}(p)$, it follows from properties *a* and *b* that $G_m(p)$ is a minimum phase. On the other hand, from the connective stability of Σ it follows that the system with transfer function $G_m(p)$ is also externally stable. Consequently, the transfer function of the system depicted in *Fig. 10.2* is the same as that depicted in *Fig. 10.3*, provided that

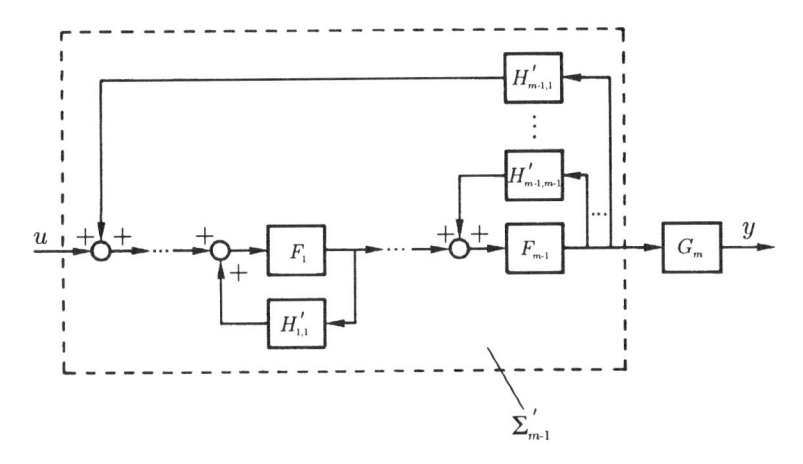

Figure 10.3 Block diagram equivalent to those of *Figs. 10.1* and *10.2*.

$$H'_{ji}(p) \doteq H_{ji}(p) \quad i \leq j \leq m-1$$
$$H'_{m-1,i}(p) = H_{m-1,i}(p) + H_{mi}(p)G_m(p) \quad i = 1, 2, \ldots, m-1$$

As is well known, these relationships (series and parallel interconnections) preserve external stability.

We can then conclude that system Σ_m in *Fig. 10.1* has the same transfer function as the system represented in *Fig. 10.3*, which is composed of the series interconnection of two subsystems: the first (Σ'_{m-1}) is structurally identical to Σ_m and the second (G_m) is externally stable and minimum phase. Moreover, properties *a* and *b* do hold for system Σ'_{m-1}. The only difference between Σ'_{m-1} and Σ_m is the length of the direct input–output path, which is $m-1$, instead of m. We can then apply to the block diagram of *Fig. 10.3* the same operations that have allowed us to transform *Fig. 10.1* into *Fig. 10.3*, and proceed like that until the whole system reduces to the series interconnection of m externally stable and minimum-phase subsystems, with transfer functions $G_i(p), i = 1, 2, \ldots, m$. Thus, the transfer function of the whole system $G(p) = \prod_{i=1}^{m} G_i(p)$ cannot have unstable zeros, since all the zeros of the transfer functions $G_i(p)$'s are stable.

$\diamondsuit$

Theorem 35 implies that the check of the minimum phase of a given system can be often performed very simply, namely, by inspection of the influence graph. For example, noting that asymptotically stable discrete-time systems that are completely reachable (see *Fig. 8.6*), or almost completely reachable with output $y = x_n$, have an influence graph with only one input–output path, we can immediately argue that such systems are minimum phase.

98 MINIMUM PHASE

EXAMPLE 15 (distillation column)
Real-time operation of distillation columns requires careful attention because such chemical plants are influenced by many parameters that randomly vary in time, like, for example, environmental temperature. In principle, one could try to control a distillation column by using standard feed-forward control schemes or adaptive feedback schemes, with time-varying gains able to adapt to the current parametric conditions. But, as already said, these control schemes are effective only if the plant is minimum phase. This is why it is important to know whether a distillation column is minimum phase or not. In this context, *Theorem 35* allows us to conclude that the simplest distillation columns, namely, the so-called plate distillation columns (see *Fig. 10.4*) are minimum phase provided that their hydraulic regime is stationary.

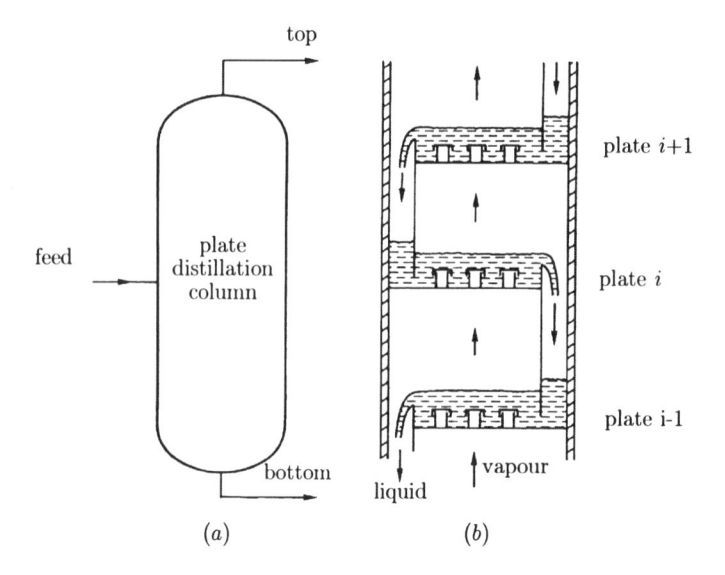

Figure 10.4 Plate distillation column: (a) feed, top and bottom; (b) direction of liquid and vapor flows in three successive plates.

In fact, in such conditions, the volume of liquid present on each plate is constant, so that the concentration $x_i(t)$ of the mixture on plate i can be described by a single differential equation (mass balance of the most volatile component), and this differential equation can be linearized around its equilibrium solution. The whole column is then modeled by a positive system in which each state variable x_i influences both the successive one (x_{i+1}) and the preceding one (x_{i-1}), because of the opposite directions of the liquid and vapor flows. Therefore, in the influence graph, which is not even worth showing, there exists a unique direct input–output path from the feed plate to the top plate of the distillation column. Since, for physical reasons, the chemical plant is obviously asymptotically stable, the conditions of *Theorem 35* are satisfied and one can conclude that a plate distillation column with a stationary hydraulic regime is a positive minimum phase system.

♣

On the basis of what was previously found, we can now make some considerations on property b relative to the existence of persistent (bounded or unbounded) hidden inputs in nonminimum phase systems. To this end assume, for simplicity, that the zeros are distinct and recall that if z_i is a zero of a continuous-time [discrete-time] linear system (not necessarily positive), the input $u(t) = e^{z_i t}$ $[u(t) = z_i^t]$ produces an output that is identically zero when applied to the system in suitable initial conditions. From linearity, we can then conclude that the output of a continuous-time [discrete-time] linear system corresponding to an initial state $x(0)$ and to an input $u(t)$, remains unchanged if we add to the input $u(t)$ the input $e^{z_i t}$ $[z_i^t]$ and we appropriately perturb the initial state. This means that the presence of the input $e^{z_i t}$ $[z_i^t]$ is not detectable by looking at the output of the system and, for this reason, such an input is called a hidden input. From the above considerations, it follows that hidden inputs associated with stable zeros tend asymptotically to zero, while the others are persistent and bounded if $z_i = 0$ $[|z_i| = 1]$ or unbounded if $z_i > 0$ $[|z_i| > 1]$. However, in view of *Theorem 34*, in the case of positive systems, persistent (bounded or unbounded) hidden inputs are, necessarily, associated with complex conjugate pairs of zeros with $\text{Re}(z_i) \geq 0$ $[|z_i| \geq 1]$ or to real zeros $z_i \leq -1$ in the case of discrete-time systems. In both cases, the hidden inputs have an oscillatory behavior. For example, in a continuous-time system with a complex zero $\alpha + i\omega$, and the hidden input $e^{\alpha t} \sin \omega t$ changes its sign each $2\pi/\omega$ units of time. We can then conclude that the nonminimum phase and stable positive systems have persistent hidden inputs that necessarily change sign over time. In other words, nonnegative persistent hidden inputs do not exist in positive systems.

The presence of hidden inputs with oscillatory behavior must be kept in mind when trying to reconstruct the input of a system from its recorded output. Normally, this operation is performed using the ARMA model of the system. For example, in the case of discrete-time systems, if β_h is the first nonzero β_i, the ARMA model

$$y(t)+\alpha_1 y(t-1)+\cdots+\alpha_n y(t-n) = \beta_h u(t-h)+\beta_{h+1} u(t-h-1)+\cdots+\beta_n u(t-n)$$

can be solved with respect to $u(t-h)$, thus yielding

$$u(t-h) = \frac{1}{\beta_h}[z(t) - \beta_{h+1} u(t-h-1) - \cdots - \beta_n u(t-n)] \qquad (10.2)$$

with $z(t) = y(t) + \alpha_1 y(t-1) + \cdots + \alpha_n y(t-n)$. If a recorded sequence $y(0), y(1), y(2)$, and so on, of the output of the system is available, it is straightforward to construct the sequence $z(n), z(n+1), z(n+2)$, and so on, and then compute the input sequence $u(n-h), u(n-h+1), u(n-h+2)$, and so on, through (10.2) provided that the first $(n-h)$ inputs $u(0), u(1), \ldots, u(n-h-1)$

are known. If these inputs are not known, the only thing we can do is to set them, in a somewhat arbitrary way, equal to some estimates $\hat{u}(0), \hat{u}(1), \ldots, \hat{u}(n-h-1)$. Thus (10.2), when used recursively, yields the following estimates of the input:

$$\hat{u}(t-h) = \frac{1}{\beta_h}[z(t) - \beta_{h+1}\hat{u}(t-h-1) - \cdots - \beta_n \hat{u}(t-n)] \quad t \geq n \quad (10.3)$$

Obviously, the estimate $\hat{u}(t-h)$ differs, in general, from the true value $u(t-h)$ of the input variable and the difference [reconstruction error $e(t-h)$] is a linear combination of all hidden inputs. Therefore, if all the zeros are stable (*i.e.*, if the system is minimum phase) the reconstruction error tends asymptotically to zero since all hidden inputs tend to zero. In contrast, if the system is a nonminimum phase, the reconstructed input $\hat{u}(t-h)$ does not converge to the real input $u(t-h)$. However, in the case of positive systems, the reconstruction dynamics will be quite peculiar. In fact, the reconstruction error $e(t-h)$ is in this case a linear combination of vanishing hidden inputs (associated with the stable zeros) and oscillatory persistent hidden inputs (associated with unstable zeros). Thus, when reconstructing the input of a nonminimum phase positive system, one should expect that, sooner or later, the reconstructed input $\hat{u}$ becomes negative. If this does not happen, one can be sure that the reconstructed input $\hat{u}$ tends asymptotically to the input of the system. Finally, it is worth noting that the rate of convergence is determined by the "dominant pole" of model (10.3), that is, by the root of maximum modulus of the equation

$$\lambda^{n-h} + \frac{\beta_{h+1}}{\beta_h}\lambda^{n-h-1} + \cdots + \frac{\beta_n}{\beta_h} = 0$$

which is the "dominant zero" of the system. In other words, if z_1 is the dominant zero, the reconstruction error tends to zero with a geometrical law of the type $|z_1|^t$.

11
Interconnected Systems

A positive system Σ with input $u(t)$ and output $y(t)$ is often composed of q interconnected positive subsystems Σ_i with input $\nu_i(t)$ and output $z_i(t)$. If the system is discrete-time, each subsystem Σ_i is described by the equations

$$x_i(t+1) = A_i x_i(t) + b_i \nu_i(t) \tag{11.1}$$

$$z_i(t) = c_i^T x_i(t) \tag{11.2}$$

where $x_i(t) \in \mathbb{R}^{n_i}, \nu_i(t) \in \mathbb{R}$ and $z_i(t) \in \mathbb{R}$.

The interconnections define the dependency of the input $\nu_i(t)$ of each Σ_i upon the outputs $z_1(t), z_2(t), \ldots, z_q(t)$ of the subsystems and upon the input $u(t)$ of Σ, as well as the dependency of the output $y(t)$ of Σ upon the outputs $z_i(t)$ of the subsystems Σ_i. In other words, letting

$$v(t) = (\nu_1(t) \quad \nu_2(t) \quad \ldots \quad \nu_q(t))^T$$
$$z(t) = (z_1(t) \quad z_2(t) \quad \ldots \quad z_q(t))^T$$

one has

$$v(t) = \Omega z(t) + \gamma u(t) \tag{11.3}$$

$$y(t) = \Psi^T z(t) \tag{11.4}$$

where Ω in a $q \times q$ matrix and γ and Ψ are q-dimensional vectors. The triplet (Ω, γ, Ψ^T) can be represented by means of a graph, called the *interconnection graph*.

102 INTERCONNECTED SYSTEMS

EXAMPLE 16 (the Italian educational system)
The Italian scholastic system is composed of four stages ["elementari" (E), "medie" (M), "superiori" (S) and "università" (U)]. The third stage has two options ["liceo" (L) and "tecnico" (T)]. Each of the five schools can be represented by a positive system with input $\nu_i(t)$ and output $z_i(t)$ given, respectively, by the number of newly enrolled individuals in years t and by the number of graduates in the same year. The structure of the scholastic system is shown in *Fig. 11.1* by means of the interconnection graph.

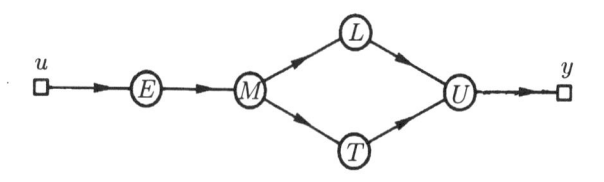

Figure 11.1 Interconnection graph of the Italian scholastic system: Each circular node represents a subsystem and the terminal nodes are input and output.

The interconnection graph shows the nonzero elements of Ω, γ, and Ψ^T regardless of their numerical value. In this particular case, the triple (Ω, γ, Ψ^T) is given by

$$\Omega = \begin{pmatrix} 0 & 0 & 0 & 0 & 0 \\ 1 & 0 & 0 & 0 & 0 \\ 0 & \alpha_{ML} & 0 & 0 & 0 \\ 0 & 0 & \alpha_{MU} & 0 & 0 \\ 0 & 0 & \alpha_{MU} & \alpha_{TU} & 0 \end{pmatrix} \quad \gamma = \begin{pmatrix} 1 \\ 0 \\ 0 \\ 0 \\ 0 \end{pmatrix}$$

$$\Psi^T = \begin{pmatrix} 0 & 0 & 0 & 0 & 1 \end{pmatrix}$$

where $\alpha_{ML} < 1$ and $\alpha_{MT} < 1$ are the fractions of "medie" graduates entering the "liceo" and the "tecnico" ($\alpha_{ML} + \alpha_{MT} \leq 1$) and, analogously, $\alpha_{LU} \leq 1$ and $\alpha_{TU} \leq 1$ are the fractions of "liceo" and "tecnico" graduates entering the university.
♣

Once the interconnection (Ω, γ, Ψ^T) and the subsystems (A_i, b_i, c_i^T) are known, it is possible to find the triplet (A, b, c^T) describing system Σ. In fact, the state $x(t)$ of Σ (of dimension $n = \sum_{i=1}^{q} n_i$) is simply

$$x(t) = \begin{pmatrix} x_1^T(t) & x_2^T(t) & \cdots & x_q^T(t) \end{pmatrix}^T$$

Equations (11.1) and (11.2) can be written in a more compact form as follows:

$$x(t+1) = [\text{diag } A_i]x(t) + [\text{diag } b_i]v(t) \qquad (11.5)$$

$$z(t) = [\text{diag } c_i^T] x(t) \qquad (11.6)$$

where $[A_i]$ is a matrix of matrices having on its diagonal the matrices $A_1, A_2, \ldots, A_q$ and zeros elsewhere, while $[\text{diag } b_i]$ and $[\text{diag } c_i^T]$ are matrices of (column and row) vectors with nonzero vectors (b_i and c_i^T) on their diagonal. If we take into account (11.3), (11.4), (11.5), and (11.6), we can obtain the state equations and the output transformation of system Σ

$$x(t+1) = Ax(t) + bu(t)$$
$$y(t) = c^T x(t)$$

and it is readily seen that

$$A = [\text{diag } A_i] + [\text{diag } b_i]\Omega[\text{diag } c_i^T]$$
$$b = [\text{diag } b_i]\gamma$$
$$c^T = \Psi^T [\text{diag } c_i^T]$$

Moreover, if the subsystems are positively interconnected, that is, if $\Omega\gamma$ and Ψ^T are nonnegative, system Σ is a positive system. In the sequel, we will refer to this class of interconnected systems.

In order to discuss the properties of an interconnected system, one could first compute the triplet (A, b, c^T), and then apply to it the methods described in the previous chapters. However, by doing so, one would not take any advantage from the specific structure of the interconnections. In some cases, it is possible to derive a number of properties of the interconnected system by studying only the analogous properties of its subsystems. When this happens, there is often a great advantage since it is generally easier to study the properties of q matrices (A_i) of dimensions $n_i \times n_i$, than the analogous property of a single matrix A of dimensions $n \times n$ with $n = \sum_{i=1}^{q} n_i$.

A first property, concerning the stability of the interconnections of positive systems, is *Theorem 36*.

THEOREM 36 (*asymptotically stable interconnected systems*)

> If an interconnected system Σ is asymptotically stable, then all its subsystems Σ_i are asymptotically stable.

Proof. The proof is a straightforward consequence of the fact, previously described, that asymptotically stable positive systems are connectively stable.

$\diamond$

It is worth noting that *Theorem 36* is false for nonpositive linear systems. For example, an unstable system can be stabilized using a feedback scheme. By contrast, an unstable positive system cannot be stabilized by positively interconnecting it with other positive systems. In particular, an unstable positive system (A, b, c^T) cannot be stabilized using an algebraic linear control law

$$u(t) = k_1 x_1(t) + k_2 x_2(t) + \cdots + k_n x_n(t)$$

if the gains k_i are constrained to be nonnegative [in order to guarantee that $u(t)$ is nonnegative].

An analogous property (*Theorem 37*) holds for excitability.

THEOREM 37 (*excitable interconnected systems*)

> If an interconnected system Σ is excitable, then all its subsystems Σ_i are excitable.

Proof. If Σ is excitable there exists a nonnegative input that brings the system from the origin $x(0) = 0$ to a strictly positive state in a finite time interval T. Since $x(T) = (x_1^T(T) \quad x_2^T(T) \quad \ldots \quad x_q^T(T))^T$, then $x_i(T) \gg 0, i = 1, 2, \ldots, q$. This implies that each subsystem Σ_i initially at rest, has been transferred to a strictly positive state in finite time. Thus, each subsystem Σ_i is excitable.

$\diamond$

For interconnected systems that are asymptotically stable and excitable (*i.e.*, standard), both *Theorems 36* and *37* hold so that *Theorem 38* holds.

THEOREM 38 (*standard interconnected systems*)

> If an interconnected system Σ is standard, then all its subsystems Σ_i are standard.

The following result (*Theorem 39*), which extends *Theorem 35*, concerns the minimum phase of interconnected positive systems.

THEOREM 39 (*condition for minimum phase*)

> An asymptotically stable interconnected system Σ with an interconnection graph containing a unique direct input–output path, is minimum phase if and only if all the subsystems Σ_i composing such a path are minimum phase.

Proof. Denote by $1, 2, \ldots, q$ the subsystems composing the unique direct input–output path of the interconnection graph and by $F_i(p)$ their transfer functions. Following the same procedure used in the proof of *Theorem 35*, it is possible to construct an interconnected system of the type shown in *Fig. 10.1* having the same transfer function of Σ. Then, all the subsystems of the block diagram in *Fig. 10.1* are externally stable, since Σ is connectively stable (being asymptotically stable), while all the subsystems composing the direct input–output path are, by assumption, minimum phase. Consequently, a and b used in the proof of *Theorem 35* are fulfilled and we can conclude, through the same reasoning, that the system is minimum phase. On the other hand, it is easy to check that if the transfer function

PROPERTIES 105

of the system has no unstable zeros, the transfer functions $F_i(p)$ must be minimum phase.

◇

Example 17 illustrates how *Theorem 39* can be fruitfully exploited to show that a system is minimum phase.

EXAMPLE 17 (reservoirs network)
Consider the reservoirs network shown in *Fig. 11.2(a)* in which $u(t)$ and $y(t)$ are the inflow and outflow rates. In such a network, α and β are the splitting coefficients of the flows at the branching points.

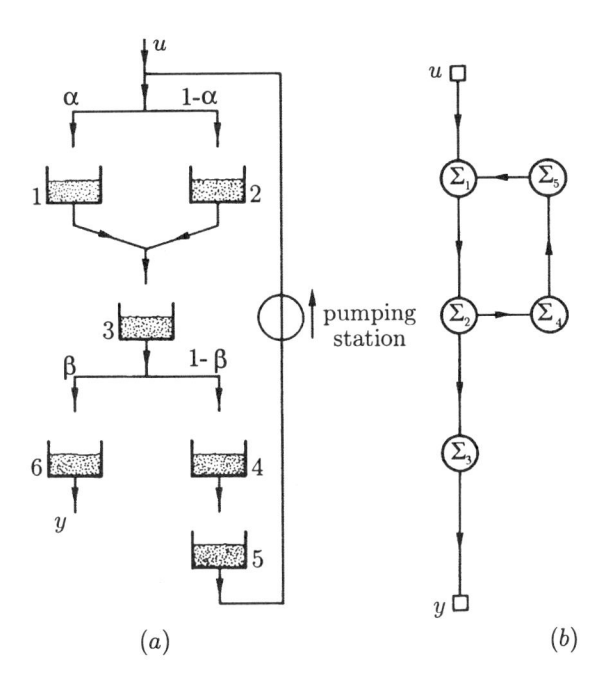

Figure 11.2 Reservoirs network: (a) links among the six reservoirs; (b) interconnection graph with the subsystems $\Sigma_1 = \{1,2\}, \Sigma_2 = \{3\} \Sigma_3 = \{6\} \Sigma_4 = \{4\} \Sigma_5 = \{5\}$.

This system has dimension 6, since the dynamics of the pumping station can be neglected, while each reservoir can be described by a single differential equation (mass balance) of the kind

$$\dot{x}(t) = -kx(t) + \nu(t)$$
$$z(t) = kx(t)$$

where $x(t)$ is the water storage and k is the ratio between outflow rate $z(t)$ and storage.

Theorem 35 is not applicable, since the influence graph of the system, which is not shown, contains two direct input–output paths (the water can flow directly from the input to the output through the reservoirs 1, 3, 6 or 2, 3, 6).

We could therefore determine the triplet (A, b, c^T) describing the whole system and explicitly evaluate its transfer function in order to check whether its zeros have a negative real part. The reader is invited to perform these calculations in order to better appreciate the following considerations, based on *Theorem 39*.

By grouping reservoirs 1 and 2 together into a single subsystem Σ_1 and considering each of the remaining reservoirs as a subsystem, the network depicted in *Fig. 11.2(a)* is described by the interconnection graph of *Fig. 11.2(b)* in which $\Sigma_1 = \{1, 2\}, \Sigma_2 = \{3\}, \Sigma_3 = \{6\}, \Sigma_4 = \{4\}, \Sigma_5 = \{5\}$. Since in such a graph there exists a unique direct input–output path and since the system is, for obvious physical reasons, asymptotically stable, we can apply *Theorem 39*, which states that the system is minimum phase if and only if the three subsystems Σ_1, Σ_2, and Σ_3 composing the direct path are minimum phase. Two of these subsystems (Σ_2 and Σ_3) are single reservoirs, that is, first-order linear systems that do not have zeros and are, therefore, minimum phase. Thus, we must only check whether subsystem Σ_1, composed of reservoirs 1 and 2, is minimum phase or not. Since the transfer function of a single reservoir is

$$G(s) = \frac{k}{s+k}$$

subsystem Σ_1, composed by the parallel connection of reservoirs 1 and 2 with splitting coefficient α, has the transfer function

$$F_1(s) = \alpha \frac{k_1}{s+k_1} + (1-\alpha)\frac{k_2}{s+k_2} = \frac{[\alpha k_1 + (1-\alpha)k_2]s + k_1 k_2}{(s+k_1)(s+k_2)}$$

The zero of the transfer function $F_1(s)$ is negative for all values of the splitting coefficient α. We can then conclude that Σ_1 is minimum phase, so that, thanks to *Theorem 39*, the whole network is minimum phase for every value of the splitting coefficients α and β.

♣

Part III

Applications

12
Input-Output Analysis

Several years before the second world war, the economist Wassily Leontief (Nobel prize winner in 1973) proposed a formal framework for the interpretation of U.S. production data. Such a framework (called afterwards *input–output analysis*) is nothing but a positive linear system at equilibrium. Thus, the model does not explain price and production dynamics but allows the evaluation of the production levels that guarantee the fulfillment of the total demand of the economic system.

Input–output analysis has been applied to a number of economic systems (national, regional, industrial) characterized by different aggregation levels of the production sectors. The approach has also been extended in various ways by Leontief himself and by others, to the dynamic case. All these studies have generated a considerable theoretical body (usually associated with the name of Leontief), which is widely used as a support for decision makers. In most cases, such models result to be positive systems.

In this chapter, we will first present input–output analysis in its simplest form and then make some considerations on production and price dynamics in an economic system. However, in order to discuss this last topic by means of positive systems, we will be obliged to make fairly strong assumptions on the mechanisms of production and price formation. We shall see, however, that, despite these simplifications, the models interpret in a rather satisfactory way the dynamics of real economic systems.

Leontief's input–output analysis is, in short, a mass balance of the goods produced and consumed in a system characterized by n sectors. If the amount of good produced in a given time period (e.g., 1 year) by sector i is denoted by x_i, the balance imposes that such a quantity covers consumption (demand) $b_i \geq 0$ and the purchase of producers of all sectors, that is

$$x_i = b_i + \sum_{j=1}^{n} z_{ij}(x_1, x_2, \ldots, x_n) \quad i = 1, \ldots, n \qquad (12.1)$$

where $z_{ij}(x_1, x_2, \ldots, x_n)$ is the amount of good i purchased by sector j. If we assume that there are no interferences in production, that is, z_{ij} depends only on x_j and that there are no economies or diseconomies of scale, the purchased quantities z_{ij} are given by

$$z_{ij} = a_{ij} x_j \qquad (12.2)$$

where a_{ij} is a technological coefficient that represents, in appropriate units, the amount of good i needed to produce one unit of good j. Obviously, $a_{ij} \geq 0$ so that the matrix A is nonnegative. By introducing the production vector $x = (x_1 \ x_2 \ \ldots \ x_n)^T$ and the consumption vector $b = (b_1 \ b_2 \ \ldots \ b_n)^T$, Eq. (12.1) and (12.2) can be written in compact form as

$$x = Ax + b \qquad (12.3)$$

In order to avoid economically meaningless cases, we will assume A and b to be positive from now on.

Input–output analysis consists of two phases. The first is the estimation, by means of all available information and data, of the technological coefficients a_{ij} and of demand b_i, while the second is the computation of the production levels by means of expression (12.3). The first phase is, by far, the most delicate and onerous one and is based, to a very large extent, on appropriate statistical analysis. In the case of very disaggregated studies, the matrix A may be very large and its identification may involve entire institutions or research groups that are often also responsible for updating the technological coefficients as well as the demands. These topics require expertise in technology innovation and socioeconomic systems and are, therefore, far beyond the aims of this book. In contrast, the second phase of the analysis is strictly related to positive systems.

Perhaps, the most fundamental question one can ask is to know whether a production system can function permanently (*i.e.*, at equilibrium) with all its sectors "alive". This is equivalent to asking whether Eq. (12.3) admits a strictly positive solution x. Moreover, one may ask how such a solution can be evaluated numerically and how much it is sensitive to variations induced by technological innovation or by socio economical changes. In order to give an answer to these questions, note that expression (12.3) can be thought of as the relation satisfied by the positive linear system

$$x(t+1) = Ax(t) + bu(t) \qquad (12.4)$$

at equilibrium $[x(t) = x(t+1) = x]$, provided the input is unitary ($u(t) = 1$). We are not allowed to imagine, however, that system (12.4) has any meaning in terms of production dynamics. Since there are no limitations on the structure (influence graph) of the production systems or on the values of the technological coefficients,

APPLICATIONS   111

system (12.4) is a generic positive system. This means that input–output analysis is nothing but, in its theoretical aspects, the issue of strict positivity of the equilibrium of a positive linear system. Since such a topic involves the stability of the system (see *Chapter 7*), the theoretical contributions to input–output analysis are to be considered as general contributions to the theory of stability and positivity of the equilibrium of positive systems. This is the reason some of the results pointed out in Chapters 5–7 have been discovered (or rediscovered) by mathematical economists. In real situations, system (12.4) is excitable, since the demands b_i are all positive. This implies (see *Theorem 21*) that in such a case Eq. (12.3) admits a strictly positive production vector if and only if the Frobenius eigenvalue λ_F of A is < 1. In other words, the production systems that can function permanently with all their sectors alive have the input–output matrix A and the demand vector b, which are a "standard" pair. Clearly, the condition $\lambda_F < 1$, implies that the solution of (12.3) can be obtained numerically by recursively using Eq. (12.4) with any initial condition $x(0) \geq 0$. Finally (*Theorem 22*), the component of maximum relative sensitivity, can be interpreted in terms of input–output analysis by saying that a perturbation of the demand of good i or of the technological coefficients $a_{i1}, a_{i2}, \ldots, a_{in}$ impacts more, in relative terms, on the production of sector i than on the production of any other sector.

Obviously, Eq. (12.3) must be modified if productions are measured in different units. For example, productions can be measured in monetary units $z_i, i = 1, \ldots, n$ provided there exists a set of prices $p_1, \ldots, p_n$, which can be used as reference.

This implies that

$$z_i = p_i x_i \qquad i = 1, \ldots, n$$

so that Eq. (12.3) becomes

$$z = A^* z + b^* \qquad (12.5)$$

with

$$a_{ij}^* = \frac{p_i}{p_j} a_{ij} \qquad b_i^* = p_i b_i$$

The interest for Eq. (12.5) lies on the fact that the sum $c_j^{+^*}$ of the elements of the jth column of the matrix A^* has a precise economical meaning. Indeed,

$$c_j^{+^*} = \sum_{i=1}^{n} a_{ij}^* = \sum_{i=1}^{n} \frac{p_i a_{ij}}{p_j} = \frac{\sum_{i=1}^{n} p_i a_{ij}}{p_j}$$

so that $c_j^{+^*}$ is the ratio between the production cost $\sum_{i=1}^{n} p_i \alpha_{ij}$ of one unit of good j and the economic value p_j of the same unit. It follows that condition

$$c_j^{+^*} < 1$$

112 INPUT–OUTPUT ANALYSIS

implies that sector j is "efficient" in the classical economic sense (revenue greater than costs). Thus, *Theorem 12* allows one to conclude that if all the sectors of an economic system are efficient, then the Frobenius eigenvalue of A^* is < 1, so that Eq. (12.5) [and, hence, Eq. (12.3)] admits a strictly positive solution.

PROBLEM 21 *(T-II) Determine whether the efficiency of all the production sectors of an economic system is a necessary condition for the existence of a solution $x \gg 0$ of (12.3). Then, give an economic interpretation of the result.*

Let us show now how it is possible to obtain formal schemes for the interpretation of the dynamics of an economic system. To this end, it is worth noting that the hypotheses one can make on production and price formation mechanisms are many and quite diversified. They range from naive ones, which correspond to models that are simple to analyze, to rather realistic ones, which correspond, however, to quite sophisticated models from which it is not possible to derive general principles concerning the dynamic behavior of economic systems. Among the simplest interpretative models, there are the two models presented below: one for price dynamics and the other for production dynamics. These two models have the nice feature of being decoupled and the advantage of referring to the theory of positive linear systems.

We start with the price model, by assuming that prices $p_1(t), \ldots, p_n(t)$ can vary over time, and that production is not instantaneous but needs (due to working and distribution times) one unit of time, say 1 year. The revenue per unit of good j produced (and sold) during year $(t + 1)$ is, therefore, $p_j(t + 1)$ while the cost of the goods needed to produce such a unit, is

$$p_1(t)a_{1j} + p_2(t)a_{2j} + \cdots + p_n(t)a_{nj}$$

since the amount a_{ij} of good i is purchased during year t at price $p_i(t)$. If the difference v_j between revenue and cost, called *added unitary value*, is assumed to be constant over time, one obtains

$$p_j(t+1) = \sum_{i=1}^{n} a_{ij}(t)p_i(t) + v_j$$

or, in matrix form,

$$p(t+1) = A^T p(t) + v \qquad (12.6)$$

where $p(t)$ and v are the vectors of the prices and of the added unitary values. Since the eigenvalues of A^T coincide with those of A, it follows that system (12.6) is asymptotically stable if and only if the Frobenius eigenvalue of A is < 1, that is, if and only if the input–output analysis (12.3) yields a solution $x \gg 0$. Moreover, if the vector v is strictly positive, system (12.6) is excitable, so that the price vector tends asymptotically toward a strictly positive equilibrium with a geometric rate of convergence (λ_F^t). *Figure 12.1* illustrates the course of the price in a hypothetical

APPLICATIONS   113

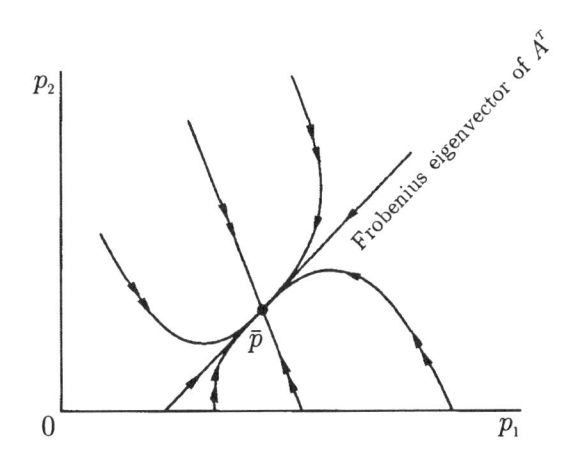

Figure 12.1 Trajectories along which the prices of system (12.6) evolve. The two straight lines are determined by the two eigenvectors of A^T.

economic system with only two sectors in the case where the input–output matrix has a positive subdominant eigenvalue λ_2.

The Frobenius eigenvector of A^T determines a straight line through the point $\bar{p} = (\bar{p}_1, \bar{p}_2)$ along which the convergence is slow (of the kind λ_F^t), while the subdominant eigenvector necessarily has components of opposite sign and determines a straight line along which the convergence is faster (of the kind λ_2^t). *Figure 12.1* clearly shows the possibility that the convergence of a price $p_i(t)$ to its equilibrium $\bar{p}_i$ is not monotone, even if all the eigenvalues of A are positive.

By using similar reasoning, we can propose a model for production dynamics. For this, let us imagine that the production $x_i(t)$ of each good i during year t fulfills the demand b_i (assumed constant over time) and the demand of the production sectors determining the production of the subsequent year, that is,

$$x_i(t) = \sum_{i=1}^{n} a_{ij} x_j(t+1) + b_i \quad i = 1, \ldots, n$$

or, in matrix form,

$$x(t) = Ax(t+1) + b \tag{12.7}$$

Equation (12.7) is not the typical state equation of a dynamic system because t and $t + 1$ are interchanged. It allows one to compute the past state from the present state. This means that, if we reorder the years with an index τ increasing as time goes back, the production would be described by the positive linear system

$$x(\tau + 1) = Ax(\tau) + b \tag{12.8}$$

which admits the equilibrium $\bar{x}$, that is, the production determined using the input–output analysis (12.3). Production dynamics is, therefore, stable with respect to τ,

114 INPUT–OUTPUT ANALYSIS

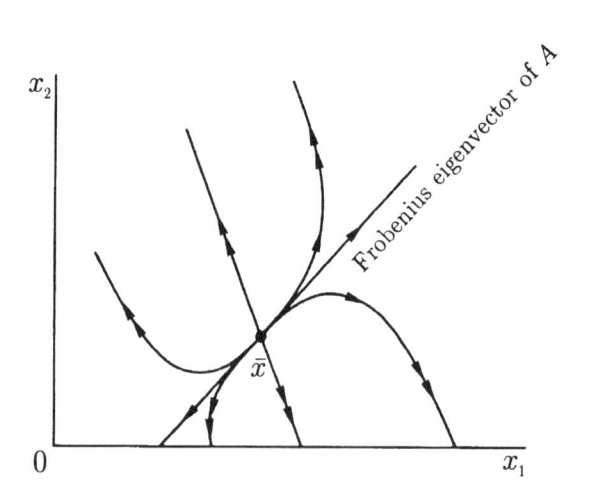

Figure 12.2 Trajectories along which the productions of system (12.7) evolve. The two straight lines are determined by the two eigenvectors of A.

that is, backward in time. The course of the trajectories $x(\tau)$ of system (12.8) is very similar to that illustrated in *Fig. 12.1* for the prices, except that the straight lines through the equilibrium $\bar{x}$ are determined by the eigenvectors of A instead of by those of A^T. By reversing the direction of the motion along trajectories, one obtains the true course of production in system (12.7). In *Fig. 12.2*, an example of a hypothetical system having only two sectors and a positive subdominant eigenvalue λ_2, is reported.

The figure clearly shows that system (12.7) is not a positive system since there exist initial states $x(0) \geq 0$, which give rise to trajectories leaving the positive orthant. Moreover, there exists only one trajectory along which productions remain positive forever, namely, the trajectory determined by the Frobenius eigenvector of A. Along this trajectory, sometimes called the *razor's edge*, the economic system develops and all the productions grow indefinitely at the same rate $(1/\lambda_F)^t$. But even small deviations from the razor's edge will, sooner or later, lead to the collapse of some production sector. Clearly, this would be true if the hypothesis supporting model (12.7) would perfectly describe the real production mechanisms. It is, however, plausible that real systems, are not so "extreme" and tend to remain in the vicinity of the razor's edge because some sort of adjustment of the production mechanism will guarantee that all sectors remain alive.

The lesson learned from models (12.6) and (12.7) is that prices converge toward an equilibrium while productions diverge at rates λ_F^t and $(1/\lambda_F)^t$, where λ_F is the Frobenius eigenvalue of the input–output matrix. Obviously, this conclusion must be considered realistic only to a limited extent, due to the serious limitations of the two models. However, many and in some sense spectacular, are the confirmations of this theory that can be found by analyzing real economic data. For purely illustrative purposes, the course of the U.S. gross national product (GNP) during

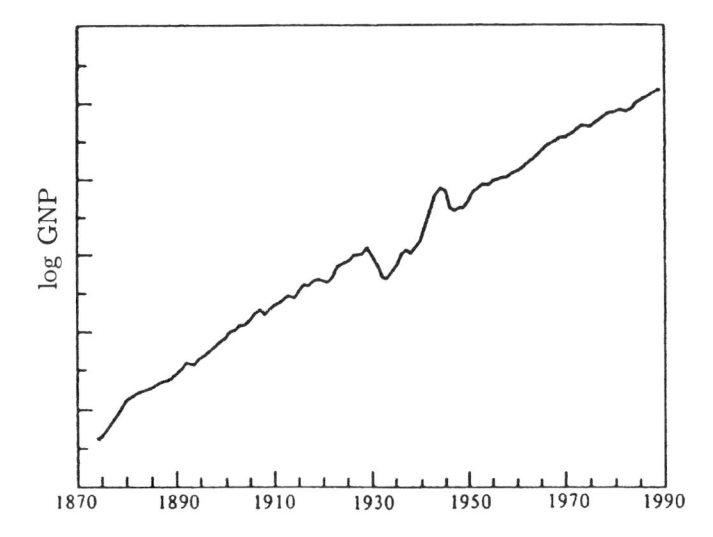

Figure 12.3 The U.S. gross national product in the last century.

the period 1870–1990, is reported in *Fig. 12.3* in appropriate units. The graph shows the logarithm of the GNP, so that a straight line denotes a geometric growth $y(t) = \lambda^t y(0)$.

The graph is, therefore, consistent with model (12.7), which actually predicts a geometric growth for the GNP. The only significant deviations from the prediction correspond to the depression of 1929 and to the second world war, two facts that, of course, could not be explained by the model.

13
Age-Structured Population Models

Plant, animal, and human population models are positive dynamic systems in which the state variables represent biomass, density, or the number of individuals of the population. Many such models, in particular those describing predation, competition, and symbiosis among different species, are nonlinear and are, therefore, out of the context of this book. An important exception is the well-known *Leslie model*, which describes the time evolution of populations in which fertility and survival rates of individuals strongly depend on their age. For this reason, such populations are called *age-structured populations*. In the Leslie model, the time t is discrete and denotes the year (or the reproduction season), while the state variables $x_1(t), x_2(t), \ldots, x_n(t)$ represent the number of females (or individuals or couples) of age $1, 2, \ldots, n$ at the beginning of year t. Assuming that there are no differences in the survival rates of males and females and that the sex ratio is balanced, one can describe the "aging" process by means of the equations

$$x_{i+1}(t+1) = s_i x_i(t) \quad i = 1, \ldots, n-1$$

where s_i is the survival coefficient at age i, that is, the fraction of females of age i that survive at least for 1 year. The first state equations take into account the *reproduction* process, and are therefore

$$x_1(t+1) = s_0(f_1 x_1(t) + f_2 x_2(t) + \cdots + f_n x_n(t))$$

where s_0 is the survival coefficient during the first year of life and f_i is the fertility rate of females of age i, that is, the mean number of females born from each female of age i. These equations, originally proposed by Leslie, are a positive linear autonomous model

118 AGE-STRUCTURED POPULATION MODELS

$$x(t+1) = Ax(t)$$

in which the matrix A, called the Leslie matrix, has

$$A = \begin{pmatrix} s_0 f_1 & s_0 f_2 & \cdots & s_0 f_{n-1} & s_0 f_n \\ s_1 & 0 & \cdots & 0 & 0 \\ 0 & s_2 & \cdots & 0 & 0 \\ \vdots & \vdots & & \vdots & \vdots \\ 0 & 0 & \cdots & s_{n-1} & 0 \end{pmatrix}$$

It is worth noting that, if the state variables denote the number of males [couples] [individuals] of each age class (instead of the number of females), the model remains unchanged since in a balanced population each male [couple] [individual] of age i generates, on average, f_i males [couples] [individuals], where f_i is the fertility of females.

Leslie models are extensively used for making demographic projections [*i.e.*, for forecasting $x(t) = A^t x(0)$, given $x(0)$]. In the case of human populations, however, the progresses in medical sciences have increased and continue to increase the survival rates (at least in rich countries), while the fluctuations of the environmental and socioeconomic conditions may produce positive and negative variations of the fertility rates. For these reasons, in the case of human populations, projections cannot be made for long periods of time and the parameters s_i and f_i must be updated each time a forecast is performed. The update of survival and fertility coefficients is done by using a number of statistical procedures that are relatively complex and sophisticated, and the scientific contribution in human demography concern, to a large extent, such statistical methods. Nevertheless, the problem of estimating survival and fertility rates from census data will not be treated here since it is not related to the topics discussed in this book.

For purely illustrative purposes, the estimated survival and fertility rates for the first 10 age groups of four animals (a fish, a bird, a deer, and a squirrel) are reported in the following table:

	s_0 / f_1	s_1 / f_2	s_2 / f_3	s_3 / f_4	s_4 / f_5	s_5 / f_6	s_6 / f_7	s_7 / f_8	s_8 / f_9	s_9 / f_{10}
Fish	$6 \cdot 10^{-5}$	0.45	0.27	0.26	0.26	0.25	0.25	0.25	0.25	0.25
	—	5000	11000	18000	24000	31000	34000	41000	45000	46000
Bird	0.50	0.80	0.36	0.37	0.38	0.39	0.39	0.38	0.38	0.37
	—	0.40	0.45	0.50	0.50	0.50	0.50	0.50	0.50	0.50
Deer	0.70	0.92	0.48	0.49	0.48	0.42	0.28	0.25	0.22	0.20
	—	—	0.10	0.40	0.50	0.50	0.50	0.50	0.45	0.400
Squirrel	0.40	0.24	0.30	0.33	0.34	0.33	0.30	0.28	0.24	0.27
	0.6	1.2	1.9	1.9	1.9	1.9	1.9	1.9	1.8	1.6

For the fish, survival rates are low, in particular at birth (s_0), but fertility rates are high and increase with age. In contrast, bird survival and fertility rates are

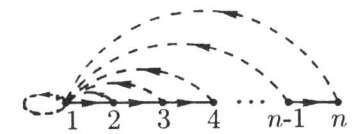

Figure 13.1 Influence graph of the Leslie model. The horizontal arcs represent aging, while the dashed arcs represent reproduction and are absent in nonfertile age classes.

almost constant after 2–3 years, while for mammals, the survival rates decrease with age.

In Leslie models, survival and fertility rates depend exclusively on age. In reality, this is true provided the individuals in each age class are not too many. In fact, as soon as the density of individuals increases, some phenomena (which, in general, reduce fertility and/or survival rates) show up. For example, finding appropriate niches for reproduction becomes more difficult if the number of fertile individuals increases; the spreading of epidemics is favored by high population densities; the search for food becomes more and more difficult as population grows, and so forth. This means that Leslie models are well suited for describing the dynamics of populations doomed to extinction, that is, characterized by small densities $x_i(t)$, for which we can suppose that survival and fertility rates remain constant over time. However, Leslie models can also be used to describe the dynamics of expanding populations provided one limits themselves to intervals of time during which the densities are not too high. For example, the survival and fertility rates reported in the above table concern expanding populations. Consequently, the corresponding Leslie model can be used for short-term forecasts, provided the initial number of individuals is not too high. Long-term forecasts ($t \to \infty$) would lead us to conclude that $\|x(t)\| \to \infty$, that is, that such populations grow indefinitely, which is obviously a biological absurdity.

The influence graph G_x of the Leslie model is depicted in *Fig. 13.1*. The survival rate $s_i, i = 1, \ldots, n-1$ is associated with each horizontal arc $(i, i+1)$, while the product of survival rate at birth s_0 times the fertility rate f_i is associated with each dashed arc $(i, 1)$.

The original Leslie model can be appropriately modified by specifying the input and the output, thus obtaining a positive linear system

$$x(t+1) = Ax(t) + bu(t)$$

$$y(t) = c^T x(t)$$

By doing so, each component $b_i u(t)$ of the vector $bu(t)$ represents the exogenous contribution of immigration or stocking on the age class i, while the output y is an aggregated population indicator. The most common choice for such an indicator, is

$$c^T = \begin{pmatrix} 1 & 1 & 1 & \cdots & 1 \end{pmatrix}$$

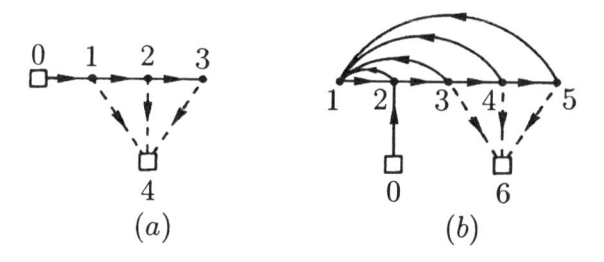

Figure 13.2 Influence graphs of two Leslie models: (a) a "rent-a-car" system; (b) a stocked fishery.

which corresponds to total population $y = x_1 + x_2 + \cdots + x_n$. Other choices allow one to specify the number of newly born individuals, the number of fertile individuals or, in the case of human populations, the number of students, of workers, of retired people, or the number of individuals of other social groups of some specific interest. As a purely illustrative example, in *Fig. 13.2* the influence graphs of two different Leslie models, are depicted.

The first graph, *Fig. 13.2(a)*, describes a rent-a-car service: $x_i(t), i = 1, 2, 3$ represents the number of cars of age i available at the beginning of year t and y (node 4) is the total number of cars available at the beginning of year t. In this case, the system is in some sense "degenerate" since all fertility rates are zero. The triple (A, b, c^T) is

$$A = \begin{pmatrix} 0 & 0 & 0 \\ s_1 & 0 & 0 \\ 0 & s_2 & 0 \end{pmatrix} \quad b = \begin{pmatrix} s_0 \\ 0 \\ 0 \end{pmatrix}$$

$$c^T = \begin{pmatrix} 1 & 1 & 1 \end{pmatrix}$$

The second graph [*Fig. 13.2(b)*] represents a fishery where an amount $u(t)$ of individuals of age 1 is stocked into the system at the beginning of year t. Thus, the number of individuals $x_2(t+1)$ of age 2 at the beginning of the subsequent year is $s_1 x_1(t) + s_1 u(t)$ and it is at this age that females become fertile. Under the assumption that fishermen are not allowed to catch nonfertile fish, the fish stock $y(t)$ that can be harvested in year t is the sum of the individuals of age 3, 4, and 5. The corresponding triplet (A, b, c^T) is

$$A = \begin{pmatrix} 0 & s_0 f_2 & s_0 f_3 & s_0 f_4 & s_0 f_5 \\ s_1 & 0 & 0 & 0 & 0 \\ 0 & s_2 & 0 & 0 & 0 \\ 0 & 0 & s_3 & 0 & 0 \\ 0 & 0 & 0 & s_4 & 0 \end{pmatrix} \quad b = \begin{pmatrix} 0 \\ s_1 \\ 0 \\ 0 \\ 0 \end{pmatrix}$$

$$c^T = \begin{pmatrix} 0 & 0 & 1 & 1 & 1 \end{pmatrix}$$

From the structure of the influence graph of the Leslie model (see *Fig. 13.1*) one can easily prove that the following three properties hold. The Leslie model is

a. *Irreducible* if and only if the last age class is fertile.

b. *Excitable* if and only if at least some of the individuals entering the system through immigration or stocking are (or will become) fertile.

c. *Nontransparent* if and only if the last age class is not fertile and does not influence the output.

More involved, but still simple, is the stability analysis of the Leslie model. To this end, let us first evaluate the characteristic polynomial of the Leslie matrix

$$\Delta_A(\lambda) = \det(\lambda I - A) = \det \begin{pmatrix} \lambda - s_0 f_1 & -s_0 f_2 & \cdots & -s_0 f_n \\ -s_1 & \lambda & \cdots & 0 \\ \vdots & \vdots & & \vdots \\ 0 & 0 & \cdots & \lambda \end{pmatrix} =$$

$$= (\lambda - s_0 f_1) \det \begin{pmatrix} \lambda & 0 & \cdots & 0 \\ -s_2 & \lambda & \cdots & \\ \vdots & \vdots & & \vdots \\ 0 & 0 & \cdots & \lambda \end{pmatrix} + s_1 \det \begin{pmatrix} -s_0 f_2 & -s_0 f_3 & \cdots & -s_0 f_n \\ -s_2 & \lambda & \cdots & 0 \\ \vdots & \vdots & & \vdots \\ 0 & 0 & \cdots & \lambda \end{pmatrix} =$$

$$= (\lambda - s_0 f_1) \lambda^{n-1} - s_0 s_1 f_2 \det \begin{pmatrix} \lambda & 0 & \cdots & 0 \\ -s_3 & \lambda & \cdots & \\ \vdots & \vdots & & \vdots \\ 0 & 0 & \cdots & \lambda \end{pmatrix} + s_1 s_2 \det \begin{pmatrix} -s_0 f_3 & -s_0 f_4 & \cdots & -s_0 f_n \\ -s_3 & \lambda & \cdots & 0 \\ \vdots & \vdots & & \vdots \\ 0 & 0 & \cdots & \lambda \end{pmatrix} =$$

$$= \lambda^n - s_0 f_1 \lambda^{n-1} - s_0 s_1 f_2 \lambda^{n-2} - s_0 s_1 s_2 f_3 \lambda^{n-3} - \cdots - s_0 s_1 \ldots s_{n-1} f_n$$

Thus, all the coefficients α_i of the characteristic polynomial $\Delta_A(\lambda) = \lambda^n + \alpha_1 \lambda^{n-1} + \cdots + \alpha_n$ are nonpositive, that is,

$$\alpha_i \leq 0 \quad i = 1, \ldots, n \tag{13.1}$$

and are given by the simple formula

$$\alpha_i = -s_0 s_1 \ldots s_{i-1} f_i$$

Since $s_0 s_1 \ldots s_{i-1}$ is the probability that a newly born female survives at least until age i, it follows that $(-\alpha_i)$ is the average contribution to total population of females of age i. In other words, on average, each female contributes to the total population with $(-\alpha_1)$ females after 1 year of life, $(-\alpha_2)$ females after 2 years, and so forth. Thus, the overall contribution of each female during her entire life is, on average, given by

$$R = \sum_{i=1}^{n}(-\alpha_i) = \sum_{i=1}^{n} s_0 s_1 \ldots s_{i-1} f_i \qquad (13.2)$$

In the demographic literature, the coefficient $(-\alpha_i)$ is called *net maternity* while R is known as *net reproduction*. This last parameter is of paramount importance for describing the dynamics of a population. In fact, taking into account that each female generates, on average, R females during her entire life, it is intuitive that the condition $R > 1$ implies that population grows (demographic explosion) while the inequality $R < 1$ characterizes populations doomed to extinction. In system theoretic terminology, this means that $R > 1$ implies instability and $R < 1$ implies asymptotic stability.

PROBLEM 22 (A-I) *Compute the net reproduction rate of the four populations (fish, bird, deer, squirrel) previously described and determine whether such populations are asymptotically stable (extinction) or unstable (growth).*

From a formal point of view, the above stability condition can be obtained using the theory presented in the previous chapters. Actually, we can show that condition

$$\sum_{i=1}^{n}(-\alpha_i) < 1$$

corresponding to $R < 1$, is a necessary and sufficient condition for asymptotic stability of any positive system satisfying inequalities (13.1). In fact, the characteristic equation $\Delta_A(\lambda) = 0$ can be written in the form

$$\lambda^n = (-\alpha_1)\lambda^{n-1} + (-\alpha_2)\lambda^{n-2} + \cdots + (-\alpha_n)$$

which, assuming that λ is positive, becomes

$$\frac{(-\alpha_1)}{\lambda} + \frac{(-\alpha_2)}{\lambda^2} + \cdots + \frac{(-\alpha_n)}{\lambda^n} = 1 \qquad (13.3)$$

Since $\alpha_i \leq 0, i = 1, \ldots, n$, the left-hand side of Eq. (13.3) decreases with λ for $\lambda > 0$ provided that at least one α_i is negative (in the trivial case, in which each α_i is zero, there are n zero eigenvalues and the system is asymptotically stable). Thus, there exists a unique real and positive eigenvalue that necessarily coincides with the Frobenius eigenvalue λ_F. The asymptotic stability condition $\lambda_F < 1$ is therefore equivalent to

$$\sum_{i=1}^{n}(-\alpha_i) < 1$$

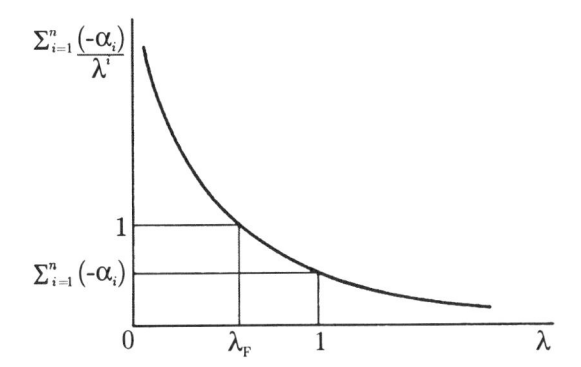

Figure 13.3 Graph of the function $\sum_{i=1}^{n}(-\alpha_i)/\lambda^i$ for $\lambda > 0$ and equivalence between condition $\lambda_F < 1$ and $\sum_{i=1}^{n}(-\alpha_i) < 1$.

as shown in *Fig. 13.3*.

Obviously, if the opposite inequality $\sum_{i=1}^{n}(-\alpha_i) > 1$ holds, the system is unstable since $\lambda_F > 1$, while if $\sum_{i=1}^{n}(-\alpha_i) = 1$ the system is simply stable.

PROBLEM 23 *(T-II) Show that condition* $\sum_{i=1}^{n}(-\alpha_i) < 1$ *is a necessary condition for the stability of any discrete-time linear system. (Hint: Use Jury's method reported in Appendix B.)*

If there is a unique dominant eigenvalue, the dynamics of the system are, in the long run, well described by the Frobenius eigenvalue λ_F, since

$$x(t+1) \cong \lambda_F x(t) \quad y(t+1) \cong \lambda_F y(t)$$

This means that the Frobenius eigenvalue λ_F is, from a biological point of view, the (asymptotic) *growth rate* of the population. In *Fig. 13.4* two examples of expanding populations are shown. One refers to a sea elephant of the Northern Pacific coast and the other is an estimate of human population. The data refer to total population [*i.e.*, $y(t) = \sum_{i=1}^{n} x_i(t)$] and are plotted on logarithmic scales, since the law $y(t) = \lambda_F^t y(0)$ [corresponding to $y(t+1) = \lambda_F y(t)$] is represented in these scales by a straight line of equation

$$\log y(t+1) = \log y(0) + (\log \lambda_F)t$$

Note that the sea elephant data lie almost on a straight line, in accordance with the Leslie model, while the human population data lie on a straight line only until two centuries ago. This is obviously due to the improvement of living conditions during the last two centuries.

It is important to distinguish between net reproduction rate R and growth rate λ_F even if both rates must be greater [less] than unity in order to have population growth [extinction]. The net reproduction rate R can be easily computed from

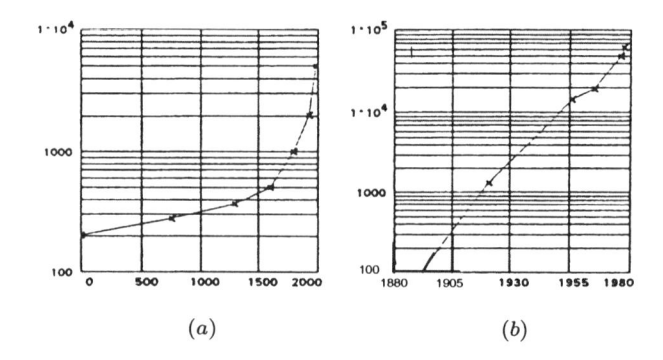

Figure 13.4 Population data in a logarithmic scales: (a) humans; (b) sea elephant.

survival and fertility data using (13.2). Conversely, the growth rate λ_F can be computed by solving equation (13.3), a task that can be relatively onerous if n is large. For this reason in many studies n is kept relatively small by defining large age classes, as in the famous work by Keyfitz and Flieger (1971) who divided the U.S. population into 10 classes of 5 years each and disregarded all individuals over 50 (see *Fig. 13.5* where the eigenvalues of the corresponding Leslie model are shown).

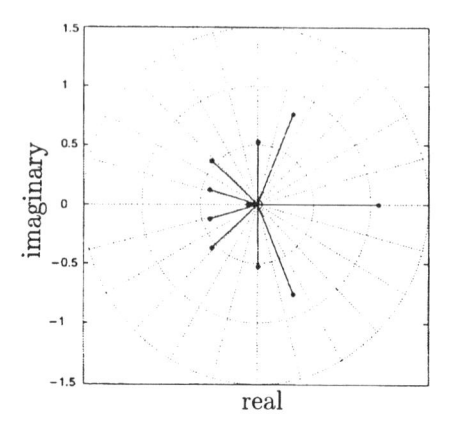

Figure 13.5 Spectrum of the Leslie matrix of the U.S. population in 1966: the Frobenius eigenvalue is the unique positive eigenvalue and it is the only one to be > 1 in modulus (from a work by Keyfitz and Flieger, 1971).

Many approximated formulas have been proposed in the literature for the computation of λ_F. *Problem 24* considers one such approximation, which can be used when the net reproduction rate R is not too far from unity.

APPLICATIONS

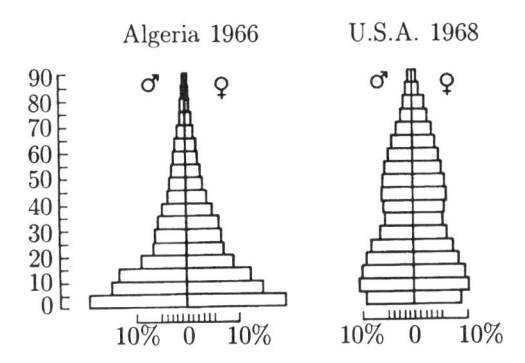

Figure 13.6 Age structure of the populations of Algeria and U.S. (from the 1968 United Nations demographic handbook).

PROBLEM 24 *(T-II)* Prove that $R \cong 1$ implies $\lambda_F \cong 1 + (R-1)/\sum_{i=1}^{n} i(-\alpha_i)$. *[Hint: Let $\lambda_F = 1 + \varepsilon$ and use (13.2) by approximating $1/\lambda_F^i$ with $1 - i\varepsilon$].*

When the dominant eigenvector is unique (thus coinciding with the Frobenius eigenvector x_F) it is possible to characterize the asymptotic behavior of the population in terms of relative abundances of the various age groups. For this, divide each component $x_i(t)$ of the state vector by the total population $\sum_{i=1}^{n} x_i(t)$ thus obtaining a normalized vector $z(t)$ called *age structure*. Since $x(t)$ tends to align, for $t \to \infty$, with the dominant eigenvector (which is unique), it follows that $z(t)$ tends, as time goes on, toward the Frobenius eigenvector x_F (more precisely toward the normalized Frobenius eigenvector). This, obviously, holds both for stable and unstable populations, so that we can state that in growing as well as in declining populations the age structure tends toward a fixed age structure, called *stable age structure*, which is nothing but the Frobenius eigenvector x_F. This result is often emphasized in textbooks of population dynamics and presented as a peculiar property of biological systems. However, this is not true, since this result is neither based on the particular structure of the Leslie model nor on the positivity of the model. Indeed, the state vector of any linear system asymptotically tends to align with its dominant eigenvector!

In *Fig. 13.6*, the age structures of two different populations (Algeria and U.S.) are shown. The different shape of the two diagrams is mainly justified by the fact that Algeria is a country with very high fertility rates and, therefore, with a high growth rate λ_F. In fact, it is easy to check (recall that $Ax_F = \lambda_F x_F$) that the Frobenius eigenvector x_F can be written in the form

$$x_F = (\lambda_F^{n-1} \quad s_1 \lambda_F^{n-2} \quad s_1 s_2 \lambda_F^{n-3} \quad \ldots \quad s_1 s_2 \ldots s_{n-2} \lambda_F \quad s_1 s_2 \ldots s_{n-1})^T$$

so that populations with a high λ_F must possess a stable age structure strongly biased toward the youngest classes.

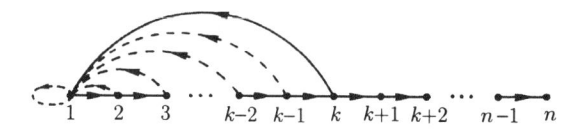

Figure 13.7 Influence graph G_x of Leslie model: k is the last fertile age class.

We still need to find out under which conditions a population has a unique dominant eigenvector. To this end, it is worth noting that the influence graph G_x of the Leslie model can be, in general, decomposed into two parts, as shown in Fig. 13.7. The first part, G'_x, contains the youngest age classes $1, 2, \ldots k$, where k is the last fertile age class. The second part, G''_x in cascade with the first, contains the remaining $(n-k)$ age classes, none of which are fertile. It is possible that the influence graph is composed only of the first part (as in most animal populations) or only of the second one (in the trivial case of sterile populations).

The corresponding state equations are

$$\begin{pmatrix} x'(t+1) \\ x''(t+1) \end{pmatrix} = \begin{pmatrix} A' & 0 \\ B & A'' \end{pmatrix} \begin{pmatrix} x'(t) \\ x''(t) \end{pmatrix}$$

where

$$x'(t) = |(x_1(t) \quad \ldots \quad x_k(t))^T \qquad x''(t) = (x_{k+1}(t) \quad \ldots \quad x_n(t))^T$$

$$A' = \begin{pmatrix} s_0 f_1 & s_0 f_2 & \ldots & s_0 f_{k-1} & s_0 f_k \\ s_1 & 0 & \ldots & 0 & 0 \\ \vdots & \vdots & & \vdots & \vdots \\ 0 & 0 & \ldots & s_{k-1} & 0 \end{pmatrix} \qquad A'' = \begin{pmatrix} 0 & 0 & \ldots & 0 & 0 \\ s_k & 0 & \ldots & 0 & 0 \\ \vdots & \vdots & & \vdots & \vdots \\ 0 & 0 & \ldots & s_{n-1} & 0 \end{pmatrix}$$

$$B = \begin{pmatrix} 0 & 0 & \ldots & s_k \\ 0 & 0 & \ldots & 0 \\ \vdots & \vdots & & \vdots \\ 0 & 0 & \ldots & 0 \end{pmatrix}$$

The matrix A'' has only zero eigenvalues, so that the dominant eigenvalue λ_d of A coincides with the dominant eigenvalue $\lambda'_d > 0$ of A'. It follows that the dominant eigenvector of A

$$x_d = \begin{pmatrix} x'_d \\ x''_d \end{pmatrix}$$

satisfying equation $A x_d = \lambda_d x_d$, has components x'_d and x''_d related one to each other by

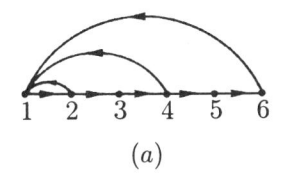

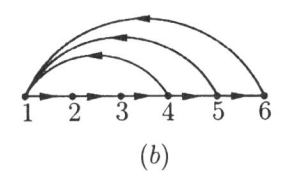

Figure 13.8 Influence graphs G'_x: (a) cyclic population with $r = 2$; (b) primitive population.

$$A'x'_d = \lambda'_d x'_d$$
$$Bx'_d + A''x''_d = \lambda'_d x''_d$$

The first one of these two equations states that x'_d is the dominant eigenvector of A' while the second one yields

$$x''_d = (\lambda'_d I - A'')^{-1} B x'_d$$

[the matrix $(\lambda'_d I - A'')$ is invertible since $\lambda'_d \neq 0 \; (= \lambda''_d)$]. The dominant eigenvector x_d is, therefore, unique if and only if the dominant eigenvector x'_d of the population composed of the first k age classes is unique. But this population is, as previously shown (see property a), irreducible. Thus (see Chapter 6), λ_d is not unique if A' is cyclic and it is unique if A' is primitive. The two cases are illustrated in *Fig. 13.8* using two different influence graphs G'_x.

The first graph [*Fig. 13.8(a)*] corresponds to a population composed of individuals reproducing every 2 years and is cyclic, with an index of cyclicity $r = 2$. Such a system has, therefore (see *Theorem 18*), two dominant eigenvalues: one positive (λ'_F) and the other one negative $(-\lambda'_F)$. There are, therefore, two distinct dominant eigenvectors. The second influence graph [*Fig. 13.8(b)*] corresponds to a population with individuals that become fertile at a certain age and, then remain such until they die. From *Theorem 6*, such a system is primitive and has, therefore, a unique dominant eigenvalue. This means that in the case shown in *Fig. 13.8(a)* the age structure of the system does not tend toward a stable age structure. In fact, if there are initially only individuals of the first age class, in the three subsequent years there will be individuals of the age classes $(2), (1,3), (2,4)$, and from the fifth year on, the population will be alternatively composed of all odd or even classes.

An interesting class of populations that do not admit stable age structures is that of *semelparous* populations whose individuals reproduce only once in their lifetime just before dying (like, e.g., *Pacific salmon, perennial bamboo, periodical cicada, lucanus cervus*, etc.).

PROBLEM 25 *(T-I)* Show that a semelparous population with k age classes has a net reproduction rate given by $R = s_0 s_1 \ldots s_{k-1} f_k$ and has no stable age structure.

Moreover, prove that its k dominant eigenvalues are regularly placed on the circle of radius R.

We consider now populations with exogenous inputs. A typical case is a population that, due to a deterioration of its environment, has a net reproduction rate $R < 1$ and is, therefore, doomed to extinction unless it is stocked from the outside. The first (indeed trivial) question concerns the possibility of sustaining this population at equilibrium in such a way that all age classes are represented. This question can be immediately answered by looking at *Theorem 21* and recalling property b. The answer is the following: A population doomed to extinction can be maintained in a strictly positive equilibrium provided it is stocked with individuals that are (or will become) fertile.

A second, definitely less trivial, question concerns the characterization of the age structures one can generate by means of an appropriate stocking strategy. Consider the case of a virgin territory $[x(0) = 0]$, where stocking at a rate $u(t)$ is possible. Then, the problem is the determination of the reachability cone $X_r(t)$. Since, in a Leslie model, condition $\alpha_i \leq 0, i = 1, \ldots, n$ holds, we can immediately conclude (see *Theorem 24*) that the reachability cone $X_r(t), t < n$ is polyhedral and characterized by t edges, which are the first t reachability vectors $b, Ab, \ldots, A^{t-1}b$. Moreover, $X_r(t)$ stops growing and becomes solid at $t = n$ if the vectors $b, Ab, \ldots, A^{n-1}b$ are linearly independent (such a condition is generally met if stocking is performed using individuals that are or will become fertile). From *Theorem 27*, it follows that not all age structures can be generated since Leslie systems, in general, are not completely reachable. A relevant exception is that of semelparous populations with stocking performed with individuals of the first age class. In fact, the reachability vectors $b, Ab, \ldots, A^{n-1}b$ of these populations, besides being linearly independent, belong to the axis of the state space.

Finally, consider the case of populations characterized by an indicator $y(t)$ linearly related to the state $x(t)$ of the system. Thus, the population is described by a positive linear system

$$x(t+1) = Ax(t) + bu(t)$$

$$y(t) = c^T x(t)$$

with a matrix A in Leslie form. Such a system has an ARMA model

$$y(t) = \sum_{i=1}^{n}(-\alpha_i)y(t-1) + \sum_{i=1}^{n}\beta_i u(t-i) \qquad (13.4)$$

with coefficients $(-\alpha_i)$ of the autoregressive part that are positive or zero (recall that $-\alpha_i$ is a maternity rate).

As stated in Chapter 10, if one is interested in reconstructing the input sequence from a recorded sequence of the output, Eq. (13.4) can be recursively solved with respect to the first input $u(t-i)$ with $\beta_i \neq 0$. The estimation error tends to zero as time goes on (*i.e.*, as the number of available data increases) provided that the system is minimum phase, that is, provided the zeros of the polynomial

$$p(z) = \beta_1 z^{n-1} + \beta_2 z^{n-2} + \cdots + \beta_n$$

are < 1 in modulus.

However, in many cases, it is not necessary to evaluate the zeros of this polynomial to ascertain whether the system is minimum phase or not. If, for example, stocking is performed using individuals of the first fertile age class, denoted by h in *Fig. 13.9*, and $y(t)$ coincides with one of the state variables, $x_i(t)$ with $i > h$, then, thanks to *Theorem 35*, we can state that the system is minimum phase provided the population is doomed to extinction if it is not stocked. In contrast, *Theorem 35* does not permit any conclusion if $i < h$ since there exist many direct input–output paths in the influence graph.

A second case in which minimum phase is guaranteed is when $y(t)$ is the total fertile population, that is, $y(t) = \sum_{i=h}^{k} x_i(t)$. In this case, the polynomial $p(z)$ is the product of two polynomials

$$p(z) = z^{h-1}(z^{k-h} + s_h z^{k-h-1} + \cdots + s_h s_{h+1} \cdots s_{k-1})$$

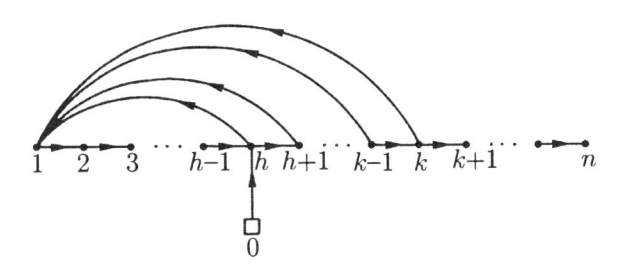

Figure 13.9 Influence graph G_{ux} of a population with stocking performed using individuals of the first fertile age class h.

The first one has $(h-1)$ trivial zeros and the second has $(k-h)$ zeros < 1 in modulus since its coefficients are positive and decreasing.

A third case of some interest is described in *Problem 26*.

PROBLEM 26 *(T-III) Prove that the population described in Fig. 13.9 (stable or unstable) is minimum phase if $y(t)$ is the total number of newly born individuals [i.e., $y(t) = \sum_{i=h}^{k} f_i x_i(t)$] and net maternity rates decrease with age. [Hint: determine the polynomial $p(z)$ and verify that such a polynomial has decreasing coefficients.]*

Clearly, there are cases in which the above conditions do not hold so that one cannot ascertain whether a system is minimum phase or not without explicitly computing its zeros. The reader should check, for example, that the above mentioned criterion (decreasing net maternity rates) is not applicable for one of the four populations (fish, bird, deer, squirrel) described at the beginning of this chapter.

14
Markov Chains

Markov processes are a well-known class of stochastic dynamical systems. For such systems, the present state and input determine the future state only in probabilistic terms. This means that, if the input and the probability distribution of the state is known at a given time, the probability distribution of the state at any subsequent time can be computed. Under very broad assumptions (known as *Markov's assumptions*) such probability distribution evolves in time, following specific rules that make the analysis elegant and effective. A particularly simple case is that in which the system has no inputs, is discrete-time, and has a finite number (n) of states. For such systems, called *Markov chains*, at each time $t = 0, 1, 2$, and so on, the transition from one state $i(= 1, 2, \ldots, n)$ to another $j(= 1, 2, \ldots, n)$ takes place with probability p_{ij}, so that the chain is identified by an $n \times n$ matrix

$$P = \begin{pmatrix} p_{11} & p_{12} & \cdots & p_{1n} \\ p_{21} & p_{22} & \cdots & p_{2n} \\ \vdots & \vdots & & \vdots \\ p_{n1} & p_{n2} & \cdots & p_{nn} \end{pmatrix}$$

called a *Markov* (or *stochastic*) *matrix*. Clearly, such a matrix is positive, since $p_{ij} \geq 0$, and each row contains at least a nonzero element, because the system must evolve from any given state. Moreover, the sum of the elements of each row equals unity so that (see *Theorem 12*) the Frobenius eigenvalue of P is unitary.

The matrix P is often represented by means of a graph, called a stochastic (or Markov) graph. Such a graph has n nodes and an arc directed from i to j if p_{ij} is positive. The stochastic graph visualizes in an effective way all the conceivable elementary transitions of the system, that is, all the state transitions that can occur

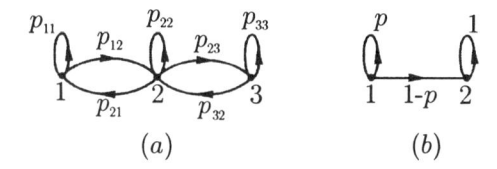

Figure 14.1 Two simple Markov graphs: (*a*) weather forecast; (*b*) Russian roulette.

in a time unit. *Figure 14.1* reports two of these graphs, which are often used as introductory examples. *Figure 14.1(a)* is the graph of the weather in which nodes $1, 2, 3$ represent, respectively, sunny, cloudy, and rainy weather. The absence of the arcs $(1, 3)$ and $(3, 1)$ shows the impossibility for the weather to go from sunny to rainy (and vice versa) without being cloudy for a certain period of time. The second example, *Fig. 14.1(b)* has only two states: one in which the system remains with probability p and another in which the system remains forever. This graph describes many situations, for example, the learning process without oblivion, the antibodies formation process (1 = absence, 2 = presence), and the notorious Russian roulette, in which $p = 5/6$ if the gun contains 6 bullets.

We now clarify in which sense we can consider Markov chains as positive linear systems. For this, denote by $x_j(t)$ the probability that the system is in state j at time t. The probability $x_i(t+1)$ is then the sum of the probabilities $p_{ji}x_j(t)$ over all states j.

Consequently, we can write

$$x_i(t+1) = \sum_{j=1}^{n} p_{ji} x_j(t)$$

or, in matrix notation,

$$x(t+1) = P^T x(t) \qquad (14.1)$$

where the vector $x(t)$, with components $x_1(t), x_2(t), \ldots, x_n(t)$ is the *state probability distribution* of the system at time t. Equation (14.1) is the state equation of an autonomous discrete-time positive system, $x(t+1) = Ax(t)$, in which the matrix A is the transpose of the Markov matrix. In conclusion, a Markov chain is a stochastic system, but its state probability distribution $x(t)$, evolves following the rules of a positive linear system. In other words, Markov chains are deterministic systems (linear and positive) in their probabilities. Moreover, the influence graph G_x of the positive system $x(t+1) = Ax(t)$ is just the Markov graph, since the arc (i, j) is present in the graph if and only if $a_{ji} (= p_{ij}) \neq 0$.

From this, it follows that Markov chains can be classified in reducible and irreducible Markov chains, where the latter can be further subdivided into primitive (also called *regular*) and cyclic. The Markov chain in *Fig. 14.1a* is, for example, irreducible (since its influence graph G_x is connected) and primitive (since G_x

APPLICATIONS 133

contains self-loops (see *Theorem 6*), while the Markov chain in *Fig. 14.1(b)* is reducible (since its influence graph G_x is not connected).

One must keep this classification in mind when studying the asymptotic behavior of a Markov chain, in particular if one likes to know whether the probability distribution $x(t)$ tends, for $t \to \infty$, to the same asymptotic distribution $\bar{x}$ regardless of the initial state, or, more generally, regardless of the initial state distribution $x(0)$. Before tackling this problem, note that the initial state $x(0)$ of system (14.1) must satisfy (being a probability distribution) the conditions

$$x_i(0) \geq 0 \qquad \sum_{i=1}^{n} x_i(0) = 1 \qquad (14.2)$$

that is, $x(0)$ must belong to the so-called simplex S, shown in *Fig. 14.2* in the three-dimensional case.

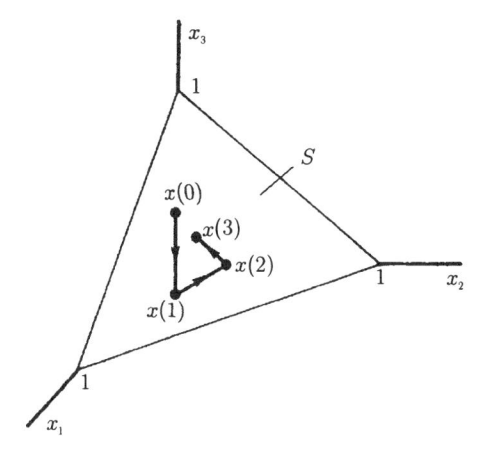

Figure 14.2 The simplex S of a three-dimensional Markov chain and the time evolution of the state vector $x(0), x(1), x(2), x(3)$, and so on.

On the other hand, from Eq. (14.1) it follows that

$$\sum_{i=1}^{n} x_i(t+1) = \sum_{i=1}^{n}\sum_{j=1}^{n} p_{ji} x_j(t) = \sum_{j=1}^{n}\sum_{i=1}^{n} p_{ji} x_j(t) = \sum_{j=1}^{n} x_j(t)$$

i.e., the sum of the components of the state vector remains unchanged over time, so that if $x(0) \in S$, then also $x(t) \in S$ for $t = 1, 2$, and so on. Thus, the time evolution of a Markov chain can be thought of (see *Fig. 14.2*) as a sequence of points $x(0), x(1), x(2)$, and so on, on the simplex S, which is the $(n-1)$-dimensional manifold on which the sum of the components of the vector x remains equal to 1. Such an invariance is a consequence of the structure of the matrix P, which has rows with unitary sum. In other words, the state $x(t)$ of system (14.1) evolves on a linear manifold since one eigenvalue of the system is unitary, so that a particular combination of the state variables remains constant in time. Moreover, the real unitary eigenvalue is the Frobenius eigenvalue, so that the remaining $n-1$ eigenvalues are ≤ 1 in modulus. These $(n-1)$ eigenvalues characterize the dynamics of the system on the simplex. In particular, if the Frobenius eigenvalue is the only dominant eigenvalue, then the system is asymptotically stable on the simplex and $x(t)$ tends, for $t \to \infty$, toward an equilibrium $\bar{x}$, which is the asymptotic state probability distribution of the Markov chain. These properties are further specified by *Theorem 40*.

THEOREM 40 *(probability distribution at equilibrium)*

> In a Markov chain, the Frobenius eigenvector x_F of the matrix P^T is an equilibrium probability distribution $\bar{x}$ (*i.e.*, $x_F = P^T x_F$). If the chain is irreducible, such a distribution is unique and strictly positive ($x_F \gg 0$). Finally, if the chain is primitive, the probability distribution $x(t)$ tends toward $\bar{x}$ for $t \to \infty$ for each $x(0)$, while if the chain is cyclic, the distribution $x(t)$ tends toward a periodic distribution $\tilde{x}(t)$, which depends on $x(0)$, has average $\bar{\tilde{x}} = x_F$ and period equal to the cyclicity index r.

Proof. Since in a Markov chain $\lambda_F = 1$, the Frobenius eigenvector x_F of P^T satisfies the relationship $x_F = P^T x_F$. It is, therefore, an equilibrium probability distribution $\bar{x}$, since it satisfies equation (14.1) with $x(t) = x(t+1)$. Uniqueness and strict positivity of $\bar{x}$ in the case of irreducible systems follow from *Theorem 17*. If the chain is primitive, then the dominant eigenvalue in unique (*Theorem 18*) so that the remaining $(n-1)$ eigenvalues are < 1 in modulus. This implies that, on the simplex, $x(t)$ tends toward $\bar{x} = x_F$ for $t \to \infty$. Finally, if the chain is cyclic with cyclicity index r, the dominant eigenvalues are the rth roots of unity, *i.e.*, $\sqrt[r]{1}$ (see *Theorem 18*), and equation (14.1) has an infinity of periodic solutions of period r. Given an initial state $x(0)$ on the simplex, either $x(0)$ belongs to one of these cycles $\tilde{x}(t)$ (this necessarily happens if $r = n$) or the trajectory starting from $x(0)$ tends toward one cycle $\tilde{x}(t)$ (quickly or slowly, depending on the modulus of the subdominant eigenvalue). Moreover, from (14.1) it follows that

$$\frac{1}{r}\sum_{t=0}^{r-1} \tilde{x}(t+1) = P^T \frac{1}{r}\sum_{t=0}^{r-1} \tilde{x}(t)$$

and, from the periodicity condition $[\tilde{x}(r) = \tilde{x}(0)]$, one can write

$$\frac{1}{r}\sum_{t=0}^{r-1} \tilde{x}(t) = P^T \frac{1}{r}\sum_{t=0}^{r-1} \tilde{x}(t)$$

Such an equation states that the average value $\bar{\tilde{x}}$ of the periodic probability distribution $\tilde{x}(t)$ is equal to x_F.

$\diamondsuit$

On the basis of *Theorem 40*, we can assert, for example, that the probability distribution of the irreducible Markov chain in *Fig. 14.1(a)* asymptotically tends toward the Frobenius eigenvector x_F of P^T independently of the initial conditions. As a useful exercise, the reader is invited to determine such a distribution, after having set the probabilities p_{ij} to some reasonable values. For the practical evaluation of x_F the reader may take advantage of the solution of *Problem 27*.

PROBLEM 27 *(T-II)* Prove that in a primitive Markov chain the matrix P^t tends for $t \to \infty$ toward a matrix with rows equal to x_F^T. This means that the Frobenius eigenvector x_F can be accurately computed by determining the matrices P^2, P^4, P^8, and so on, and stopping when the difference between two successive matrices is sufficiently small. Prove also that this numerical method is more effective than pure simulation [consisting of a recursive use of (14.1) starting from any initial state $x(0)$] if the modulus of the subdominant eigenvalue which determines the rate of convergence to x_F on the simplex, is close to 1.

In order to illustrate the last part of *Theorem 40*, we analyze two simple cyclic Markov chains in Example 18.

EXAMPLE 18 (cyclic Markov chains)
Consider the two Markov chains represented by means of their Markov graph in *Fig. 14.3*.

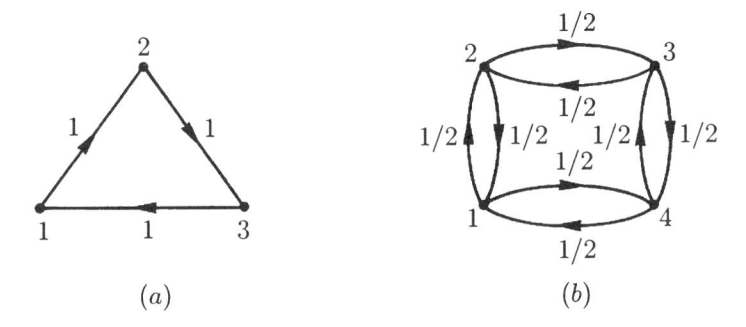

Figure 14.3 Two cyclic Markov chains: (a) $n = r = 3$; (b) $n = 4, r = 2$.

The chain in *Fig. 14.3(a)* is a degenerate "deterministic" Markov chain. If the system is initially in state 1, then it goes through states 2, 3, and then back to 1, thus closing the cycle composed of three vertices of the simplex. But even starting from any other point on the simplex one obtains a cycle of period 3

$$x(0) = (x_1(0) \quad x_2(0) \quad x_3(0))^T$$
$$x(1) = (x_2(0) \quad x_3(0) \quad x_1(0))^T$$
$$x(2) = (x_3(0) \quad x_1(0) \quad x_2(0))^T$$
$$x(3) = x(0)$$

The unique equilibrium solution is then given by the center of the simplex, which coincides with the eigenvector x_F of P^T. This is in agreement with *Theorem 40*, since the system has a cyclicity index $r = 3$.

The Markov chain in *Fig. 14.3(b)* is also cyclic but its cyclicity index is 2 (the greatest common divisor M of the lengths of all simple cycles in the graph is $= 2$). This means that there are two dominant eigenvalues ($\sqrt[2]{1}$), that is, $\lambda'_{\text{dom}} = \lambda_F = 1$ and $\lambda''_{\text{dom}} = -1$. These two eigenvalues with unitary modulus imply the existence of cycles of period 2, for example, of the kind

$$x(0) = \left(\frac{\alpha}{2} \quad \frac{1-\alpha}{2} \quad \frac{\alpha}{2} \quad \frac{1-\alpha}{2}\right)^T$$
$$x(1) = \left(\frac{1-\alpha}{2} \quad \frac{\alpha}{2} \quad \frac{1-\alpha}{2} \quad \frac{\alpha}{2}\right)^T$$
$$x(2) = x(0)$$

Since $r < n$, many trajectories starting from initial states $x(0)$ on the simplex tend toward such cycles. For example, if

$$x(0) = (\alpha \quad 1-\alpha \quad 0 \quad 0)^T$$

one gets

$$x(1) = \left(\frac{1-\alpha}{2} \quad \frac{\alpha}{2} \quad \frac{1-\alpha}{2} \quad \frac{\alpha}{2}\right)^T$$

so that, after only one transition the state of the system is on a cycle. This is due to the fact that the two subdominant eigenvalues are zero (as the reader can easily check).

◇

In order to check what she/he has learned up to now, the reader is invited to solve *Problem 28*, concerning a stochastic model proposed at the beginning of the century by *Ehrenfest* to explain diffusion phenomena.

PROBLEM 28 *(A-III)* Suppose that $n-1$ balls numbered from 1 to $n-1$ are kept in two urns A and B. At regular intervals of time, $t = 0, 1, 2$, and so on, an integer number between 1 and $n-1$ (including the extremals) is randomly chosen, the corresponding ball is identified and moved into the other urn. Describe this (diffusion) process using a Markov chain with n states in such a way that the chain is in state $i (= 1, 2, \ldots, n)$ when there are $i - 1$ balls in urn A. Check that such a Markov chain is not primitive (as someone may think) but cyclic, with cyclicity index $r = 2$. Then, prove that the probability distribution at equilibrium is given by

$$\bar{x}_i = \frac{1}{2^n} \binom{n}{i}$$

and identify the cycles of the probability distribution, at least in the case where there are three balls. Finally, find out what would happen if an integer number between 1 and n (not $n-1$) is randomly chosen and no swop is performed if the extracted number is n.

Since *Theorem 40* is concerned with the asymptotic behavior of irreducible Markov chains, we must now deal with reducible Markov chains. For this, recall (see Chapter 4) that the states of a reducible system $x(t+1) = Ax(t)$, can be renumbered and partitioned into k classes $C_1, \ldots, C_k$, which give rise to a block triangular matrix A

$$A = \begin{pmatrix} A_{11} & 0 & \ldots & 0 \\ A_{21} & 0 & \ldots & 0 \\ \vdots & \vdots & & \vdots \\ A_{k1} & A_{k2} & \ldots & A_{kk} \end{pmatrix}^T$$

Since in a Markov chain $A = P^T$, it follows that the Markov matrix is upper triangular, that is,

$$P = \begin{pmatrix} P_{11} & P_{12} & \ldots & P_{1k} \\ 0 & P_{22} & \ldots & P_{2k} \\ \vdots & \vdots & & \vdots \\ 0 & 0 & \ldots & P_{kk} \end{pmatrix}^T$$

This chain can be described by a graph with k nodes in which each node is a class C_i and the arc (C_i, C_j) is present if and only if $P_{ij} \neq 0$. *Figure 14.4* illustrates the case of a Markov chain with six classes.

If for $t = 0$ the chain is in a state of the class C_1 then, sooner or later, it will be in a state of the class C_3 or C_4 and will never go back to C_1. For this reason, the class C_1 is called *transient*. This means that the probability $x_i(t)$ of being in a state i of the class C_1 tends to zero as $t \to \infty$. In contrast, if at a given time the state of the chain is in a state of the class C_5, it will remain in that class forever. For this reason, such a class is called *absorbing* (or *terminal*). In the most general case, a chain is composed of many transient and absorbing classes (the latter are the classes C_i with $P_{ij} = 0 \ \forall j \neq i$). The Markov chain of *Fig. 14.4* has, for example,

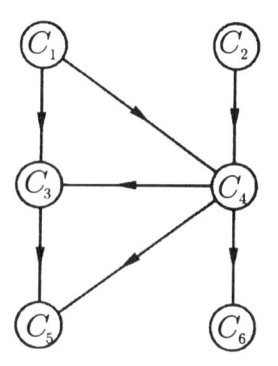

Figure 14.4 A reducible Markov chain: C_5 and C_6 are absorbing classes, while $C_1, \ldots, C_4$ are transient classes.

four transient classes and two absorbing classes. Obviously, each absorbing class is an irreducible chain for which *Theorem 40* holds. In conclusion, the probability distribution $x(t)$ of a reducible Markov chain has all the components relative to the transient classes that tend to zero for $t \to \infty$. The other components of the vector $x(t)$ tend toward an equilibrium or toward a cycle, depending on the initial distribution $x(0)$. If, for example, the chains corresponding to the absorbing classes C_5 and C_6 in *Fig. 14.4* were, respectively, primitive and cyclic, and the chains were initially in a state of class C_3, we would have

$$\lim_{t\to\infty} x_i(t) = 0 \qquad i \notin C_5$$
$$\lim_{t\to\infty} x_i(t) = \bar{x}_i > 0 \quad i \in C_5$$

where $\bar{x}_i$ is the asymptotic probability distribution of the primitive chain corresponding to class C_5. Conversely, if the chain were initially in a state of class C_1, we would have

$$\lim_{t\to\infty} x_i(t) = 0 \qquad i \notin C_5 \cup C_6$$
$$\lim_{t\to\infty} x_i(t) = \alpha \bar{x}_i > 0 \quad i \in C_5$$

with $0 < \alpha < 1$ while $x_i(t), i \in C_6$, would tend to $(1-\alpha)\tilde{x}_i(t)$ where $\tilde{x}_i(t)$ is one of the cycles characterizing the cyclic chain corresponding to class C_6 (see *Theorem 40*). Obviously, once the system is "settled", at each time t

$$\alpha \sum_{i \in C_5} \bar{x}_i + (1-\alpha) \sum_{i \in C_6} \tilde{x}_i(t) = 1$$

It is clear that the parameter α depends on the initial condition $x(0)$ and can be interpreted as the probability [given $x(0)$] that the chain enters, sooner or later, class C_5.

Clearly, the analysis is simplified when there is a single absorbing class, since the asymptotic behavior of the chain is easily determined by applying *Theorem 40* to the irreducible chain describing the absorbing class. However, in many applications one has to deal with reducible chains with multiple absorbing classes, as shown in all the examples and problems that follow. This is true in particular in many important applications in genetics that aim to study the evolution of the species. The fundamental law, in this context, is *Mendel's law*, which allows one to determine, in probabilistic terms, the genotypes (aa, Aa or AA) of each individual given those of the parents ($aa - aa, aa - Aa, \ldots, AA - AA$). Such a law simply states that each individual randomly inherits one allele (a or A) from the father and one from the mother. Thus, for example, individuals born from a father Aa and a mother AA cannot be of genotype aa and will have equal probability ($1/2$) to be of genotype AA or Aa. The following table reports the probability of occurences in all conceivable cases.

Mendel's Table

Offspring genotypes	Parents genotypes					
	$aa - aa$	$aa - Aa$	$aa - AA$	$Aa - Aa$	$Aa - AA$	$AA - AA$
aa	1	1/2	0	1/4	0	0
Aa	0	1/2	1	1/2	1/2	0
AA	0	0	0	1/4	1/2	1

For predicting the evolution of the genetic features of a population, it is then sufficient to know if mating between males and females is independent of their genotypes (random mating) or is influenced, for natural or controlled selection reasons, by their genotypes. Mating rules and Mendel's law can be easily embedded in a Markov chain, which is a genuine genetic model that can be used to predict the evolution of the genetic features of a population. As a purely illustrative example, we show in *Example 19* how the Markov chain corresponding to a particular mating rule can be derived.

EXAMPLE 19 (nonrandom mating)
Suppose that mating among individuals of a population happens only between individuals of the same genotype. In this case, one is interested only in three columns of Mendel's table: the first ($aa - aa$), the fourth ($Aa - Aa$), and the sixth ($AA - AA$). Denote, then, by 1, 2, and 3 the genotypes Aa, aa, and AA and by $x_i(t)$ the probability that an individual of the tth generation is of genotype i. From Mendel's table and from the mating rule (and also from the hypothesis that the sex and the survival and fertility rates do not depend on genotypes), it follows that

$$x_1(t+1) = \tfrac{1}{2} x_1(t)$$
$$x_2(t+1) = \tfrac{1}{4} x_1(t) + x_2(t)$$
$$x_3(t+1) = \tfrac{1}{4} x_1(t) + x_3(t)$$

140   MARKOV CHAINS

These three state equations obviously describe a Markov chain $x(t+1) = P^T x(t)$ with

$$P = \begin{pmatrix} 1/2 & 1/4 & 1/4 \\ 0 & 1 & 0 \\ 0 & 0 & 1 \end{pmatrix}$$

The corresponding influence graph, depicted in *Fig. 14.5(a)*, is reducible (the first variable is not influenced by the remaining two). The eigenvalue corresponding to such a variable is $1/2$, while the remaining two eigenvalues, corresponding to the absorbing classes 2 and 3, are unitary.

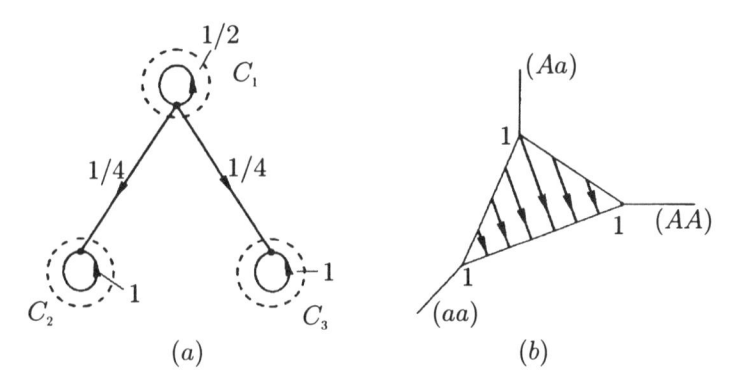

Figure 14.5 Example of a genetic model: (a) Markov graph and its absorbing C_2 and C_3 and transient (C_1) classes; (b) trajectories on the simplex evidencing the extinction of the heterozygotes (Aa).

Consequently, the probability vector $x(t)$ asymptotically tends toward the edge of the simplex corresponding to $x_1 = 0$, as shown in *Fig. 14.5(b)*. This means that the heterozygotes of this population (*i.e.*, the individuals of genotype Aa) are doomed to extinction (they are halved at each generation). Moreover, the law of variation of the two groups of homozygotes (aa and AA) is very simple: They undergo an equal increase at each generation [see the term $(1/4)x_1(t)$ in the second and third state equation]. This is an obvious consequence of the multiplicity of the Frobenius eigenvalue, which renders the trajectories of the system parallel one to each other, as shown in *Fig. 14.5(b)*.

♣

As a second useful exercise on genetic models, the reader is invited to solve *Problem 29*.

PROBLEM 29 (A-III) *Prove that if the females of a given species are fertile only if they belong to one of the two homozygote groups, then, individuals of the other homozygote group would not exist and heterozygotes would be doomed to extinction.*

When a Markov chain is initially in a transient class, it is often useful to derive information on its short-term behavior. For example, one would like to answer the following two questions: For how many transitions will the chain remain in transient classes? and What is the probability that a chain enters an absorbing class visiting first a specified state? One can simply answer these and other questions if the chain has been decomposed into two chains, one corresponding to all transient classes and one corresponding to all absorbing classes. If this decomposition is performed, the Markov matrix P turns out to be

$$P = \begin{pmatrix} P_{tr} & R \\ 0 & P_{abs} \end{pmatrix} \quad (14.3)$$

where the matrix P_{tr} (of dimension $m \times m$, if m is the number of transient states) specifies the behavior of the chain within the transient classes and the matrix P_{abs} [of dimension $(n-m) \times (n-m)$] specifies the behavior of the chain within the absorbing classes. The matrix P_{tr} has the property

$$\lim_{t \to \infty} (P_{tr})^t = 0$$

since the probabilities associated with the transient states must tend to zero a $t \to \infty$. Therefore, P_{tr} is an asymptotically stable positive matrix, which implies (easy to prove) that the matrix $(I - P_{tr})$ has an inverse. Such an inverse matrix, denoted by F, is the so-called *fundamental matrix*:

$$F = (I - P_{tr})^{-1} \quad (14.4)$$

It is easy to check that

$$F = I + P_{tr} + P_{tr}^2 + P_{tr}^3 + \cdots \quad (14.5)$$

so that $F > 0$. We are now ready to prove *Theorem 41*.

142 MARKOV CHAINS

THEOREM 41 *(fundamental matrix)*

> The element f_{ij} of the fundamental matrix F is the average number of times in which the chain passes through the transient state j if initially in the transient state i. Thus, the sum r_i of the elements of the ith row of F is the average number of transitions necessary to enter an absorbing class when starting from the transient state i. Finally, the element g_{ij} of the matrix $G = FR$ is the probability that a chain initially in the ith transient state, enters one of the absorbing class visiting first the jth state of those classes.

Proof. The probability that a Markov chain, initially in the transient state i, is after t transitions in the transient state j is given by the element (i,j) of the matrix $(P_{\text{tr}})^t$ [note that $(P_{\text{tr}})^0 = I$]. The element (i,j) of the matrix $(I + P_{\text{tr}} + P_{\text{tr}}^2 + \ldots)$ is, therefore, the average number of times in which a chain, initially in state i, will visit state j. In view of Eq. (14.5), the first statement of the theorem is proven.

The second statement is a trivial consequence of the first, since a chain leaves forever the transient classes when it enters an absorbing class.

Finally, noting that a chain can go from transient state i to state j of the absorbing classes in just one step with probability r_{ij} [see (14.3)] or in more steps visiting other transient states, we can write

$$g_{ij} = r_{ij} + \sum_{h=1}^{m} p_{ih} g_{hj}$$

In matrix notation, such an expression becomes

$$G = R + P_{\text{tr}} G$$

so that

$$G = (I - P_{\text{tr}})^{-1} R$$

which, in view of (14.4), completes the proof.

$\diamond$

We show now (*Example 20*) a typical application of *Theorem 41*.

EXAMPLE 20 (production process)
Consider the two stage production process described by the Markov graph in *Fig. 14.6*.

Nodes 1 and 2 represent the first and the second operations to be performed sequentially on one item. The probability that the i-th operation is correctly performed is p_i, while the probability that the operation has not been correctly performed but can be repeated is r_i. Obviously,

$$p_i + r_i + s_i = 1 \quad i = 1, 2$$

APPLICATIONS 143

where, s_i is the probability that the item must be eliminated after the operation. States 1 and 2 of the chain are transient states, while states 3 and 4 are absorbing states. The Markov matrix is

$$P = \begin{pmatrix} r_1 & p_1 & s_1 & 0 \\ 0 & r_2 & s_2 & p_2 \\ \hline 0 & 0 & 1 & 0 \\ 0 & 0 & 0 & 1 \end{pmatrix}$$

and is already in the upper triangular form [see (14.3)], with

$$P_{tr} = \begin{pmatrix} r_1 & p_1 \\ 0 & r_2 \end{pmatrix} \quad R = \begin{pmatrix} s_1 & 0 \\ s_2 & p_2 \end{pmatrix} \quad P_{abs} = \begin{pmatrix} 1 & 0 \\ 0 & 1 \end{pmatrix}$$

Thus, the fundamental matrix F is

$$F = (I - P_{tr})^{-1} = \frac{1}{(1-r_1)(1-r_2)} \begin{pmatrix} 1-r_2 & p_1 \\ 0 & 1-r_1 \end{pmatrix}$$

Thus, the average number of times one item goes through the second operation is

$$f_{12} = \frac{p_1}{(p_1 + s_1)(p_2 + s_2)}$$

while the average number of operations performed on a single item is

$$f_{11} + f_{12} = \frac{p_1 + p_2 + s_2}{(p_1 + s_1)(p_2 + s_2)}$$

The probabilities that one item is eliminated or accepted can be read, instead, on the first row of the matrix

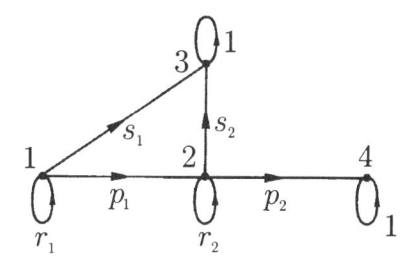

Figure 14.6 Production process with two stages: 3 = elimination and 4 = acceptance of the item.

144   MARKOV CHAINS

$$G = FR = \begin{pmatrix} \dfrac{s_1 s_2 + p_1 s_2 + p_2 s_1}{(p_1 + s_1)(p_2 + s_2)} & \dfrac{p_1 p_2}{(p_1 + s_1)(p_2 + s_2)} \\ \\ \dfrac{s_1 s_2 + p_1 s_2}{(p_1 + s_1)(p_2 + s_2)} & \dfrac{p_1 p_2 + p_2 s_1}{(p_1 + s_1)(p_2 + s_2)} \end{pmatrix}$$

♣

We end this chapter by proposing a problem that can be solved using *Theorem 41*.

PROBLEM 30 *(A-III) Describe the evolution of the score of a tennis game between two players A and B with a Markov chain. The chain must have 17 states: state 1 is the initial state and states 16 and 17 are the only absorbing states, corresponding to the game won by player A and by player B, respectively. Pay attention to the "equivalence" of 30 − 30, 40 − 40, and deuce. Suppose player A is better but more emotional than B, that is, suppose that the probability for A to win a single point is $p' = 1/2 - \varepsilon$ in the case of a "game ball" and $p'' = 1/2 + \delta$ otherwise. Which relationship must exist between ε and δ for the match to be well balanced?*

15
Compartmental Systems

Compartmental systems are systems composed of interconnected reservoirs. They are frequently used in hydrology to describe natural and artificial networks of reservoirs designed for electric power supply, flood control, or water supply to agricultural users. Models of the same type are used by biologists and bioengineers to describe the transportation, accumulation, and drainage processes of many elements and compounds such as hormones, glucose, insulin, and metals in the human body. In this case, the reservoirs are organs or tissues and are called *compartments*. Other examples of compartmental systems are stocking and industrial systems containing chemical reactors, heat exchangers, distillation columns, and other processes.

We shall limit ourselves to the case of linear, continuous-time, single input–single output systems. Thus, a compartmental system is composed of n interconnected compartments (reservoirs) described by a first-order differential equation and an output transformation

$$\dot{x}_i(t) = -\alpha_i x_i(t) + w_i(t) \tag{15.1}$$

$$y_i(t) = \beta_i x_i(t) \tag{15.2}$$

where $x_i(t)$ is the amount of resource present in the ith compartment at time t, w_i is the inflow, y_i is the outflow, and $\alpha_i x_i$ is the sum of the outflow and the losses of resource in the compartment. All the parameters and variables in (15.1), (15.2), and (15.4) are nonnegative and

$$\alpha_i \geq \beta_i \tag{15.3}$$

where the equality holds when there are no losses. In the case of lakes or artificial reservoirs, Eq. (15.3) holds with the strict inequality sign because there are losses due to evaporation. The transfer function of each compartment is

$$G_i(s) = \frac{\mu_i}{1 + sT_i}$$

where the gain μ_i and the time constant T_i are related to the parameters α_i and β_i in (15.1), (15.2), and (15.4) by the equations

$$\mu_i = \frac{\beta_i}{\alpha_i} \qquad T_i = \frac{1}{\alpha_i}$$

Condition (15.3) is therefore equivalent to $\mu_i \leq 1$.

The compartmental system is said to be *acyclic* when there are no loops of compartments (as in irrigation systems) and *cyclic* in the opposite case (as in chemical processes with recycles). The interconnections among the n compartments are described by the dependency of each inflow w_i on the outflows $y_1, y_2, \ldots, y_n$, and on the external inflow $u(t)$, which feeds the whole system,

$$w_i(t) = \sum_{j=1}^{n} d_{ji} y_j(t) + b_i u(t) \tag{15.4}$$

The parameter b_i is the fraction of input directly feeding the ith compartment, so that

$$0 \leq b_i \leq 1 \qquad \sum_{i=1}^{n} b_i = 1 \tag{15.5}$$

while the parameters d_{ji} satisfy the inequalities

$$0 \leq d_{ji} \leq 1 \qquad \sum_{i=1}^{n} d_{ji} \leq 1 \tag{15.6}$$

since d_{ji} is the fraction of the outflow of the jth compartment that feeds the ith compartment. The last relationship is a strict inequality if the jth compartment discharges some resource outside the system, as in the case of a reservoir that supplies water to an agricultural user. All the previous expressions state that resource can only be transported, stored, and discharged but not "created". This means that biological systems characterized by reproductive processes cannot, in general, be described by compartmental models.

Equations (15.1), (15.2), and (15.4) are the state equations of the compartmental system and can be written in the more compact form

$$\dot{x}(t) = Ax(t) + bu(t)$$

where matrix A is a Metzler matrix, since (15.5) and (15.6) imply $a_{ij} = \beta_j d_{ji} \geq 0$ for $i \neq j$, and $b \geq 0$. A compartmental system is, therefore, a positive system. One of its properties is to have nonpositive entries on the diagonal of its matrix A. In fact, $a_{ii} = -\alpha_i + \beta_i d_{ii}$ is nonpositive due to (15.3) and (15.6). More precisely, the sum c_i^+ of the elements of the ith column of A is

$$c_i^+ = -\alpha_i + \beta_i \sum_{j=1}^{n} d_{ij}$$

so that, in view of (15.3) and (15.6), we have

$$c_i^+ \leq 0 \quad i = 1, \ldots, n \tag{15.7}$$

The dominant eigenvalue λ_F is, therefore, zero or negative (see *Theorem 12*) and, in the latter case, the system is asymptotically stable. The case $\lambda_F = 0$ can be excluded provided there are no "degenerate" cycles of compartments, that is, cycles in which a constant amount of resource can indefinitely circulate when the system is not supplied from outside [$u(t) \equiv 0$]. This degeneration occurs when all the compartments composing a cycle have no losses and do not supply external users. Since, for obvious reasons, this is never the case in applications, one can conclude that compartmental systems are asymptotically stable positive linear systems satisfying constraint (15.7) (with the strict inequality for at least one i). It is, however, possible to show that constraint (15.7) is not very important. In fact, any asymptotically stable positive system

$$\dot{x}(t) = Ax(t) + bu(t)$$

not satisfying condition (15.7) can be transformed, through a change of coordinates

$$z = Tx$$

where T is a diagonal matrix with positive entries on the diagonal, into a new system

$$\dot{z} = A^* z(t) + b^* u(t)$$

with $A^* = TAT^{-1}$ satisfying constraint (15.7). This means that any asymptotically stable positive system can be regarded as equivalent to a compartmental system and that the transformation $z = Tx$ realizing such an equivalence is nothing but a change of the units measuring the amount of resource stored in each reservoir.

PROBLEM 31 *(T-III) Give a formal proof of the last statement.*

Compartmental systems, like any positive system, can be profitably represented by their influence graphs. However, they are more often represented by a diagram, called a *compartmental network*, which shows the flow of resource among the reservoirs composing the system. The three elements of compartmental networks are depicted in *Fig. 15.1*: (*a*) the compartment with its inflow and outflow; (*b*)

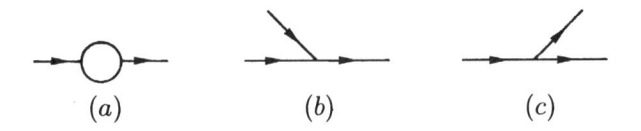

Figure 15.1 Elements of a compartmental network: (a) compartment; (b) confluence point; (c) branching point.

the confluence point where two flows are summed; (c) the branching point where the flow is split into two parts (not necessarily equal).

As for the influence graph, the compartmental network also describes only the structure of the system, namely, the interconnections among the compartments, but does not give any information about the magnitude of gains and time constants.

EXAMPLE 21 (discrete model of an aquifer)
The groundwater flow in an aquifer is described by means of partial differential equations, usually solvable only numerically provided the boundary and initial conditions are specified. For example, in the case of an isotropic, homogeneous, and unidimensional aquifer, the pressure head $h(l,t)$ is described by the partial differential equation

$$k\frac{\partial^2 h}{\partial l^2} = \sigma\frac{\partial h}{\partial t} \qquad t \geq 0 \quad 0 \leq l \leq L \tag{15.8}$$

with boundary conditions

$$-k\left[\frac{\partial h}{\partial l}\right]_{l=0} = \sigma u(t) \qquad -k\left[\frac{\partial h}{\partial l}\right]_{l=L} = \sigma y(t) \quad t \geq 0$$

and initial conditions

$$h(l,0) = h_0(l) \qquad 0 \leq l \leq L$$

The coefficients k (transmissivity) and σ (storage coefficient) are positive and depend on the type of soil.

For obtaining an easily tractable model, one can perform a spatial discretization by dividing the interval $[0, L]$, which determines the aquifer, in n elementary stretches of length $\Delta = L/n$. By setting

$$x_i(t) = h\left(i\Delta - \frac{\Delta}{2}, t\right) \qquad i = 1, \ldots, n$$

and considering that, for n sufficiently large,

$$\frac{\partial^2 h}{\partial l^2}\left(i\Delta - \frac{\Delta}{2}, t\right) \cong \frac{1}{\Delta^2}(x_{i+1}(t) - 2x_i(t) + x_{i-1}(t))$$

APPLICATIONS 149

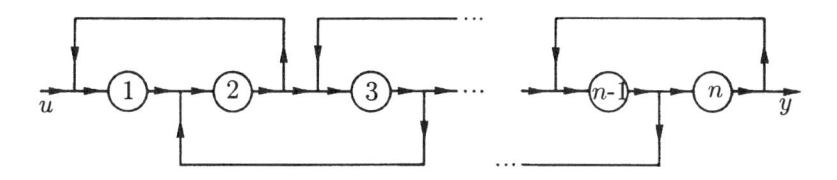

Figure 15.2 Compartmental network of the aquifer described by Eq. (15.8).

Eq. (15.8) can be approximated by the following n ordinary differential equations

$$\dot{x}_1(t) = -\frac{k}{\sigma\Delta^2}x_1(t) + \frac{k}{\sigma\Delta^2}x_2(t) + \frac{1}{\Delta}u(t)$$
$$\dot{x}_i(t) = \frac{k}{\sigma\Delta^2}x_{i-1}(t) - \frac{2k}{\sigma\Delta^2}x_i(t) + \frac{k}{\sigma\Delta^2}x_{i+1}(t) \quad i = 2,\ldots,n-1$$
$$\dot{x}_n(t) = \frac{k}{\sigma\Delta^2}x_{n-1}(t) - \frac{2k}{\sigma\Delta^2}x_n(t)$$

The compartmental network of *Fig. 15.2*, composed of a chain of n reservoirs with recycles, interprets these equations. The time constant of the first reservoir is $T_1 = \sigma\Delta^2/k$, while all remaining reservoirs have a time constant $T_2 = \sigma\Delta^2/2k$.

♣

PROBLEM 32 *(T-III) Consider the two communicating tanks shown in Fig. 15.3(a) and suppose that the exchange flow between the two tanks is proportional to the pressure difference at their bottoms. Then, write the state equations of the system and check that the corresponding compartmental network is the one depicted in Fig. 15.3b. In view of this result, give an interpretation of the compartmental network in Fig. 15.2.*

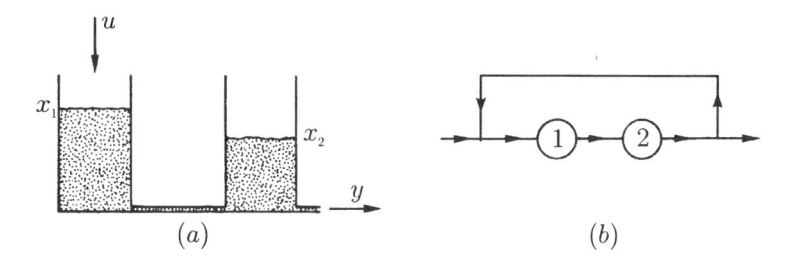

Figure 15.3 Communicating tanks (a) and corresponding compartmental network (b).

Theorems 35 and *39*, which are useful to prove that a system is minimum phase, can be applied to compartmental systems because such systems are asymptotically

150 COMPARTMENTAL SYSTEMS

stable. For example, from *Theorem 35* it follows that the aquifer (15.8) is minimum phase since its compartmental network (see *Fig. 15.2*) clearly shows that the input–output path is unique. This means that the inflow $u(t)$ can be computed from an outflow record $y(t)$ provided the parameters k and σ of the aquifer are known. From *Theorem 39*, one can state that a system composed of two compartmental subsystems connected in series is minimum phase if and only if the two subsystems are such. For feedback interconnections, the same theorem gives the following result: A compartmental system composed of two compartmental subsystems connected in feedback is minimum phase if and only if the subsystem in the forward direction (from input to output) is minimum phase.

We now give some extra results on the minimum phase of compartmental systems.

THEOREM 42 *(parallel interconnection of compartments)*

> A compartmental system composed of n compartments connected in parallel is minimum phase. Moreover, its transfer function $G^{(n)}(s)$ has alternate zeros and poles.

Proof. First, enumerate the compartments in decreasing order of their time constant T_i, so that $-p_1 \geq -p_2 \geq \cdots \geq -p_n$, where $-p_i = -1/T_i$ is the pole of the ith compartment. Second, note that the theorem holds for $n = 2$, since

$$G^{(2)}(s) = \frac{\rho_1}{s+p_1} + \frac{\rho_2}{s+p_2} = \rho^{(2)} \frac{s+z^{(2)}}{(s+p_1)(s+p_2)}$$

with

$$z^{(2)} = \frac{\rho_2}{\rho_1+\rho_2} p_1 + \frac{\rho_1}{\rho_1+\rho_2} p_2$$

Finally, note that *Theorem 42* holds for $n = k+1$ if it holds for $n = k$. In fact, the zeros of the transfer function $G(s) + H(s)$ of two systems connected in parallel coincide with the poles of the system obtained by connecting system $G(s)$ in (negative) feedback with the system with transfer function $1/H(s)$. Therefore, the zeros of the parallel connection of system

$$G^{(k)}(s) = \sum_{i=1}^{k} \frac{\rho_i}{s+p_i} = \rho^{(k)} \frac{\prod_{i=1}^{k-1}(s+z_i^{(k)})}{\prod_{i=1}^{k}(s+p_i)}$$

with the $(k+1)$-st compartment with transfer function $H_{k+1}(s) = \rho_{k+1}/(s+p_{k+1})$, can be found by drawing the root locus of the open loop transfer function $G^{(k)}(s)/H_{k+1}(s)$, which has k poles $(-p_1, -p_2, \ldots, -p_k)$ and k zeros $(-z_1^{(k)}, -z_2^{(k)}, \ldots, -z_{k-1}^{(k)}, -p_{k+1})$ alternating on the negative real axis, as shown in *Fig. 15.4*.

APPLICATIONS 151

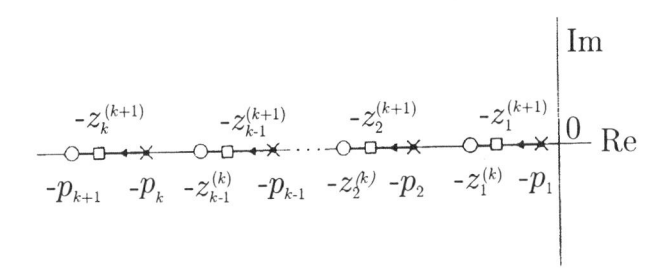

Figure 15.4 The root locus of $G^{(k)}(s)/H_{k+1}(s)$.

In this case, the root locus is composed of k segments of the real axis connecting each pole of $G^{(k)}(s)/H_{k+1}(s)$ with the nearest zero on its left. On each one of these segments, there is one zero $-z_i^{(k+1)}$ of the transfer function $G^{(k+1)}(s)$. Since all these zeros are negative, the system is minimum phase. Moreover, the zeros $(-z_1^{(k+1)}, -z_2^{(k+1)}, \ldots, -z_k^{(k+1)})$ and the poles $(-p_1, -p_2, \ldots, -p_{k+1})$ of $G^{(k+1)}(s)$ alternate. Therefore, the theorem is proved by induction.

◇

Since n compartments connected either in series or in parallel are minimum phase, it is natural to conjecture that all acyclic compartmental systems are minimum phase. Unfortunately, this is false (see, e.g., *Problem 33*), though a weaker form of this conjecture is valid, as shown by the following theorem, in which the notion of *topological diversity D* of a compartmental network is used. Topological diversity is defined as

$$D = \max_h\{n_h\} - \min_h\{n_h\}$$

where n_h is the number of compartments composing the hth direct input–output path. The topological diversity is therefore the difference, measured in compartment numbers, between the longest and the shortest direct input–output path. For example, the compartmental networks denoted by (*a*) and (*b*) in *Fig. 15.5* have two direct input–output paths and their topological diversity is equal to 2 and 1, respectively, while the whole network has topological diversity equal to 3.

THEOREM 43 *(acyclic interconnection of compartments)*

A compartmental system with more than one direct input–output path is minimum phase if

a. It is acyclic.

b. It has topological diversity < 3.

c. The time constants of the compartments are not "too different".

Proof. Suppose that an acyclic system has n compartments, all with the same time constant $T_i = T = 1/p$ and that each direct input–output path contains $k, k+1$ or $k+2$ compartments, so that $D < 3$. Since the system is acyclic, its transfer function is the sum of the transfer functions of all direct input–output paths, that is,

$$G(s) = \frac{P}{(s+p)^k} + \frac{Q}{(s+p)^{k+1}} + \frac{R}{(s+p)^{k+2}} =$$
$$= \frac{Ps^2 + (2Pp+Q)s + Pp^2 + Qp + R}{(s+p)^{k+2}}$$

where P, Q, and R are nonnegative and at least one of them is positive. If $PQR > 0$, the transfer function $G(s)$ has two zeros with a negative real part, since the coefficients of the second-order polynomial are positive. It is also straightforward to check that in the case $PQR = 0$, the zeros of $G(s)$ (if they exist) have a negative real part. Therefore, the system is minimum phase.

To complete the proof, we must check that a sufficiently small perturbation of the time constants of the compartments, say from T to $T + \varepsilon_i, i = 1, 2, \ldots, n$, does not generate zeros with a nonnegative real part. For this, note that, in general, $k + 2 < n$ since, by hypothesis, there exist at least two direct input–output paths. Therefore, the unperturbed system is not completely reachable and observable (in the usual sense, see Appendix B) since its transfer function has $< n$ poles. A perturbation of the time constants give rise, in such a case, to a transfer function with a higher number of poles and zeros. However, if the ε_i's are sufficiently small, two of those zeros are, by continuity, close to the two roots of the above mentioned second-order polynomial and therefore have a negative real part. The remaining zeros are close to the eigenvalues of the unreachable and/or unobservable part of the unperturbed system (they coincide with them for $\varepsilon_i = 0, i = 1, \ldots, n$). But such eigenvalues are all equal to $-1/T$ since the system is acyclic. In conclusion, all the zeros have a negative real part if the time constants T_i of the compartments are not too different.

◇

Theorem 43 states that acyclic compartmental networks are minimum phase if they have low "total diversity", that is, if they have a low mix of topological diversity and dynamic diversity $\max_{ij}(T_i - T_j)$. If the system has a very low topological diversity ($D = 0$ or $D = 1$), it may happen that the system is minimum phase even if its dynamic diversity is high. One example is the parallel connection of one compartment with the series interconnection of two compartments [system (*b*) with $D = 1$ in *Fig. 15.5*], which is minimum phase for every value of the three time constants (easy to check). Another example (see *Theorem 42*) is the parallel interconnection of n compartments ($D = 0$). *Problem 33* shows, however, that an acyclic system with $D = 2$ may be a nonminimum phase when condition *c* is not met, that is, when dynamic diversity is too large.

PROBLEM 33 *(T-II)* Show that the compartmental system denoted by (a) in *Fig. 15.5* can be a nonminimum phase. (Hint: Use the root locus technique as in the proof of Theorem 42 .)

The properties we have just presented are often useful for reducing the analysis of the nonminimum phase property of a complex compartmental system to the study of the sign of the real part of the zeros of a low-order polynomial. For example, the system in *Fig. 15.5*, which has 19 compartments and, consequently, might have 18 zeros, can be analyzed by studying only the second-order polynomial defining the zeros of subsystem (a) (see *Problem 33*). More precisely, if such zeros have a negative real part, that is, if system (a) is minimum phase, such is the whole system, even if the acyclic subsystem denoted by (c), which has $D = 3$, is not minimum phase. This follows from the result we have previously pointed out concerning the nonminimum phase of series and feedback interconnections.

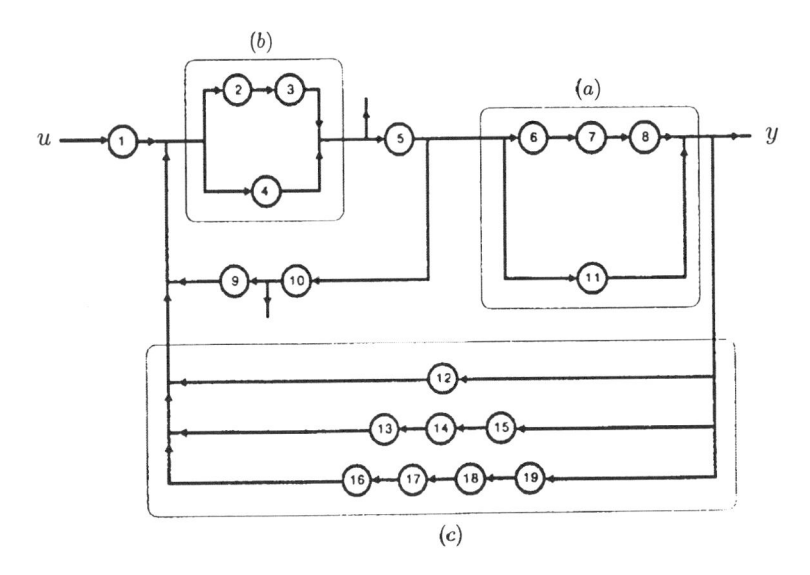

Figure 15.5 Compartmental network of a cyclic system with four direct input–output paths and topological diversity $d = 3$.

16

Queueing Systems

Systems in which an operation (service) is performed on individuals, files, machines, or other matters, are frequently encountered in applications. Such systems are generally composed of two parts, as shown in *Fig. 16.1*

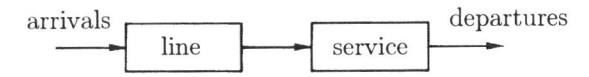

```
arrivals ──→ [ line ] ──→ [ service ] ──→ departures
```

Figure 16.1 Structure of a queueing system.

The first part is a kind of reservoir where the users wait before they are served. In the second part, often composed of parallel channels (multiserver), a *service* is performed in a certain time. Users *arrivals* and *departures* are, at least to some extent, random and characterize the first and second part of the system, respectively. An airport with a queue of airplanes ready for landing, an office where the papers for passport renewal are processed, or the waiting room of a medical center are typical examples of systems characterized by a waiting time followed by a service.

Such systems can be modeled when the statistics of the arrivals, the rule used for selecting the next user to be served, and the statistics of the service times are specified. We will assume that the selection rule is the "first in–first out" rule, characterizing systems with users waiting in line. This choice is, however, not relevant for our scope, which is the description of the time evolution of the number of users waiting to be served. In fact, any other selection rule, like the " last in–last out" or the random one, would not vary the dynamics of the number of waiting users but only change the waiting time of each user. Moreover, we shall assume that the service has limited capacity (s servers working in parallel) and that there are no users in queue if some server is inactive. Arrivals and departures will be assumed to be random processes characterized by the property that the probability of one arrival or departure during an infinitesimal interval of time dt is proportional to dt itself. Processes of this kind are called *Poisson's processes*. The proportionality coefficients, denoted by λ and μ, respectively, depend, in general, on the total number i of users present in the system (viz., the users in queue and under service). Thus, $\lambda_i dt$ [$\mu_i dt$] is the probability of one arrival [departure] during the interval $(t, t+dt)$ if there are i users in the system. In other words, λ_i and μ_i are the arrival and departure frequencies and $1/\lambda_i$ and $1/\mu_i$ are the time intervals separating, on average, two successive arrivals and departures when there are i users in the system. Such hypotheses, called *Markovian hypotheses*, imply that the future behavior of the system is determined (obviously, in probabilistic terms) once the initial state of the system (probability distribution of the number of users present in the system at the initial time) is known. Queueing systems are, therefore, the continuous-time version of Markov chains (see Chapter 14).

We are now ready to show how queueing systems can be modeled. For this, denote by $x_i(t)$ the probability that i users are in the system at time t, and let us compute the probability $x_i(t+dt)$ as a function of $x_j(t)$. We begin with $i = 0$, that is with the case of an empty system, which is, in some sense, a degenerate situation. Since, from the hypotheses, the probability that two or more arrivals or departures occur during the interval of length dt is negligible with respect to dt, we can write

$$x_0(t+dt) = x_0(t)(1 - \lambda_0 dt) + x_1(t)\mu_1 dt(1 - \lambda_1 dt) \qquad (16.1)$$

Such an expression states that at time $t+dt$ there are no users in the system if one of the following facts happen:

1. There are no users in the system at time t and no user arrives during the interval $(t, t+dt)$.

2. There is one user in the system at time t and such a user leaves the system while no other user arrives during the interval $(t, t+dt)$.

For dt going to zero, Eq. (16.1) becomes

$$\dot{x}_0(t) = -\lambda_0 x_0(t) + \mu_1 x_1(t) \qquad (16.2)$$

which is the first state equation of the queueing system. If $i > 0$, we can write

$$x_i(t+dt) = x_{i-1}(t)\lambda_{i-1} dt(1 - \mu_{i-1} dt) + x_i(t)(1 - \lambda_i dt)(1 - \mu_i dt)$$
$$+ x_{i+1}(t)\mu_{i+1} dt(1 - \lambda_{i+1} dt)$$

which has three terms instead of two at the right-hand sides, since in this case there is also the possibility that the number of users decreases by one unit during the time interval $(t, t+dt)$. As $dt \to 0$, one obtains the generic state equation

$$\dot{x}_i(t) = \lambda_{i-1}x_{i-1}(t) - (\lambda_i + \mu_i)x_i(t) + \mu_{i+1}x_{i+1}(t) \qquad (16.3)$$

Equations (16.2) and (16.3) with $i = 1, 2$, and so on, are the model of the system when the queue is not bounded. Such equations are those of an infinite-dimensional linear system. In contrast, if the servers are s and the maximum number of users waiting in queue is c, we get a system of dimension $n + 1$, where

$$n = c + s$$

In fact, Eq. (16.3) holds for $i = 1, 2, \ldots, n - 1$ while, for $i = n$, we have

$$\dot{x}_n(t) = \lambda_{n-1}x_{n-1}(t) - \mu_n x_n(t) \qquad (16.4)$$

which is Eq. (16.3) with $\lambda_n = \mu_{n+1} = 0$. Equations (16.2) and (16.3) with $i = 1, 2, \ldots, n-1$, and (16.4) are then the state equations of a system with bounded queue and a finite number of services.
Such a system is of the kind

$$\dot{x}(t) = Ax(t) \qquad (16.5)$$

where the state vector $x = \begin{pmatrix} x_0 & x_1 & \ldots & x_n \end{pmatrix}^T$ is $(n+1)$-dimensional and satisfy the equation $\sum_{i=0}^{n} x_i(t) = $ constant since $\sum_{i=0}^{n} \dot{x}_i(t) = 0$. This means that, if the initial state $x(0)$ is on the simplex $\sum_{i=0}^{n} x_i = 1$ (as it must be, x_i being a probability), the state $x(t)$ remains on the simplex forever. The matrix A, of dimensions $(n+1) \times (n+1)$, is given by

$$A = \begin{pmatrix} -\lambda_0 & \mu_1 & 0 & 0 & \ldots & 0 & 0 \\ \lambda_0 & -\lambda_1 - \mu_1 & \mu_2 & 0 & \ldots & 0 & 0 \\ 0 & \lambda_1 & -\lambda_2 - \mu_2 & \mu_3 & \ldots & 0 & 0 \\ \vdots & \vdots & \vdots & \vdots & & \vdots & \vdots \\ 0 & 0 & 0 & 0 & \ldots & \lambda_{n-1} & -\mu_n \end{pmatrix}^T \qquad (16.6)$$

and has negative elements on its diagonal, positive elements on the two pseudodiagonals, and zero elements elsewhere. It is therefore a Metzler matrix. In other words, system (16.5), (16.6) is a continuous-time positive linear system and its state vector x is the probability distribution of the number of users in the system. The influence graph of system (16.5), (16.6) is connected, as shown in *Fig. 16.2*, so that the system is irreducible. From this, it follows (see *Theorem 17*) that the Frobenius eigenvector x_F is unique and strictly positive. Moreover, the Frobenius eigenvalue λ_F is zero, since all the columns of A have a zero sum, so that the eigenvector x_F satisfies the equation

$$Ax_F = 0 \qquad (16.7)$$

158 QUEUEING SYSTEMS

Figure 16.2 Influence graph of the queueing system (16.5), (16.6).

In conclusion, $x_F \gg 0$ is an equilibrium state of the system. Moreover, the probability distribution $x(t)$ asymptotically tends toward x_F since the system is asymptotically stable on the simplex (all the eigenvalues of A, except λ_F, have a negative real part). This means that for computing the asymptotic probability distribution $\bar{x} = x_F$ it is sufficient to integrate Eqs. (16.5), (16.6) from any initial condition $x(0)$ on the simplex. This allows one to obtain useful information on the rate of convergence of the probability distribution $x(t)$ toward $\bar{x}$. Obviously, such a rate of convergence is related to the real part of the subdominant eigenvalue of matrix A.

PROBLEM 34 (A-II) *A small data-processing center is open 10 h/day for 24 days/month, contains 5 computers, and serves a community of 350 students. Each student goes randomly to the center 3 times per month on average. The student works for 40 min, if one of the computers is free and otherwise goes away without waiting. Knowing that the center opens at 9.00 a.m., find out at what time the center is practically in steady conditions (obviously in a statistical sense). Moreover, determine the average number of computers used in steady conditions and the probability to find all computers busy at* 9.30, 10.30, 11.30, ..., 6.30pm. *(Hint: The time needed to reach, from a practical point of view, the equilibrium can be assumed to be equal to 5 times the subdominant time constant.)*

The asymptotic probability distribution $\bar{x}$ of a queueing system can be easily evaluated because the matrix A [see (16.6)] is almost diagonal. In fact, Eqs. (16.6), (16.7) at equilibrium give

$$\bar{x}_{i+1} = \frac{\lambda_i}{\mu_{i+1}} \bar{x}_i \quad i = 0, 1, 2, \ldots \quad (16.8)$$

so that all the components of the vector $\bar{x}$ can be recursively computed starting from the first component $\bar{x}_0$ (probability that no user is in the system). From Eq. (16.8), we obtain

$$\bar{x}_i = \frac{\lambda_0 \lambda_1 \ldots \lambda_{i-1}}{\mu_1 \mu_2 \ldots \mu_i} \bar{x}_0 \quad i = 1, \ldots, n$$

which can be written as

$$\bar{x}_i = \phi_i \bar{x}_0 \quad i = 1, \ldots, n \quad (16.9)$$

where the parameter ϕ_i is given by

$$\phi_i = \frac{\lambda_0 \lambda_1 \ldots \lambda_{i-1}}{\mu_1 \mu_2 \ldots \mu_i} \qquad i = 1, \ldots, n \qquad (16.10)$$

By recalling that the probabilities $\bar{x}_i$ must have a unitary sum, the probability $\bar{x}_0$ can be evaluated by using the formula

$$\bar{x}_0 = \frac{1}{1 + \sum_{i=1}^{n} \phi_i} \qquad (16.11)$$

Once $\bar{x}_0$ and, hence, $\bar{x}_1, \bar{x}_2$, and so on, are known, one can compute various statistical indicators measuring the system performance at equilibrium. Among them, we can recall the mean number of users in the system

$$\bar{n} = \sum_{i=0}^{n} i \bar{x}_i$$

and in queue

$$\bar{c} = \sum_{i=s+1}^{n} (i-s) \bar{x}_i$$

the average frequency of arrivals

$$\bar{\lambda} = \sum_{i=0}^{n} \lambda_i \bar{x}_i$$

the fraction of time during which the system is not used

$$u = 1 - \bar{x}_0 = \sum_{i=1}^{n} \bar{x}_i$$

the probability that a user cannot be served on arrival

$$p_{\text{busy}} = \sum_{i=s}^{n} \bar{x}_i = 1 - \sum_{i=0}^{s-1} \bar{x}_i$$

and many others, as the average waiting time $\bar{T}_w$, the average service time $\bar{T}_s$, and the average response time $\bar{T} = \bar{T}_w + \bar{T}_s$. Note that all these indicators can be referred to unsteady situations by replacing $\bar{x}_i$ with $x_i(t)$ and that they are linear combinations of the components of the state vector $x(t)$. Thus, they can be expressed with the usual output transformation $y(t) = c^T x(t)$ of a linear system. In other words, a queueing system and its associated performance indicator are formally an autonomous positive linear system (A, c^T).

Clearly, some indicators are related one to each other, often in a quite intuitive manner. For example, the famous *Little's formula* states that the average frequency of arrivals λ is the ratio between the average number of users in queue $\bar{c}$ and the average waiting time $\bar{T}_w$ or between the average number of users in the system $\bar{n}$ and the average response time $\bar{T}$, that is,

$$\bar{c} = \lambda \bar{T}_w \quad \bar{n} = \lambda \bar{T}$$

Explicit formulas for the computation of the asymptotic probability distribution $\bar{x}$ and performance indicators are available for a large number of specific systems of practical interest and represent the kernel of queueing theory. It is worth noting that, nowadays, such a theory has lost part of its importance since the computation of probability distributions and performance indicators can be performed by numerically integrating Eqs. (16.5) on a computer, thus obtaining additional information on the transient behavior of the queue. However, we report next some of the most known formulas for systems without queue, with bounded queue, and with unbounded queue.

Services without queue

This is the case of a telephone system with calls that are accepted if at least one line is free and rejected otherwise. Thus, the frequency of arrivals λ is independent of the number of busy lines if at least one line is free. Each accepted call engages a line for an average time $1/\mu$, which is independent of the number of busy lines. A system of this type is therefore characterized by $c = 0$, $s = n$, and

$$\lambda_i = \lambda \quad i = 0, 1, \ldots, n-1$$
$$\lambda_n = 0$$
$$\mu_i = i\mu \quad i = 1, 2, \ldots, n$$

so that [see (16.10)]

$$\phi_i = \left(\frac{\lambda}{\mu}\right)^i \frac{1}{i!} \quad i = 1, \ldots, n$$

From (16.11), it follows that

$$\bar{x}_0 = \frac{1}{1 + \sum_{i=1}^{n} \left(\frac{\lambda}{\mu}\right)^i \frac{1}{i!}} = \frac{1}{\sum_{i=0}^{n} \left(\frac{\lambda}{\mu}\right)^i \frac{1}{i!}}$$

and from (16.9)

$$\bar{x}_i = \left(\frac{\lambda}{\mu}\right)^i \frac{1}{i!} \bar{x}_0 \quad i = 1, \ldots, n$$

The most interesting performance indicator for such a system is the rejection probability $p_r = x_n$, which can be given the form

$$p_r = \frac{\left(\dfrac{\lambda}{\mu}\right)^n \dfrac{1}{n!}}{\sum_{i=0}^{n}\left(\dfrac{\lambda}{\mu}\right)^i \dfrac{1}{i!}}$$

Such a formula, known as *Erlang's B formula* is used for the computation of the minimum number of lines that can guarantee that the rejection probability is smaller than an acceptable standard. Moreover, since $\lambda_n = 0$, the average arrival frequency $\bar{\lambda}$ is given by

$$\bar{\lambda} = \lambda \bar{x}_0 + \lambda \bar{x}_1 + \cdots + \lambda \bar{x}_{n-1} = \lambda(1 - p_r)$$

so that, from Little's formula $\bar{n} = \bar{\lambda}\bar{T}$ with $\bar{T} = 1/\mu$, it follows that the average number of busy lines is simply given by

$$\bar{n} = \frac{\lambda}{\mu}(1 - p_r)$$

In conclusion, the frequency of the calls (accepted or rejected) is λ while the frequency of the accepted calls is $\bar{\lambda} = \bar{n}/\bar{T} = \bar{n}\mu = \lambda(1 - p_r)$.

PROBLEM 35 *(A-I) Consider the system described in Problem 34 and check that such a system is a service without queue with $n = s = 5$. . Then, evaluate the rejection probability p_r using Erlang's B formula and compare such a result with the solution of Problem 34.*

Services with bounded queue

This is the case of a service with s servers and a queue of finite capacity c. If the arrival frequency and the time needed to perform the service are independent of the number of users in the system, we have

$$\lambda_i = \lambda \quad i = 0, 1, \ldots, n - 1$$
$$\lambda_n = 0$$
$$\mu_i = i\mu \quad i = 1, \ldots, s$$
$$\mu_i = s\mu \quad i = s + 1, \ldots, n$$

Let us now apply Eqs. (16.9), (16.10), and (16.11) for the computation of $\bar{x}$ in the particular case of a single server ($s = 1$). Thus, $\mu_i = \mu$ for $i = 1, \ldots, n$ so that

$$\phi_i = \left(\frac{\lambda}{\mu}\right)^i \quad i = 1, \ldots, n$$

and

$$\bar{x}_o = \frac{1}{1+\sum_{i=1}^{n}\phi_i} = \frac{1}{\sum_{i=0}^{n}\left(\frac{\lambda}{\mu}\right)^i} = \frac{1-\left(\frac{\lambda}{\mu}\right)}{1-\left(\frac{\lambda}{\mu}\right)^{n+1}}$$

while the rejection probability is

$$p_r = \bar{x}_n = \left(\frac{\lambda}{\mu}\right)^n \bar{x}_0$$

Finally, the average number of users in queue is

$$\bar{c} = \sum_{i=2}^{n}(i-1)\bar{x}_i$$

Since in this case

$$\bar{\lambda} = \lambda\bar{x}_0 + \lambda\bar{x}_1 + \cdots + \lambda\bar{x}_{n-1} = \lambda(1-\bar{x}_n) = \lambda(1-p_r)$$

the average waiting time $\bar{T}_w$ can be obtained by means of Little's formula $\bar{c} = \bar{\lambda}\bar{T}_w$.

PROBLEM 36 *(A-I)* A medical team for heart transplants accepts no more than 10 people on its waiting list. If we take into account the scarcity of donors, each operation keeps the team busy for 10 days. If the center serves a population of 1,000,000 people with 70 years average life and the need for heart transplants is 1 each 100 individuals, determine the average waiting time $\bar{T}_w$ and the rejection probability p_r.

Services with unbounded queue
In this case the system is described by Eqs. (16.2) and (16.3) with $i = 1, 2$, and so on, that is, by an infinite number of differential equations. However, formulas (16.9), (16.10), and (16.11) still hold provided n is replaced with ∞. The most easily tractable case is that of "nondiscouraged" arrivals, that is,

$$\lambda_i = \lambda \quad i = 0, 1, 2, \text{ and so on}$$

and servers not influenced by the number of users waiting in queue, that is,

$$\mu_i = i\mu \quad i = 1, \ldots, s$$

$$\mu_i = s\mu \quad i = s+1, s+2, \text{ and so on}$$

Under these hypotheses, the recurrent equation

$$\bar{x}_{i+1} = \frac{\lambda_i}{\mu_{i+1}}\bar{x}_i$$

remains unchanged for $i \geq s$, that is,

$$\bar{x}_{i+1} = \frac{\lambda}{s\mu} \bar{x}_i \quad i \geq s$$

This means that

$$\lambda < s\mu$$

must hold for the queue to possess the asymptotic property

$$\lim_{i \to \infty} \bar{x}_i = 0 \tag{16.12}$$

If we assume that this property is satisfied and we use the procedure previously outlined, we obtain

$$\bar{x}_0 = \frac{1}{\dfrac{\left(\dfrac{\lambda}{\mu}\right)^s}{s!\left(1 - \dfrac{\lambda}{s\mu}\right)} + \sum_{i=0}^{s-1} \left(\dfrac{\lambda}{\mu}\right)^i \dfrac{1}{i!}}$$

so that it is possible to recursively compute [see Eq. (16.8)] the components $\bar{x}_i$ of the asymptotic probability distribution, and all the related performance indicators. For example, the probability p_{busy} of finding the service busy is given by the following formula, known as *Erlang's C formula*

$$p_{\text{busy}} = \left(\frac{\lambda}{\mu}\right)^s \frac{\bar{x}_0}{s!\left(1 - \dfrac{\lambda}{s\mu}\right)} \tag{16.13}$$

The derivation of this formula is left to the reader as an exercise.

The particular case of a single server ($s = 1$) is perhaps the best known and used in queueing theory. In this case, Eqs. (16) and (16.13) become

$$\bar{x}_0 = 1 - \frac{\lambda}{\mu}$$

$$p_{\text{busy}} = \frac{\lambda}{\mu}$$

Moreover,

$$\bar{n} = \sum_{i=0}^{\infty} i\bar{x}_i = \sum_{i=0}^{\infty} i\left(\frac{\lambda}{\mu}\right)^i \left(1 - \frac{\lambda}{\mu}\right) = \left(1 - \frac{\lambda}{\mu}\right) \sum_{i=0}^{\infty} i\left(\frac{\lambda}{\mu}\right)^i$$

Then, setting $\rho = (\lambda/\mu)$ one obtains

$$\bar{n} = (1-\rho)\sum_{i=0}^{\infty} i(\rho)^i = \rho(1-\rho)\sum_{i=0}^{\infty} \frac{\partial}{\partial \rho}(\rho)^i = \rho(1-\rho)\frac{\partial}{\partial \rho}\sum_{i=0}^{\infty}(\rho)^i$$
$$= \rho(1-\rho)\frac{\partial}{\partial \rho}\frac{1}{(1-\rho)} = \rho(1-\rho)\frac{1}{(1-\rho)^2} = \frac{\rho}{1-\rho}$$

so that, in the end,

$$\bar{n} = \frac{\frac{\lambda}{\mu}}{1 - \frac{\lambda}{\mu}}$$

From Little's formula $\bar{n} = \bar{\lambda}\bar{T}$ with $\bar{\lambda} = \lambda$, it follows that the average response time is

$$\bar{T} = \frac{\frac{1}{\lambda}}{1 - \frac{\lambda}{\mu}}$$

EXAMPLE 22 (communication system)
Consider a communication system that first divides the message to be transmitted into segments and then sends the first segment waiting in queue to the first free transmission line. Let λ be the arrival frequency of the segments and $1/\mu$ the average transmission time of each single segment. If there is only one line with transmission time $1/\mu'$, the previously considered formulas hold, that is,

$$\bar{n}' = \frac{\frac{\lambda}{\mu'}}{1 - \frac{\lambda}{\mu'}} \qquad \bar{T}' = \frac{\bar{n}'}{\lambda}$$

We can now determine the analogous formulas for a system with two transmission lines ($s = 2$) and denote by $\bar{n}''$ and $\bar{T}''$ the corresponding performance indicators. Since, in this case, $\mu_1 = \mu''$ and $\mu_i = 2\mu''$ for $i = 2, 3$, and so on, we have

$$\phi_1 = \frac{\lambda}{\mu''} \quad \phi_2 = \frac{\lambda}{\mu''}\frac{\lambda}{2\mu''} \quad \phi_3 = \frac{\lambda}{\mu''}\frac{\lambda}{2\mu''}\frac{\lambda}{2\mu''} \text{ and so on}$$

so that

$$\phi_i = 2\left(\frac{\lambda}{2\mu''}\right)^i \quad i = 1, 2, \text{ and so on}$$

From (16.11) (with n replaced with ∞) it follows that

$$\bar{x}_0 = \frac{1}{1 + 2\sum_{i=1}^{\infty}\left(\frac{\lambda}{2\mu''}\right)^i} = \frac{1 - \frac{\lambda}{2\mu''}}{1 + \frac{\lambda}{2\mu''}}$$

and

$$\bar{x}_i = 2\left(\frac{\lambda}{2\mu''}\right)^i \bar{x}_0$$

The average number $\bar{n}''$ of segments in the system is

$$\bar{n}'' = \sum_{i=0}^{\infty} i\bar{x}_i = 2\frac{\lambda}{2\mu''} \frac{1 - \frac{\lambda}{2\mu''}}{1 + \frac{\lambda}{2\mu''}} \sum_{i=0}^{\infty} i\left(\frac{\lambda}{2\mu''}\right)^{i-1} = \frac{2\frac{\lambda}{2\mu''}}{1 - \left(\frac{\lambda}{2\mu''}\right)^2}$$

We can now compare the performances of systems with one or two lines. Clearly, if the transmission time of each segment is the same in both systems, the comparison is in favor of the system with two lines. In fact, if $\mu' = \mu'' = \mu$ one has

$$\frac{\bar{n}''}{\bar{n}'} = \frac{1 - \frac{\lambda}{\mu}}{1 - \left(\frac{\lambda}{2\mu}\right)^2} < 1$$

so that $\bar{n}'' < \bar{n}'$ and, consequently,

$$\bar{T}'' < \bar{T}'$$

that is, the response time is smaller in the system with two lines. In contrast, if we assume $\mu' > \mu''$, then the comparison can be in favor of the system with a single line. For example, if $\mu' = 2\mu''$ (so that $\lambda/2\mu'' = \lambda/\mu'$) one obtains (recall the condition $\lambda < s\mu$, which, in this case, becomes $\lambda < \mu'$)

$$\frac{\bar{n}''}{\bar{n}'} = \frac{2}{1 + \frac{\lambda}{\mu'}} > 1$$

so that

$$\bar{T}'' > \bar{T}'$$

This result can be justified by noting that in the case of a single line, a segment is transmitted with a higher speed if it is the only one to be present in the system.

♣

Conclusions

In this book, we have collected and synthesized all the results (to the best of our knowledge) regarding positive linear systems. For the sake of simplicity, we have considered only the case of time invariant finite-dimensional systems. However, we have dealt with both the continuous-time and the discrete-time case, unifying the treatment when possible. The emphasis is on the structural properties of positive systems, that is, to those properties exclusively linked to the topology of the influence graph. The existence of such structural properties is, basically, the reason for the conceptual interest and the success in applications of this class of dynamical systems. There are three peculiar aspects of our treatment. First, the presentation of a unified theory synthesizing results known for a long time, in many different areas such as economy (input–output analysis), demography (Leslie models), genetics (Markov chains), hydraulics, chemistry and bioengineering (compartmental models), computer sciences, and telecommunications (queueing theory). Second, the presentation of innovative concepts and results (at least in the just cited applicative areas) such as excitability, transparency, realization and minimum phase. Third, the demonstration of the usefulness of the theory, given by a large number of examples treating problems with an applicative flavor.

The theory presented in this book can certainly be extended, completely or partly depending on the topic, to general classes of positive systems such as periodic systems (those in which A, b, and c vary periodically over time), time-variant systems (those in which A, b, and c vary over time in a generic manner), or infinite dimensional systems (described by partial differential equations). In fact, the core of the theory presented in this book, namely, the Frobenius theorem, stated here in the case of constant matrices of dimension $n \times n$, can be extended to various classes of positive linear operators, able to describe periodical, time invariant, and infinite-dimensional systems. However, the most attractive and significant extensions are certainly the ones concerning *nonlinear* systems. To this end, we recall that in the chapters devoted to applications, we have repeatedly insisted on the need for more realistic nonlinear models. Considering the importance of those applications, and the indubitable attractiveness of the theory, it is therefore reasonable to forecast an increasing interest in the years to come for positive nonlinear systems. But, before tackling this hard task, it surely makes sense to acquire a sound background on positive *linear* systems.

Annotated Bibliography

Chapter 1 (*Introduction*)

- There are no books, to date, treating in a general way positive linear systems while a general review has been given very recently by Anderson (1996). In contrast, numerous are the books on special classes of positive systems, for example, compartmental systems [Anderson (1983)] and Markov chains [Seneta (1981)]. A few extensions of the positivity concept can also be found, in particular in Berman *et al.* (1989), where the basic mathematical tool is convex analysis, and in Krasnoselskij *et al.* (1989) who approaches the topic through functional analysis. Other interesting extensions are discussed by Krause (1998). The properties of 2D positive systems are studied by Fornasini and Valcher (1996, 1997), Valcher and Fornasini (1995), Valcher (1997). Finally, the *behavioral* approach to positivity is presented by Nieuwenhuis (1998) and Valcher (2000).

- In this book, we have restricted ourselves to the case of time-invariant finite-dimensional systems. The basic mathematical tools are those of linear systems theory, which are nowadays well known to most researchers in any field of science. The approach is similar to that of Luenberger (1979), who in one chapter presents some theoretical results on positive linear systems as well.

- For nonnegative matrices and Metzler matrices, the reference book is Gantmacher (1959). But other more recent books dealing with such matrices are also available, such as Berman and Plemmons (1994), Minc (1988), Graham

(1987) and Horn and Johnson (1985). Theoretical results of some interest can be also found in the already cited book by Anderson (1983) and in Varga (1962) who discuss, in particular, matrix iterative methods used in numerical analysis. A very simple but interesting survey paper on the role of nonnegative matrices in modeling, is Rumchev and James (1990).

Chapter 2 (*Definitions and Conditions of Positivity*)

- Notation and jargon on positive matrices are extremely different in the literature. In some books [e.g., Minc (1988) and Gantmacher (1959)], strictly positive matrices are called positive, while other authors [e.g., Graham (1987)] call positive the nonnegative matrices. The notation adopted in this book is the most common one in the applicative literature [see, e.g., Berman and Plemmons (1994) and Berman et al. (1989)].

- The definition of *external positivity* is original. The need for distinguishing between *internal* and external positivity will be clear in Chapter 9, which is devoted to realization.

- The definition of positive linear systems is taken from Luenberger (1979). The extension to the case in which the invariant is a generic cone, not necessarily the nonnegative orthant, has been treated by numerous authors and the most important contributions are due to Birkhoff (1967), Vandergraft (1968) and Barker (1972), who extended the results of Perron (1907) and Frobenius (1912) on nonnegative matrices. Other related results can be found in Farina and Valcher (1998). A different definition of positive systems, requiring only the nonnegativity of the state variables for some appropriate feasible control, has been given by Berman et al. (1989) and d'Alessandro and De Santis (1994). It has been proved by Berman and Stern (1987) that such systems can be made positive in the classical sense by means of a static state feedback so that, in practice, one is brought to study nonnegative (or Metzler) matrices in any case. Positive nonlinear systems have been studied by Krause (1992) and Gouzé (1988, 1990, 1991, 1995).

- Metzler matrices are called *essentially nonnegative* by some authors [e.g., Berman et al. (1989) and Varga (1962)]. It is worth noting that a matrix A where $-A$ is a Metzler matrix with eigenvalues with a negative real part is often called a *nonsingular M matrix* (or *Leontief matrix*). In the book by Berman and Plemmons (1994), 50 conditions equivalent to "A is a nonsingular M matrix" are listed!

Chapter 3 (*Influence Graphs*)

- *The influence graph* is sometimes called an *interaction graph* or *Coates graph* and the corresponding binary matrix of zeros and ones is also called an *adiacency matrix* or *structure matrix*. A classical textbook on oriented graphs and their applications is Harary *et al.* (1965). A survey on the properties of influence graphs in positive systems can be found in Farina (1992).

- A survey of the properties of the influence graphs related to dynamical systems (not necessarily positive) can be found in Reinschke (1988), while a complete treatment of the properties of binary matrices and of their associated graphs is given in Cvetkovic *et al.* (1980).

Chapter 4 (*Irreducibility, Excitability, and Transparency*)

- The main results of this chapter can be found in Minc (1988), where an entire paragraph is devoted to the structural properties of nonnegative matrices, and in Varga (1962), where such properties are used in the analysis of algorithms for numerical analysis. *The index of cyclicity* is sometimes called the *imprimitivity index* and, *Theorem 7* is sometimes used as a definition of primitivity so that the last statement of *Definition 4* becomes a theorem.

- The definition of *excitability* (or *invasion*) can be found in Muratori and Rinaldi (1990, 1991a), while the definition of *transparency* is original.

Chapter 5 (*Stability*)

- The first work on strictly positive matrices is a paper by Perron, dated 1907. The most important results, however, are due to the German mathematician Georg Frobenius (1849–1917), who is known, among other things, for having introduced the notion of rank of a matrix. In 1912, Frobenius published the paper "*Über Matrizen aus Nicht Negativen Elementen*" in which many properties are proved, such as those concerning the positivity of the dominant eigenvalue and the spectrum of an irreducible nonnegative matrix.

- There are plenty of papers about upper and lower bounds of the Frobenius eigenvalue (sometimes called *Perron root*, Song (1992)): some of them are reported in Horn and Johnson (1985) and Minc (1988). The interested reader can look at the journal *Linear Algebra and its Applications* to find some of these bounds.

- *Theorems 13* and *14* are based on results reported in Berman and Plemmons (1994). In the same book, the history of the contributions on the theory of M matrices is also reconstructed.

- The existence of a strictly positive diagonal solution of the Liapunov equation has been studied in the context of robustness of linear systems. Systems whose associated Liapunov equation has a strictly positive diagonal solution are called *diagonally stable* or *Volterra–Liapunov stable*. This property holds for positive systems, but it may hold also for other classes of systems. Robust stability of positive systems is also studied in Hinrichsen and Son (1996, 1998).

- The notion of D stability was introduced during the 1950s by Arrow and McManus (1958) in the course of their work on price equilibrium in economic systems. The extension to the case of discrete-time systems considered in this book is original. Kaszkurewicz and Bhaya (1993) proposed a different type of perturbation that, however, appears to be rather unnatural in the context of positive systems.

- The regulation problem for positive systems has been studied by Heemels *et al.* (1998), Heemels (1995), Kaczorek (1998) and Rumchev and James (1994, 1995) and the concept of connective stability has been presented by Siljak (1974).

Chapter 6 (*Spectral Characterization of Irreducible Systems*)

- All the results contained in this chapter are due to Perron (1907) and Frobenius (1912). A more detailed treatment of the same topic can be found in Minc (1988).

Chapter 7 (*Positivity of Equilibria*)

- The link between positivity of equilibria and stability is reported in Luenberger (1979), while *Theorems 20* and *21* are presented in Muratori and Rinaldi (1990, 1991a).

- *Theorem 22* is due to Morishima (1964). Some interesting applications can be found in Luenberger (1979), while a recent application to the study of interpersonal relationships can be found in Rinaldi (1998).

Chapter 8 (*Reachability and Observability*)

- *Theorem 27* is due to Maeda and Kodama (1981) who gave a slightly cryptic proof. Later, Murthy (1986) proved again the sufficiency and, in 1989, Rumchev and James (1989) and Fanti *et al.* (1989) gave new proofs of necessity. The result has been extended to systems with multiple inputs by Rumchev and James (1989) and Fanti *et al.* (1990). Also, the case of almost complete reachability (*Theorem 28*) has been solved by Maeda and Kodama (1981), while extension to the multiple inputs case has been carried out by Valcher (1996). Other results on the reachability cone can be found in Farina (1996a) and Farina and Benvenuti (1998).

- Extension of the reachability conditions to the periodic case are given in Bru *et al.* (1998), to the 2D case, in Kaczorek (1996), and results on the observability problem are given in Dautrebande and Bastin (1998).

Chapter 9 (*Realization*)

- One of the first contributions to the positive realization problem can be found in Maeda *et al.* (1977), where compartmental systems are addressed. The interest in this problem is motivated by our need to obtain information on the number of compartments in the model.

- Necessary and sufficient geometric conditions for the existence of a positive realization of the transfer function of a continuous-time system have been given in Ohta *et al.* (1984). Necessary conditions that make use of the impulse response of the system are given in Farina (1995).

- The discrete-time case has been considered by Maeda and Kodama (1981), where the focus is on the properties of all the nonnegative solutions of an autoregressive model. Other results on the discrete-time case can be found in Wende and Daming (1989) and in Farina (1994).

- Some preliminary results for the solution of the positive realization problem can be found in Nieuwenhuis (1982), Farina and Benvenuti (1995). *Theorems 31* and *33* are due to Anderson *et al.* (1996). The most general case has been solved by Farina (1996b) and Kitano and Maeda (1998) and the problem of finding the minimal order of a positive realization is addressed in Farina (1996c), Benvenuti and Farina (1998a), and Benvenuti *et al.* (1998). Other interesting results can be found in Kaczorek (1997a, 1997b), Hadjicostis (1999), Foerster and Nagy (1998a, 1998b), Benvenuti and Farina (1999) and Benvenuti *et al.* (1999). A general review on the realization problem is given by Farina (1998).

- A different, but important, approach to the realization and minimality problem using positive linear algebra and shift invariant polyhedral cones has been considered by Picci (1978), Picci and van Schuppen (1992), Picci et al. (1994), van den Hof and van Schuppen (1994), and Van den Hof (1996, 1997a, 1997b).

- References on Charge Routing Networks can be found in Benvenuti and Farina (1996a, 1996b) and Benvenuti (1998).

Chapter 10 (*Minimum Phase*)

- *Theorem 34* is due to Farina (1995).

- The other results of this chapter are original, even if *Theorem 35* is a refinement of a similar theorem presented in Muratori et al. (1994).

Chapter 11 (*Interconnected Systems*)

- The results presented in this chapter are refinements of analogous results proved in Muratori et al. (1994).

Chapter 12 (*Input–Output Analysis*)

- The input–output model at the equilibrium has been presented by Leontief (1941, 1951) together with an estimate of the 10×10 dynamic matrix A referring to the U.S. economy.

- The dynamic input–output model has been proposed by Leontief both in its continuous-time version [Leontief (1953)] and in the discrete-time version [Leontief (1970)]. The model considered in this chapter is drawn from a particular case of the second version.

- There are plenty of publications on input–output analysis. With regard to the economical aspects and the estimation problems, we suggest the book by Pasinetti (1975), while for the analytical aspects one can refer to Takayama (1985).

Chapter 13 (*Age-Structured Population Models*)

- The paper usually cited as the origin for the studies on age-structured populations models is Leslie (1945). Actually, the same model was presented for the first time by Bernardelli (1941) and, afterward, by Lewis (1942). Numerous and interesting are the extensions to the case of populations with vital stages, for which the reader can refer to Caswell (1989).

- Reachability and minimum phase of age-structured population models with immigration and stocking has been studied by Muratori and Rinaldi (1990, 1991b). The nonlinear case of age-structured populations subject to harvesting is discussed in Muratori and Rinaldi (1988).

- With regard to the availability of demographical data and the associated estimation methods, the interested reader can refer to Harper (1977) for plant populations, to Seber (1982) for animal populations, and to Keyfitz and Flieger (1971) for human populations.

Chapter 14 (*Markov Chains*)

- A general introduction to Markov chains can be found in Bhat (1972) or, in the more recent text by Seneta (1981). A more advanced treatment of the topic can be found in Feller (1968).

- Some of the applicative examples presented in this chapter have been derived from Luenberger (1979) and Graham (1987).

Chapter 15 (*Compartmental Systems*)

- One of the most important compartmental models used in hydrology has been proposed by Nash (1957) who analyzes a river stretch viewing it as a network of equivalent reservoirs. A classical book in this field is Moran (1959).

- Applications in biology and medicine can be found in Sheppard (1962), Jacquez (1985), Anderson (1983), and Kajiya *et al.* (1984).

- The equivalence between compartmental systems and asymptotically stable systems is mentioned in Maeda *et al.* (1977), while the results on minimum phase can be found in Muratori *et al.* (1992).

- A topic not addressed in this book, but of particular relevance in medicine, is the parameter identification problem from a set of input–output measures. The interested reader can make reference to Cobelli and Di Stefano (1981) and van den Hof (1996, 1998).

- For extensions to nonlinear models, the reader can refer to Maeda *et al.* (1978) and Jacquez and Simon (1993).

- The realization problem for compartmental systems is discussed in Farina (1997).

Chapter 16 (*Queueing Systems*)

- The core of queuing systems theory is the model presented in this chapter, also known as the "death and birth process". Such a model has found interesting applications in physics and biology and has been also useful in engineering, in particular for the analysis and design of communication systems and information processing.

- The first books on this topic, among which we cite Syski (1960), Saaty (1961), and Riordan (1962), were published in the 1960s and mainly focus on statistical aspects. They deal, however, exclusively with the determination of the equilibrium distribution of the queue without giving any information on the dynamical aspects. During the 1970s the theory was considerably extended in order to include more complex systems such as telecommunication and computer networks. Traces of this effort can be found in Svobodova (1977), Ferrari (1978), and Kobayashi (1978).

Bibliography

d'Alessandro P., de Santis E., "Positiveness of dynamic systems with non positive coefficient matrices", *IEEE Trans. Automatic Control*, **39** 131–134, (1994).

Anderson B.D.O., Deistler M., Farina L., Benvenuti L., "Nonnegative realization of a linear system with nonnegative impulse response", *IEEE Trans. Circuits Systems, I: Fundamental Theory Appl.*, **43** 134–142, (1996).

Anderson B.D.O., "New Developments in the theory of positive systems", Conference on Mathematical Theory of Networks and Systems, 1996, *Systems and Control in the 21st Century*, C.I. Byrnes, B.N. Datta, D.S. Gilliam, C.F. Martin, Eds., Birkhauser, Boston, pp. 17–36, (1996).

Anderson D.H., "Compartmental modeling and tracer kinetics ", *Systems and Control in the 21st Century, C.I. Byrnes*, Springer-Verlag, New York, 1983.

Arrow K.J., McManus M., "A note on dynamic stability ", *Econometrica* , **26** 448–454, (1958).

Barker G.P., " On matrices having an invariant cone", *Czech. Math. J.*, **22** 49–68, (1972).

Benvenuti L., " Positive linear filters with charge routing networks", *Mathematical Theory of Networks and Systems—MTNS98, Padova* , July 6–11, (1998).

Benvenuti L., Farina L., Anderson B.D.O., De Bruyne F., "Minimal discrete-time positive realizations of transfer functions with positive real poles", *Mathematical Theory of Networks and Systems—MTNS98*, Padova, July 6–11, (1998).

Benvenuti L., Farina L., "Discrete-time filtering using charge routing networks", *Signal Proc.*, **49** 207–215, (1996a).

Benvenuti L., Farina L., "On the class of linear filters attainable with charge routing networks", *IEEE Trans. Circuits Syst., II: Analog Digital Signal Proc.*, **43** 618–622, (1996b).

Benvenuti L., Farina L., "A note on minimality of positive realizations", *IEEE Trans. Circuits and Syst., I: Fundamental Theory Appl.*, **45** 676–678, (1998a).

Benvenuti L., Farina L., "An example of how positivity may force realizations of 'large' dimensions", *Syst. Control Lett.*, **36** 261–266 (1999).

Benvenuti L., Farina L., Anderson B.D.O. "Filtering through combination of positive filters", *IEEE Trans. Circuits Systems, I: Fundamental Theory Appl.*, **46** 1431–1440 (1999).

Berman A., Neumann M., Stern R., *Nonnegative matrices in dynamic systems*, Wiley, New York, 1989.

Berman A., Plemmons R.J., *Nonnegative matrices in the mathematical sciences*, SIAM Press, Philadelphia, 1994.

Berman A., Stern R., "Linear feedback, irreducibility and M-matrices", *Linear Algebra Appl.*, **97** 1–11 (1987).

Bernardelli H., "Population waves", *J. Burma Res. Soc.*, **31** 1–18 (1941).

Bhat U.N., *Elements of applied stochastic processes*, Wiley, New York, 1972.

Birkhoff G., "Linear transformations with invariant cones", *Am. Math. Mon.*, **72**, 274–276 (1967).

Bru R., Romero S., Sanchez E., "On reachability and controllability of nonnegative discrete-time linear periodic systems", *Mathematical Theory of Networks and Systems—MTNS98*, Padova, July 6–11 (1998).

Caswell H., *Matrix population models: construction, analysis and interpretation*, Sinaner Associates, Inc. Publishers, Sunderland, MA, 1989.

Cobelli C., Di Stefano III J.J., "Parameter and structural identifiability concepts and ambiguities: A critical review and analysis", *Am. J. Physiol.*, **239** R7–R24 (1981).

Cvetkovic D.M., Doob M., Sachs H., *Spectra of graphs*, Academic Press, New York, 1980.

Dautrebande N., Bastin G., *Positive linear observers for positive linear systems*, CESAME Technical Report 98.85, (1998).

Fanti M.P., Maione B., Turchiano B., "Controllability of linear single-input positive discrete time systems", *Int. J. Control*, **50** 2523–2542 (1989).

Fanti M.P., Maione B., Turchiano B., "Controllability of multi-input positive discrete time systems", *Int. J. Control*, **51** 1295–1308 (1990).

Farina L., "Positive linear systems: a contribution to the study of structural properties", Master's Thesis in Electrical Engineering, University of Rome "La Sapienza" (in Italian) (1992).

Farina L., "A note on discrete-time positive realizations", *Syst. Control Lett.*, **22** 467–469 (1994).

Farina L., "Necessary conditions for positive realizability of continuous-time linear systems", *Syst. Control Lett.*, **25** 121–125 (1995).

Farina L., Benvenuti L. "Positive realizations of linear systems", *Syst. Control Lett.*, **26** 1–9 (1995).

Farina L., "Finite-time reachability for a class of input constrained linear systems", *Proceedings of the 35th Conference on Decision and Control*, Kobe, Japan, 3078–3079 (1996a).

Farina L., "On the existence of a positive realization", *Syst. Control Lett.*, **28** 219–226 (1996b).

Farina L.,"Minimal order realizations for a class of positive linear systems", *J. Franklin Inst.*, **333B** 893–900 (1996c).

Farina L., "Is a system representable as a compartmental system?", *European Control Conference—ECC97*, Bruxelles, July 1–4 (1997).

Farina L., Benvenuti L., "Polyhedral reachable set with positive controls", *Math. Control, Signals Syst.*, **10** 364-380 (1998).

Farina L., Valcher M.E., "An algebraic approach to the construction of polyhedral invariant cones", *Mathematical Theory of Networks and Systems—MTNS98, Padova*, July 6–11 (1998).

Farina L., "Positive systems and the realization problem", PhD Thesis, University of Rome "La Sapienza" (in Italian) (1998).

Feller W., *An introduction to probability theory and its applications*, Wiley, New York, 1968.

Ferrari D., *Computer systems performance evaluation*, Prentice-Hall, Englewood Cliffs, NJ, 1978.

Foerster K.H., Nagy B., "Spectral properties of rational matrix functions with nonnegative realizations", *Linear Algebra and its Applications* **275-276** 189–200 (1998a).

Foerster K.H., Nagy B., "On nonnegative realizations of rational matrix functions and nonnegative input-output systems", *Operator Theory: Advances and Applications* **103** 89–104 (1998b).

Fornasini E., Valcher M.E., "On the spectral and combinatorial structure of 2D positive systems", *Linear Algebra and its Applications* **245** 224–258 (1996).

Fornasini E., Valcher M.E., "Recent developments in 2D positive system theory", *J. of Applied Mathematics and Computer Science* **7** 101–123 (1997)

Frobenius G., 'On matrices with nonnegative elements", *S.-B. Deutsch. Akad. Wiss. Berlin, Math.-Nat.* **Kl**, 456–477 (1912) (in German).

Gantmacher F.R., *The theory of matrices*, Chelsea, New York, 1959.

Gouzé J. L., "A criterion for global convergence to equilibrium for differential systems", *Rapport INRIA n. 894* (1988).

Gouzé J. L., "On the global behaviour of polynomial differential systems in the positive orthant", *Rapport INRIA n. 1345* (1990).

Gouzé J. L., "Transformation of polynomial differential systems in the positive orthant", H. Kimura, S. Kodama Ed., *Proceedings of the International Symposium, MTNS-91*, pp. 87–93, 1991.

Gouzé J. L., "Positivity, space scale and convergence towards the equilibrium", *J. Biol. Syst.*, **3** 613–620 (1995).

Graham A. *Nonnegative matrices and applicable topics in linear algebra*, Ellis Horwood Limited, Chichester, 1987.

Hadjicostis C., "Bounds on the size of minimal nonnegative realizations for discrete time LTI systems", *Syst. Control Lett.* **37** 39-43 (1999).

Harary F, Norman R.Z., Cartwright D., *Structural models: an introduction to the theory of directed graphs*, Wiley, New York, 1965.

Harper J.L. *Population biology of plants*, Academic, London, 1977.

Heemels W.P.M.H., van Eijndoven S.J.L., Stoorvogel A.A., "Linear quadratic regulator problem with positive controls", *Int. J. Control* **70** 551–578 (1998).

Heemels W.P.M.H., "Optimal positive control and optimal control with positive state entries", *Master's Thesis, Eindhoven University of Technology* (1995).

Hinrichsen D., Son N.K., "Robust stability of positive continuous-time systems", *Numer. Funct. Anal. Opt.* **17** 649–659 (1996).

Hinrichsen D., Son N.K., "μ-analysis and robust stability of positive linear systems", *Appl. Math. Computer Sci.* **8** (1998).

van den Hof J.M., van Schuppen J.H., "Realization of positive linear systems using polyhedral cones", *Proceedings 33rd Conference on Decision and Control*, 3889–3893 (1994).

van den Hof J.M., "System theory and system identification of compartmental systems", PhD Thesis, Groningen University, Germany (1996).

van den Hof J.M., "Realization of continuous-time positive linear systems", *Syst. Control Lett.* **31** 243–253 (1997a).

van den Hof J.M., "Realizations of positive linear systems", *Linear Algebra and its Applications* **256** 287–308 (1997b).

van den Hof J.M., "Structural identifiability from input–output observations of linear compartmental systems", *IEEE Trans. Automatic Control*, **43** 800–818 (1998).

Horn R.A., Johnson C.R., *Matrix analysis*, Cambridge University Press, London and New York, 1985.

Jacquez J.A., *Compartmental analysis in biology and medicine*, 2nd ed., University of Michigan Press, Ann Arbor, MI, 1985.

Jacquez J.A., Simon C.P. "Qualitative theory of compartmental systems", *SIAM Rev.* **35** 43–79 (1993).

Kaczorek T., "Realization problem for discrete-time positive linear systems", *Appl. Math. Computer Sci.* **7** 117–124 (1997a).

Kaczorek T., "Positive realisations of improper transfer matrices of discrete-time linear systems", *Bull. Pol. Acad. Sci., Tech. Sci.* **15** 277–286 (1997b).

Kaczorek T., "Reachability and controllability of nonnegative 2D Roesser type models", *Bull. Pol. Acad. Sci., Tech. Sci.* **44** 405–410 (1996).

Kaczorek T., "Stabilization of positive linear systems", *Bull. Proceedings of the 37-th IEEE Conf. on Decision and Control*, Tampa, FL (1998).

Kajiya F., Kodama S., Abe H., *Compartmental analysis*, Karger, Basel, 1984.

Kaszkurewicz E., Bhaya A., "On discrete-time diagonal and D-stability", *Linear Algebra Appl.*, **187** 87–104 (1993).

Keyfitz N., Flieger W., *Population: facts and methods of demography*, W.H. Freeman, San Francisco, 1971.

Kitano T., Maeda H., "Positive realization of discrete-time system by geometric approach", *IEEE Trans. Circuits Syst.,I: Fundamental Theory Appl.* **45** 308–311 (1998).

Kobayashi H., *Modeling and analysis: an introduction to system performance evaluation methodology*, Addison-Wesley, Reading, MA, 1978.

Krasnoselskij M.A., Lifshits J.A., Sobolev A.V., *Positive linear systems: the method of positive operators*, Heldermann-Verlag, Berlin, 1989.

Krause U., "Positive nonlinear systems", *Proceedings of the First World Congress of Nonlinear Analysts*, Tampa, FL, August 19–26, (1992).

Krause U., "Positivity in systems theory", *Mathematical Theory of Networks and Systems—MTNS98*, Padova, July 6–11, (1998).

Leontief W.W., *The structure of the American economy 1919–1929*, Harvard University Press, Cambridge, 1941.

Leontief W.W., *The structure of the American economy 1919–1939*, Oxford University Press, New York, 1951.

Leontief W.W., *Dynamic analysis*, in Studies in the structure of the American economy, W.W. Leontief et al., eds., Oxford University Press, New York, 1953.

Leontief W.W,. *The dynamic inverse*, in Contributions to Input–Output analysis, A.P. Carter, A. Brody, eds., North Holland, Amsterdam, 1970.

Leslie P.H., "On the use of matrices in certain population mathematics", *Biometrika*, **35** 183–212 (1945).

Lewis E.G., "On the generation and growth of a population, *Sankhya*", *Indian J. Statistics* **6** 93–96 (1942).

Luenberger D. G., *Introduction to dynamic systems*, Wiley, New York, 1979.

Maeda H., Kodama S., "Positive realization of difference equations", *IEEE Trans. Circuits Syst.*, **28** 39–47 (1981).

Maeda H., Kodama S., Kajiya F., "Compartmental system analysis: realization of a class of linear systems with physical constraints", *IEEE Trans. Circuits Syst.*, **24** 8–14 (1977).

Maeda H., Kodama S., Ohta Y., "Asymptotic behaviour of nonlinear compartmental systems: nonoscillation and stability", *IEEE Trans. Circuits Syst.*, **25** 372–378 (1978).

Minc H., *Nonnegative matrices*, Wiley, New York, 1988.

Moran P.A.P., *The theory of storage*, Methuen, London, 1959.

Morishima M., *Equilibrium, stability and growth. A multisectorial analysis*, Oxford at the Clarendon Press, Ely House, London, 1964.

Muratori S., Rinaldi S., "Structural properties of controlled population models", *Syst. Control Lett.*, **10** 147–153 (1988).

Muratori S., Rinaldi S., "Equilibria,stability, and reachability of Leslie systems with nonnegative inputs", *IEEE Trans. Autom. Contr.*, **35** 1065–1068 (1990).

Muratori S., Rinaldi S., "Excitability, stability and sign of equilibria in positive linear systems", *Syst. Control Lett.*, **16** 59–63 (1991a).

Muratori S., Rinaldi S., "On the invertibility of Leslie systems", *Appl. Math. Comp.*, **41** 179–187 (1991b).

Muratori S., Piccardi C., Rinaldi S., "On the minimum phase of compartmental systems", *Int. J. Control*, **56** 23–34 (1992).

Muratori S., Piccardi C., Rinaldi S., "On the zeroes of interconnected positive linear systems", *Int. J. Syst. Sci.*, **25** 2255–2263 (1994).

Murthy D.N.P., "Controllability of a linear positive dynamic system", *Int. J. Syst. Sci.*, **17** 49–54 (1986).

Nash J.E., "The form of the instantaneous unit hydrograph", *I.A.S.H.*, **45** (1957).

Nieuwenhuis J.W., "About nonnegative realizations", *Syst. Control Lett.*, **1** 283–287 (1982).

Nieuwenhuis J.W., "When to call a linear system nonnegative", *Linear Algebra and its Applications*, **281** 43–58 (1998).

Ohta Y., Maeda H., Kodama S., "Reachability, observability and realizability of continuous time positive systems", *SIAM J. Control Opt.*, **22** 171–180 (1984).

Pasinetti L., *Lezioni di teoria della produzione*, Il Mulino, Bologna, 1975.

Perron O., "Zur Theorie der Matrizen", *Math. Ann.*, **64** 248–263 (1907).

Picci G., "On the internal structure of finite-state stochastic processes", in *Lecture Notes in Economics and Mathematical Systems*, Springer-Verlag, Berlin, Vol. 162, pp. 288–304, 1978.

Picci G., van Schuppen J.H., "Stochastic realizations of finite-valued processes and primes in the positive matrices", in H. Kimura and S. Kodama, eds., *Proceedings of the International Symposium, MTNS-91*, pp. 227–232, 1992.

Picci G., van den Hof J.M., van Schuppen J.H., "Positive linear algebra for stochastic realizations of finite-valued processes", in U. Helmke, R. Mennicken and J. Saurer eds., *Proceedings of the International Symposium, MTNS-93*, pp. 425–428, 1994.

Reinschke K.J., *Multivariable control: a graph-theoretic approach*, Springer-Verlag, Berlin, 1988.

Rinaldi S., "Love dynamics: the case of linear couples", *Appl. Math. Comp.*, **95** 181–192 (1998).

Riordan J., *Stochastic service systems*, Wiley, New York, 1962.

Rumchev V.G., James D.J.G., "Controllability of positive linear discrete-time systems", *Int. J. Control*, **50** 845–857 (1989).

Rumchev V.G., James D.J.G., "Oscillatory behaviour of controllable single-input single-output positive discrete time systems", *IMA J. of Mathermatical Control and Information*, **12** 235–243 (1994).

Rumchev V.G., James D.J.G., "Spectral characterization and pole assignment for positive linear discrete time systems", *Int. J. Systems Science*, **26** 295–312 (1995).

Rumchev V.G., James D.J.G., "The role of nonnegative matrices in discrete-time mathematical modelling", *Int. J. Math. Educ. Sci. Technol.*, **21** 169–182 (1990).

Saaty T.L. *Elements of queueing theory*, McGraw-Hill, New York, 1961.

Sanchez E., Bru R., Hernandez V., "Nonnegative periodic realizations of non-negative discrete-time linear periodic systems", *European Control Conference—ECC93* (1993).

Seber G.A.F., *The estimation of animal abundance and related parameters*, Charles Griffin & Company LTD, London, 1982.

Seneta E., *Nonnegative matrices and Markov chains*, Springer-Verlag, New York, 1981.

Siljak D.D., "Connective stability of complex ecosystems", *Nature (London)*, **249** 280–286 (1974).

Sheppard C.W., *Basic principles of the tracer method*, Wiley, New York, 1962.

Son K.S., "On the real stability radius of positive linear discrete time systems", *Numer. Funct. Anal. and Optimiz.*, **16** 1067–1085 (1995).

Song Y., "Lower bounds for Perron root of a nonnegative matrix", *Linear Algebra Appl.*, **169** 269–278 (1992).

Svobodova L., *Computer performance measurements and evaluation methods: analysis and applications*, Elsevier, New York, 1977.

Syski R., *Introduction to congestion theory in telephone systems*, Oliver and Boyd, London, 1960.

Takayama A., *Mathematical Economics*, 2nd ed., Cambridge University Press, New York, 1985.

Valcher M.E., "Controllability and reachability criteria for discrete-time positive systems", *Intern. J. Control*, **65** 511–536 (1996).

Valcher M.E., Fornasini E., "State models and asymptotic behaviour of 2D positive systems", *IMA J. of Math.Control and Information*, **12** 17–36 (1995).

Valcher M.E., "On the internal stability and asymptotic behavior of 2D positive systems", *IEEE Trans. on Circ. and Sys., Part I*, **44** 602–613, (1997).

Valcher M.E., "Nonnegative linear systems in the behavioral approach: The autonomous case", *Preprint* (2000).

Vandergraft J.S., "Spectral properties of matrices which have invariant cones", *SIAM J. Appl. Math.*, **16** 1208–1222 (1968).

Varga R.S, *Matrix iterative analysis*, Prentice Hall, New Jersey, 1962.

Wende C., Daming L., "Nonnegative realizations of systems over nonnegative quasi-fields", *Acta Math. Appl. Sinica*, **5** 252–261 (1989).

Appendix A: Elements of Linear Algebra and Matrix Theory

A.1 REAL VECTORS AND MATRICES

An n-dimensional real *vector* a is an ordered set of n real numbers $a_1, a_2, \ldots, a_n$ that, conventionally, are written in column as follows

$$a = \begin{pmatrix} a_1 \\ a_2 \\ \vdots \\ a_n \end{pmatrix}$$

A real *matrix* of dimension $m \times n$ is, a set of mn real numbers a_{ij}, $i = 1, \ldots, m$, $j = 1, \ldots, n$ ordered by rows and columns

$$A = \begin{pmatrix} a_{11} & a_{12} & \ldots & a_{1n} \\ a_{21} & a_{22} & \ldots & a_{2n} \\ \vdots & \vdots & & \vdots \\ a_{m1} & a_{m2} & \ldots & a_{mn} \end{pmatrix}$$

Two matrices [vectors] of the same dimension can be summed by summing the corresponding elements. Analogously, a matrix [vector] can be multiplied by a real number by multiplying all the elements of the matrix [vector] by such a number. If $m = n$, the matrix A is said to be *square*. If $n = 1$, the matrix A is a *column vector*, while if $m = 1$ the matrix A is a *row vector*. A matrix A can be transposed, by exchanging the rows with the columns: The matrix A^T thus obtained has dimension $n \times m$ and is called the *transposed* matrix of A and is given by

$$A^T = \begin{pmatrix} a_{11} & a_{21} & \cdots & a_{m1} \\ a_{12} & a_{22} & \cdots & a_{m2} \\ \vdots & \vdots & & \vdots \\ a_{1n} & a_{2n} & \cdots & a_{mn} \end{pmatrix}$$

The vector a^T is, therefore, the row vector

$$a^T = \begin{pmatrix} a_1 & a_2 & \cdots & a_n \end{pmatrix}$$

Two matrices A and B can be multiplied provided that the number of columns of A coincides with the number of rows of B. In other words, if A is of dimension $m \times n$ and B is of dimension $p \times q$, the product A times B is possible if and only if $n = p$. The result is a matrix $C = AB$ of dimension $m \times q$ whose element c_{ij} can be obtained by multiplying the ith row of A by the jth column of B, that is,

$$c_{ij} = \sum_{h=1}^{n} a_{ih} b_{hj}$$

Obviously, it is possible to multiply a matrix A of dimension $m \times n$ by a vector b of dimension n, the result being a vector $c = Ab$ of dimension m whose ith element c_i is given by

$$c_i = \sum_{h=1}^{n} a_{ih} b_h$$

If we use the above rules, it is straightforward to see that

$$(AB)^T = B^T A^T$$

The class of square $n \times n$ matrices is particularly interesting. The *identity matrix* I has unitary elements on its diagonal and is zero elsewhere, that is,

$$I = \begin{pmatrix} 1 & 0 & \cdots & 0 & 0 \\ 0 & 1 & & 0 & 0 \\ \vdots & & \ddots & & \vdots \\ 0 & 0 & & 1 & 0 \\ 0 & 0 & \cdots & 0 & 1 \end{pmatrix}$$

If the product AB of two square matrices is the identity matrix, that is, $AB = I$, B is called the inverse of A and denoted by A^{-1}. Moreover, the inverse matrix A^{-1} of a matrix A is unique if it exists and $A^{-1}A = AA^{-1}(= I)$. In this regard, it is worth recalling that a matrix A is invertible if and only if its *determinant* is nonzero, where the determinant of a matrix $n \times n$ is defined as follows:

$$n = 1 \quad : \det A = a_{11}$$
$$n = 2, 3, \ldots \quad : \det A = \sum_{i=1}^{n} (-1)^{i+1} a_{i1} \det A_i$$

where A_i is the $(n-1) \times (n-1)$ matrix obtained by deleting the ith row and the first column of A. Two square matrices A and B such that $AB = BA$ are said to commute. From the previous statement, a matrix commutes with its own inverse (if it exists).

Matrices and vectors are particularly useful to represent problems in a compact form. For example, a system of n linear equations with n unknowns

$$a_{11}x_1 + a_{12}x_2 + \cdots + a_{1n}x_n = b_1$$
$$a_{21}x_1 + a_{22}x_2 + \cdots + a_{2n}x_n = b_2$$
$$\vdots$$
$$a_{n1}x_1 + a_{n2}x_2 + \cdots + a_{nn}x_n = b_n$$

becomes, in vector notation,

$$Ax = b$$

and its solution, provided the inverse matrix A^{-1} exists, can be synthetically expressed as

$$x = A^{-1}b \tag{A.1}$$

This does not mean that for computing of the solution of a system of n equations with n unknowns it is convenient to compute the matrix A^{-1} and use formula (A.1).

A.2 VECTOR SPACES

We now give the definition of vector space, but we warn the reader that in *Definition 1*, when no ambiguities are possible, the term *vector* is omitted.

DEFINITION 1 *(space)*

A set X is called a *space* if any element x of X can be multiplied by a scalar a and the product ax is an element of X, if any pair (x, y) of elements of X can be summed up and the term $z = x + y$ is an element of X, and if the following properties hold

$$x + y = y + x$$
$$(x + y) + z = x + (y + z)$$
$$0 + x = x$$
$$x + (-x) = 0$$
$$1 \cdot x = x$$
$$0 \cdot x = 0$$
$$a(x + y) = a \cdot x + a \cdot y$$
$$(a + b) \cdot x = a \cdot x + b \cdot x$$
$$a \cdot (b \cdot x) = (a \cdot b) \cdot x$$

The third and fourth properties state that in a space there exists the additive identity (zero element or vector) and that each element x has its opposite $-x$.

The simplest example of a space is the set of real numbers endowed with the usual arithmetic operations. Such a space, denoted by $\mathbb{R}$, can be geometrically represented as an oriented straight line. An obvious extension is the space $\mathbb{R}^n$ composed of the n-tuples of real numbers. The set of real matrices of dimensions $m \times n$ is also a space since all the axioms required in *Definition 1* hold, provided that one uses as zero matrix the matrix composed of zero elements. Other examples of spaces are the set of rational functions with real or complex coefficients, the set of polynomials with real or complex coefficients, the set of polynomials with degree smaller than n and real or complex coefficients, the set of real and continuous [integrable] [differentiable] functions in an interval $[a, b]$.

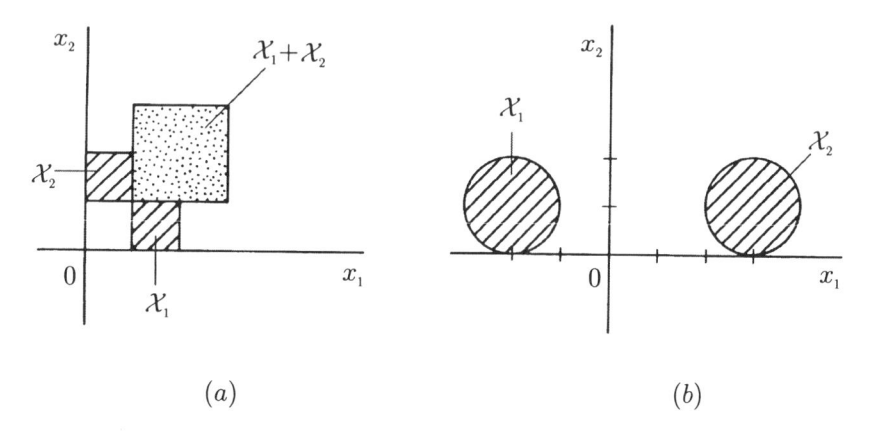

Figure A.1 Examples of sum of two sets in $\mathbb{R}^2$: (*a*) sum of two squares; (*b*) sum of two circles (to be completed by the reader).

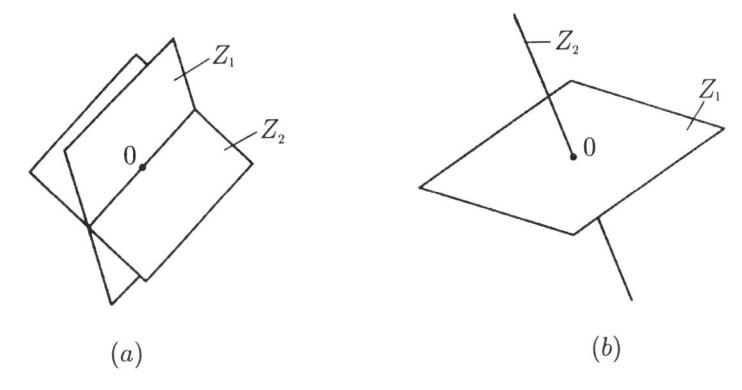

Figure A.2 The space $\mathbb{R}^3$ is the sum of Z_1 and Z_2 but in (a) it is not their direct sum while in (b) it is.

DEFINITION 2 (*sum of sets*)

> Given two sets $\mathcal{X}_1$ and $\mathcal{X}_2$ contained in a space X, the set $\mathcal{X} = \mathcal{X}_1 + \mathcal{X}_2$ composed of all the elements $x = x_1 + x_2$ with $x_1 \in \mathcal{X}_1$ and $x_2 \in \mathcal{X}_2$ is said to be the *sum of the sets* $\mathcal{X}_1$ and $\mathcal{X}_2$.

An example is illustrated in *Fig. A.1(a)*, where the space X is $\mathbb{R}^2$. In contrast, in *Fig. A.1(b)* only the sets $\mathcal{X}_1$ and $\mathcal{X}_2$ are shown and the reader is invited to determine the set $\mathcal{X}_1 + \mathcal{X}_2$.

Other important operations on sets are the well–known union ($\cup$) and intersection ($\cap$), which, however, are not typical of vector spaces since no algebraic structure is required to perform them.

DEFINITION 3 (*subspace*)

> A set $Z \subseteq X$ is a *subspace* of the space X if Z itself is a space.

The zero element of a space, as well as the whole space, satisfy *Definition 3* and are therefore subspaces. They are called *not proper*, while any other subspace is called *proper*. Every proper subspace in $\mathbb{R}^3$ is geometrically represented by a straight line or by a plane passing through the origin (being a space, the subspace must contain the zero element). Other examples of proper subspaces are the set of differentiable functions on the interval $[a, b]$ within the space of continuous functions in $[a, b]$ and the set of the polynomials of a degree smaller than n in the space of the polynomials.

THEOREM 1 (*sum and intersection of two subspaces*)

> The sum [intersection] of two subspaces is a subspace.

Obviously, the union of two subspaces is not, in general, a subspace.

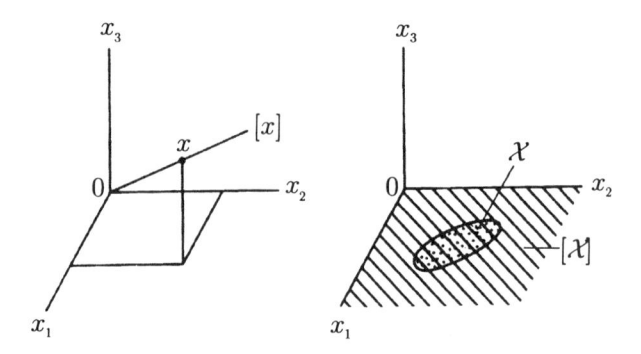

Figure A.3 Examples of subspaces generated in $\mathbb{R}^3$: (a) subspace generated by one element x; (b) subspace generated by a subset $\mathcal{X}$ of a plane.

DEFINITION 4 *(direct sum)*

> A space X, is called the *direct sum* of two subspaces Z_1 and Z_2, and we write
>
> $$X = Z_1 \oplus Z_2 \qquad (A.2)$$
>
> when Z_1 and Z_2 are disjoint (*i.e.*, they have only the zero vector in common) and $X = Z_1 + Z_2$.

Definition 4 is illustrated in *Fig. A.2*.

THEOREM 2 *(direct sum and decomposition)*

> Every element x of a vector space $X = Z_1 \oplus Z_2$ can be uniquely decomposed into the sum of two elements z_1 and z_2, belonging to Z_1 and Z_2, respectively.

DEFINITION 5 *(linear manifold)*

> Given a vector x and a subspace Z in a space X, the set $x + Z$ is called a *linear manifold*.

A straight line in a plane, is, therefore, a linear manifold in $\mathbb{R}^2$, just like the class of polynomials $p(\lambda) = \lambda^5 + a_2\lambda^2 + a_1\lambda + a_0$ is a linear manifold in the space of polynomials.

DEFINITION 6 *(subspace generated by a set)*

> Let $\mathcal{X}$ be a subset of a vector space. The smallest subspace containing $\mathcal{X}$ (*i.e.*, the intersection of all the subspaces containing $\mathcal{X}$) is called the *subspace generated* by $\mathcal{X}$ and is denoted by $[\mathcal{X}]$.

Two examples are illustrated in *Fig. A.3*.

Often, the subspace generated by a set is defined in a different but equivalent way (see *Theorem 3*) and *Definition 6* is, then, presented as a theorem.

A.3 DIMENSION OF A VECTOR SPACE

DEFINITION 7 *(linear combination)*

A vector x of a vector space X is a *linear combination* of an n-tuple of vectors $x^1, x^2, \ldots, x^n$ if there exists an n-tuple of real numbers $a_1, a_2, \ldots, a_n$ such that

$$x = \sum_{i=1}^{n} a_i x^i \qquad (A.3)$$

From this definition, it follows that the zero vector is a linear combination of any n-tuple of vectors, since it suffices to set $a_1 = a_2 = \cdots = a_n = 0$ for (A.3) to hold. Such a linear combination is called *not proper* while any linear combination with at least one nonzero a_i is called *proper*. The above definition, referring to an n-tuple $(x^1, \ldots, x^n)$, can be extended in the following way:

DEFINITION 8 *(linear combination)*

A vector x of a vector space is a *linear combination* of a set $\mathcal{X}$ when it is a linear combination of an n-tuple of vectors belonging to $\mathcal{X}$.

One can prove that the set of all linear combinations of a set $\mathcal{X}$ is a subspace and that such a subspace coincides with the subspace $[\mathcal{X}]$ previously defined (see *Definition 6*). Therefore, *Theorem 3* holds.

THEOREM 3 *(subspace generated by a set)*

Given a set $\mathcal{X}$, the subspace $[\mathcal{X}]$ generated by $\mathcal{X}$ is the set of all linear combinations of $\mathcal{X}$.

DEFINITION 9 *(linear independence)*

The vectors $x^1, \ldots, x^n$ are said to be *linearly independent* if the zero vector is not a proper linear combination of them, that is, if the equation

$$\sum_{i=1}^{n} a_i x^i = 0$$

holds only for $a_1 = a_2 = \ldots = a_n = 0$.

An analogous definition (*Definition 10*) holds for the linear independence of vectors of a set $\mathcal{X}$.

DEFINITION 10 *(linear dependence)*

The vectors $x^1, \ldots, x^n$ are said to be *linearly dependent* if they are not linearly independent.

DEFINITION 11 *(basis of a space)*

> A set $\mathcal{X}$ of vectors is a basis of a vector space X if every element of X can be uniquely written as a linear combination of the elements of $\mathcal{X}$.

The set

$$\mathcal{X} = (1, 1+\lambda, 1+\lambda+\lambda^2, \ldots, 1+\lambda+\ldots+\lambda^{n-1})$$

is, for example, a basis of the space of polynomials with a degree $< n$, while the set

$$\mathcal{X} = (1, 1+\lambda, \ldots, 1+\lambda+\ldots+\lambda^k, \ldots)$$

is a basis of the space of polynomials.

Due to the uniqueness property required by *Definition 11*, the following important property holds:

THEOREM 4 *(linear independence of the elements of a basis)*

> The vectors of a basis are linearly independent vectors.

In general, the basis of a vector space is not unique; for example, the set $\mathcal{X} = (1, \lambda, \ldots, \lambda^{n-1})$, which differs from the previously cited one, is another basis of the space of polynomials with a degree $< n$. However, *Theorem 5* holds.

THEOREM 5 *(invariance of the number of elements of a basis)*

> If a vector space has a basis composed of a finite number of elements, any other basis is composed of the same number of elements.

This last result justifies *Definition 12*.

DEFINITION 12 *(finite dimensional and functional spaces)*

> A vector space with a basis composed of n elements is called *finite dimensional* (the dimension of the space is n). All other spaces are called *functional* (or infinite dimensional) spaces.

Examples of finite dimensional spaces are the space $\mathbb{R}^n$, the space of polynomials with a degree $< n$, and the space of $p \times q$ matrices (the latter having dimension $n = pq$). Examples of functional spaces are the space of polynomials and the space of continuous functions.

Obviously, the concept of dimension can be extended to subspaces. The set of the polynomials is, for example, an infinite dimensional subspace of the space of the continuous functions, while the set of real numbers is a subspace of finite dimension of the space of polynomials, which is infinite dimensional.

A.4 CHANGE OF BASIS

Given a basis $\mathcal{X} = \{e^1, \ldots, e^n\}$ of a vector space X of dimension n, each vector x of X can be uniquely written as a linear combination of the basis, that is,

$$x = \sum_{i=1}^{n} a_i e^i$$

The *abstract* vector x is, therefore represented by the *vector* $a \in \mathbb{R}^n$ given by

$$a = \begin{pmatrix} a_1 & a_2 & \ldots & a_n \end{pmatrix}^T$$

If the basis is changed, one obtains a different representation of the vector x. In fact, if one considers a new basis $\bar{\mathcal{X}} = \{\bar{e}^1, \ldots, \bar{e}^n\}$, then

$$x = \sum_{i=1}^{n} \bar{a}_i \bar{e}^i$$

and the abstract vector x is represented by the new vector $\bar{a} \in \mathbb{R}^n$ given by

$$\bar{a} = \begin{pmatrix} \bar{a}_1 & \bar{a}_2 & \ldots & \bar{a}_n \end{pmatrix}^T$$

It is, therefore, important to know how to obtain $\bar{a}$ from a, once the change of basis is specified. To this end, the following important result holds.

THEOREM 6 *(change of basis)*

> Let a and $\bar{a}$ be the representations of the same vector x, in the bases $\mathcal{X}$ and $\bar{\mathcal{X}}$, respectively. Then,
>
> $$\bar{a} = Ta \qquad (A.4)$$
>
> where T is the $n \times n$ square matrix whose ith column is the representation of the ith element of the basis $\mathcal{X}$ in the basis $\bar{\mathcal{X}}$.

To prove this theorem, it suffices to note that a relationship of the kind (A.4) must hold for each vector $x \in X$. Thus, consider as vector x the ith vector e^i of the basis X, which is represented in the basis X itself by a vector with zero entries except the ith one, which is equal to 1. Hence, by performing the matrix product one finds out that the representation of e^i in the basis $\bar{X}$ is the ith column of T.

An obvious but important property of matrix T is that it is composed of linearly independent vectors (since the elements of a basis are such). Moreover, from *Theorem 6* it follows that matrix T is invertible and that the ith column of matrix T^{-1} is the representation of the ith element of the basis $\bar{\mathcal{X}}$ in the basis $\mathcal{X}$.

A.5 LINEAR TRANSFORMATIONS AND MATRICES

DEFINITION 13 *(linear transformation)*
> Given two vector spaces X and Y, a transformation $L : X \to Y$ is said to be *linear* if
> $$L(ax^1 + bx^2) = aLx^1 + bLx^2 \qquad (A.5)$$

Note that in (A.5) the two + symbols work in different spaces: the first one works on elements of X, while the second works on elements of Y.

Examples of linear transformations are the integration over an interval $[a, b]$, which transforms elements of the space of the integrable functions into the space of real numbers, the approximation of a polynomial by truncation, which transforms elements of the space of the polynomials into elements of the space of the polynomials with a degree $< n$, and the derivation, which transforms the space of the polynomials with a degree $< n$ into the space of the polynomials with a degree $< n - 1$.

Every transformation $y = Lx$ admits a particular representation as a basis $\mathcal{X}$ of X and a basis $\mathcal{Y}$ of Y are chosen. A remarkable case is the one in which X and Y have finite dimensions, since the transformation can be represented by a matrix. In fact, *Theorem 7* holds.

THEOREM 7 *(matrices for the representation of transformations)*
> Let L be a linear transformation $X \to Y$ between finite dimensional spaces and let $\mathcal{X} = \{x^1, \ldots, x^n\}$ and $\mathcal{Y} = \{y^1, \ldots, y^m\}$ be bases of X and Y respectively. Let a be the representation of x in the basis $\mathcal{X}$ and b be the representation of Lx in the basis $\mathcal{Y}$. Then,
> $$b = Aa$$
> where A is the $m \times n$ matrix whose ith column is the representation of Lx^i in the basis $\mathcal{Y}$.

The proof of this theorem is analogous to that of *Theorem 6*.

EXAMPLE 1 (differentiation of polynomials)
Consider the differentiation D of polynomials x with a degree $< n$. Let $\mathcal{X} = \{1, \lambda, \ldots, \lambda^{n-1}\}$ and $\mathcal{Y} = \{1, \lambda, \ldots, \lambda^{n-2}\}$ be the bases of the spaces of the polynomials with a degree $< n$ and $n-1$, respectively. In these bases, the differentiation $y = Dx$ is represented by the $(n-1) \times n$ matrix

$$A = \begin{pmatrix} 0 & 1 & 0 & 0 & \cdots & 0 \\ 0 & 0 & 2 & 0 & \cdots & 0 \\ 0 & 0 & 0 & 3 & \cdots & 0 \\ \vdots & \vdots & \vdots & \vdots & & \vdots \\ 0 & 0 & 0 & 0 & \cdots & n-1 \end{pmatrix}$$

LINEAR TRANSFORMATIONS AND MATRICES

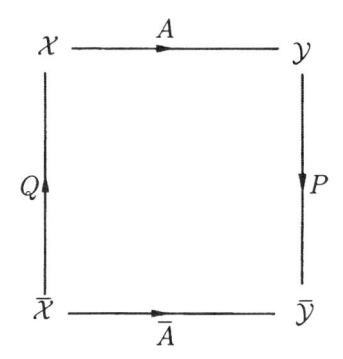

```
    𝒳 ─────A────→ 𝒴
    │             │
   Q│             │P
    │             │
    ↓             ↓
    𝒳̄ ─────Ā────→ 𝒴̄
```

Figure A.4 Illustration of $\bar{A} = PAQ$.

♣

Obviously, when the bases change from $(\mathcal{X}, \mathcal{Y})$ to $(\bar{\mathcal{X}}, \bar{\mathcal{Y}})$ the matrix representing the transformation changes from A to $\bar{A}$. If P and Q are the matrices that describe the change of basis from $\mathcal{Y}$ to $\bar{\mathcal{Y}}$ and from $\bar{\mathcal{X}}$ to $\mathcal{X}$, so that $\bar{b} = Pb$ and $a = Q\bar{a}$, the diagram in *Fig. A.4* clearly shows that

$$\bar{A} = PAQ \tag{A.6}$$

The matrices P and Q are easily computed (see *Theorem 6*) and, therefore, expression (A.6) can be actually used for the calculation of matrix $\bar{A}$.

EXAMPLE 2

Continuing with *Example 2*, suppose one wants to find the representation $\bar{A}$ of the transformation $y = Dx$ when the bases in the two spaces are

$$\bar{\mathcal{X}} = \{1, 1+\lambda, \ldots, 1+\lambda+\ldots+\lambda^{n-1}\} \quad \bar{\mathcal{Y}} = \{1, \lambda, \ldots, \lambda^{n-2}\}$$

The matrix $\bar{A}$ can be obtained directly through *Theorem 7*, or, alternatively, by using (A.6) and the representation A of *Example 1*. Following this last possibility, since $\mathcal{Y} = \bar{\mathcal{Y}}$ we have $P = I$, while from *Theorem 6* it follows that

$$Q = \begin{pmatrix} 1 & 1 & 1 & \cdots & 1 \\ 0 & 1 & 1 & \cdots & 1 \\ 0 & 0 & 1 & \cdots & 1 \\ \vdots & \vdots & \vdots & & \vdots \\ 0 & 0 & 0 & & 1 \end{pmatrix}$$

Therefore,

$$\bar{A} = PAQ = AQ =$$

$$= \begin{pmatrix} 0 & 1 & 0 & 0 & \ldots & 0 \\ 0 & 0 & 2 & 0 & \ldots & 0 \\ 0 & 0 & 0 & 3 & \ldots & 0 \\ \vdots & \vdots & \vdots & \vdots & & \vdots \\ 0 & 0 & 0 & 0 & \ldots & n-1 \end{pmatrix} \begin{pmatrix} 1 & 1 & 1 & \ldots & 1 \\ 0 & 1 & 1 & \ldots & 1 \\ 0 & 0 & 1 & \ldots & 1 \\ \vdots & \vdots & \vdots & & \vdots \\ 0 & 0 & 0 & \ldots & 1 \end{pmatrix} = \begin{pmatrix} 0 & 1 & 1 & 1 & \ldots & 1 \\ 0 & 0 & 2 & 2 & \ldots & 2 \\ 0 & 0 & 0 & 3 & \ldots & 3 \\ \vdots & \vdots & \vdots & \vdots & & \vdots \\ 0 & 0 & 0 & 0 & \ldots & n-1 \end{pmatrix}$$

♣

All the matrices representing the same transformation L are called *equivalent*. A particular but interesting case is that of the transformation of a space X into itself, that is, $L : X \to X$. In this case, it is often appropriate to choose the same basis for the space X, considered as the set of the elements to be transformed and as the set of the transformed elements. In such a case, $Q = P^{-1}$ so that (A.6) becomes

$$\bar{A} = PAP^{-1} \tag{A.7}$$

Expression (A.7) is frequently used in linear algebra, where the two matrices A and $\bar{A}$ are called *similar*.

A.6 IMAGE AND NULL SPACE

In this section, we present the definitions and the main properties of the image and null space of a linear transformation.

DEFINITION 14 *(image and null space of a transformation)*

> The *image* (or range) of a linear transformation $L : X \to Y$ is the set $\mathcal{I}(L)$ of all the elements $y \in Y$ such that $y = Lx$ for some $x \in X$. The *null* space (or kernel) of the transformation is the set $\mathcal{N}(L)$ of all the elements $x \in X$ such that $Lx = 0$.

THEOREM 8 *(property of the image and null space)*

> The image and the null space of a linear transformation $L : X \to Y$ are two subspaces of Y and X, respectively.

DEFINITION 15 *(rank and nullity)*

> The dimensions of the image and null space of a linear transformation, called rank and nullity, are denoted by $\rho(L)$ and $\nu(L)$, that is,
>
> $$\rho(L) = \dim \mathcal{I}(L) \quad \nu(L) = \dim(L)$$

If n and m are the dimensions of X and Y, then

$$\rho(L) \leq \min(m, n) \quad \nu(l) \leq n$$

An important relationship between rank and nullity of a linear transformation is given by *Theorem 9*.

THEOREM 9 *(relationship between rank and nullity)*

Let L be a linear transformation $X \to Y$ between an n-dimensional space X and a space Y. Then
$$\rho(L) + \nu(L) = n$$

According to the previous remarks on vectors and transformations, it is important to know how the image and null space of a transformation can be represented and how the rank and nullity can be obtained from the matrix representing the transformation. To this end, *Theorem 10* can be used.

THEOREM 10 *(representation of the image)*

The image of a transformation represented in a given basis by a matrix A is the subspace generated by the column vectors of the matrix A.

EXAMPLE 3
Let,

$$A = \begin{pmatrix} 0 & 1 & 1 \\ 0 & 1 & 1 \\ 1 & 0 & 1 \end{pmatrix}$$

In $\mathbb{R}^3$, the subspace $\mathcal{I}(A)$ is represented by the plane passing through the origin and determined by the three vectors

$$a_1 = \begin{pmatrix} 0 \\ 0 \\ 1 \end{pmatrix} \quad a_2 = \begin{pmatrix} 1 \\ 1 \\ 0 \end{pmatrix} \quad a_3 = \begin{pmatrix} 1 \\ 1 \\ 1 \end{pmatrix}$$

as shown in *Fig. A.5*. This subspace is two dimensional, thus $\rho(A) = 2$ e $\nu(A) = 1$

♣

As a consequence of *Theorem 10* and of the definition of rank, one obtains the following criterion (*Theorem 11*) for the determination of the rank of a transformation.

THEOREM 11 *(property of the rank)*

The rank of a transformation L is given by the maximum number of linearly independent columns of any of its representations A.

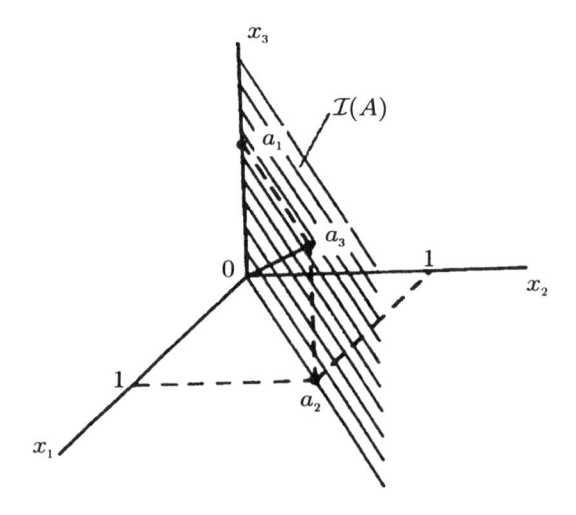

Figure A.5 Representation of the image of a matrix.

One can also prove that the number of linearly independent columns of a matrix A is equal to the number of linearly independent rows. Therefore, the rank of a transformation L can be simply determined by checking the linear independence of the columns or rows of any representation of L. If matrix A is square and the columns (and, hence, the rows) of A are linearly independent one can say that the transformation L is nonsingular (or invertible) in accordance with *Definition 16*.

DEFINITION 16 *(nonsingular transformation)*

A linear transformation $L : X \to X$ is called *nonsingular* (or invertible) if

$$\mathcal{I}(L) = X$$

For each nonsingular transformation, it is possible to define an inverse transformation L^{-1} as the one for which

$$L^{-1}(Lx) = x$$

Therefore, the transformation $L^{-1}L$ is the identity transformation that leaves all the elements of X unchanged. It is easy to prove also that LL^{-1} is the identity transformation and that if L is represented by A, the representation of L^{-1} is the matrix A^{-1}.

As for the rank of the transformation $L = L_1 L_2$, where $L_1 = Y \to Z$ and $L_2 = X \to Y$, the following result (*Theorem 12*) holds:

THEOREM 12 *(rank of the product of transformations)*

> If $L = L_1 L_2$, then
> $$\rho(L) \leq \min[\rho(L_1), \rho(L_2)]$$
> Moreover, if L_1 is nonsingular $\rho(L) = \rho(L_2)$ and if L_2 is nonsingular, $\rho(L) = \rho(L_1)$.

A consequence of *Theorem 12* is that the product $L = L_1 L_2$ of two nonsingular transformations is nonsingular. It can also be proved that

$$(L_1 L_2)^{-1} = L_2^{-1} L_1^{-1}$$

A.7 INVARIANT SUBSPACES, EIGENVECTORS, AND EIGENVALUES

DEFINITION 17 *(transformed set)*

> Given a set $\mathcal{X}$ of space X and a linear transformation $L : X \to Y$, the set
> $$L\mathcal{X} = \{y : y = Lx, x \in \mathcal{X}\}$$
> is called the *transformed set* of $\mathcal{X}$

Obviously, the transformed set LZ of a subspace Z is a subspace and, in particular, the image $\mathcal{I}(L)$ of a transformation is the transformed set of the whole space.

DEFINITION 18 *(invariant set)*

> A set $\mathcal{X} \subset X$ is said to be invariant with respect to the linear transformation $L : X \to X$ if $L\mathcal{X} \subset \mathcal{X}$.

Definition 18 also clearly applies to the subspaces Z, which are therefore called invariant when the transformation L transforms them into subspaces LZ contained in or coinciding with Z. Particularly important in applications are the one-dimensional invariant subspaces that are related with the notions of eigenvector and eigenvalue, as pointed out by *Definition 19*.

DEFINITION 19 *(eigenvector and eigenvalue)*

> Let $L : X \to X$ be a linear transformation. A nonzero vector $x \in X$ is called an *eigenvector* of L if there exists a scalar λ such that
> $$Lx = \lambda x$$
> The scalar λ associated with the eigenvector x is called *eigenvalue* of L.

Obviously, the expression $Lx = \lambda x$ holds for every representation A of L, so that

$$Aa = \lambda a$$

where a is the representation of the eigenvector.

An eigenvector x determines a one-dimensional invariant subspace (the subspace $[x]$ generated by x) and each nonzero vector of such a subspace is an eigenvector. Thus, an infinite number of eigenvectors is associated with each eigenvalue λ: clearly, what is important is the one-dimensional invariant subspace associated with the eigenvalue. However, it is worth noting that different one-dimensional invariant subspaces can be associated with the same eigenvalue λ. A trivial example is represented by the identity transformation in which each vector is an eigenvector associated with the unitary eigenvalue. On the other hand, there exist transformations that do not admit eigenvectors (and, consequently, eigenvalues). For example, the clockwise rotation in $\mathbb{R}^2$ of an angle α is a linear transformation and if $\alpha \neq n\pi$ no one-dimensional invariant subspace exists. Finally, an example in which there is only one-dimensional invariant subspace and, hence, only one eigenvalue is given by the differentiation in the space of the polynomials. Indeed, the one-dimensional subspace of constant polynomials is transformed into the origin and, is therefore invariant; but since there are no other one-dimensional invariant subspaces there exists only one eigenvalue ($\lambda = 0$).

From the definition of an eigenvector, it follows that a scalar λ is an eigenvalue if and only if

$$\exists x \neq 0 : (\lambda I - L)x = 0$$

that is, if and only if the null space of the transformation $(\lambda I - L)$ has dimension $\nu \geq 1$. Therefore, if A is a representation of L and I is the identity matrix, λ is an eigenvalue of L if and only if it satisfies

$$\det(\lambda I - A) = 0$$

DEFINITION 20 (*characteristic polynomial*)

Let $L : X \to X$ be a linear transformation represented by a matrix A in a basis $\mathcal{X}$ of X. The polynomial $\Delta_L(\cdot)$ given by

$$\Delta_L(\lambda) = \det(\lambda I - A)$$

is called a *characteristic polynomial* of the transformation L.

Definition 20 makes sense only if the polynomial $\Delta_L(\cdot)$ is independent of the basis. But this is actually the case since a change of basis from $\mathcal{X}$ to $\bar{\mathcal{X}}$ implies that the representation changes from A to $\bar{A}$ with

$$\bar{A} = TAT^{-1}$$

and T nonsingular, so that

$$\det(\lambda I - \bar{A}) = \det(\lambda I - TAT^{-1}) = \\ \det T(\lambda I - A)T^{-1} = \det T \det(\lambda I - A) \det T^{-1} = \det(\lambda I - A)$$

DEFINITION 21 *(characteristic equation and characteristic roots)*
> Given a linear transformation $L : X \to X$, the equation
> $$\Delta_L(\lambda) = 0$$
> is called a *characteristic equation*. The roots of this equation in the scalar field of the vector space are called *characteristic roots*.

In view of the previous statements, *Theorem 13* holds.

THEOREM 13 *(eigenvalues and characteristic roots)*
> The eigenvalues of a linear transformation $L : X \to X$ coincide with its characteristics roots.

Theorem 13 justifies the "algebraic" definition of an eigenvalue that is often given instead of the "geometric" definition adopted here. It is worth noting, however, that the "algebraic" definition (an eigenvalue is a solution of the characteristic equation) can easily induce the mistake of considering as eigenvalues scalar values that do not belong to the field over which the vector space X is defined. To better clarify this point, consider *Example 4*.

EXAMPLE 4 (differentiation of polynomials)
Consider the transformation $L : \mathbb{R}^2 \to \mathbb{R}^2$ that rotates in counterclockwise direction each vector of the plane by an angle $\pi/2$. Obviously, this transformation does not admit eigenvectors and, hence, eigenvalues. However, the transformation L is represented by the matrix
$$A = \begin{pmatrix} 0 & -1 \\ 1 & 0 \end{pmatrix}$$
so that
$$\Delta_L(\lambda) = \det \begin{pmatrix} \lambda & 1 \\ -1 & \lambda \end{pmatrix} = \lambda^2 + 1$$
and the characteristic equation is $\lambda^2 + 1 = 0$, whose solutions are $\lambda = \pm i$. However, such (imaginary) solutions should not be regarded as characteristic roots since they do not belong to the field of real numbers.

♣

In the case where scalars are complex numbers, the solutions over the complex field of the characteristic equation are the eigenvalues of the transformation L. In the sequel, we shall always refer to such a case; in particular, if the matrix representing a transformation is real, we shall consider it as belonging to the set of matrices with complex entries. In this way, one can state that the transformation $L : X \to X$, represented by a real matrix, has a number of eigenvalues equal to the dimension of the space X, since a polynomial equation of degree n with real coefficients has n roots over the complex field. Clearly, the n eigenvalues are not always distinct. Thus, one must distinguish between simple and multiple eigenvalues and introduce the notion of algebraic multiplicity of an eigenvalue. In general, there are $k(\leq n)$ *distinct* eigenvalues $\lambda_1, \lambda_2, \ldots, \lambda_k$, and each one of them has an *algebraic multiplicity* $a_i, i = 1, \ldots, k$ (the eigenvalues with unitary algebraic multiplicity are called *simple*). Thus, the characteristic polynomial can be factorized as follows:

$$\Delta_L(\lambda) = (\lambda - \lambda_1)^{a_1}(\lambda - \lambda_2)^{a_2} \ldots (\lambda - \lambda_k)^{a_k}$$

Finally, it is worth noting that complex eigenvalues are always present in an even number because if $\lambda = a + ib$ is a complex eigenvalue, its conjugate $\bar{\lambda} = a - ib$ is also an eigenvalue.

DEFINITION 22 *(generalized eigenvector)*

> Given a linear transformation $L : X \to X$ represented by a matrix A, a vector $x \neq 0$ is a *generalized eigenvector of order k* associated with the eigenvalue λ if
> $$(A - \lambda I)^{k-1}x \neq 0 \quad (A - \lambda I)^k x = 0$$

Note that the generalized eigenvectors of order 1 are the eigenvectors considered in *Definition 19*, since $(A - \lambda I)^0 = I$.

Once a generalized eigenvector of order k has been found, it is possible to construct a set of k linearly independent generalized eigenvectors, as shown by *Theorem 14*.

THEOREM 14 *(chain of generalized eigenvectors)*

> Let x be a generalized eigenvector of order k associated with an eigenvalue λ. Then, the k vectors
> $$\begin{aligned} x^k &= x \\ x^{k-1} &= (A - \lambda I)x^k \\ x^{k-2} &= (A - \lambda I)x^{k-1} \\ &\vdots \\ x^1 &= (A - \lambda I)x^2 \end{aligned} \quad (A.8)$$
> are linearly independent generalized eigenvectors (x^i is of order i).

The fact that the vectors x^i given by (A.8) are generalized eigenvectors of order i, can be easily checked, since

$$(A - \lambda I)^i x^i = (A - \lambda I)^{i+1} x^{i+1} = \cdots = (A - \lambda I)^k x^k = 0$$

while

$$(A - \lambda I)^{i-1} x^i = (A - \lambda I)^{k-1} x^k \neq 0$$

since $x^k = x$ is an eigenvector of order k. Moreover,

$$Ax^1 = \lambda x^1$$

since x^1 is an eigenvector of order 1, while for $i \geq 2$ the following relationship holds [obtained from (A.8)]:

$$Ax^i = x^{i-1} + \lambda x^i$$

The linear independence of the generalized eigenvectors can, then, be proved by contradiction.

But more can be said on generalized eigenvectors as shown by *Theorem 15*.

THEOREM 15 *(independence of chains of generalized eigenvectors)*

> The generalized eigenvectors corresponding to distinct eigenvalues are linearly independent. On the other hand, if x and y are two generalized eigenvectors of order h and k associated with the same eigenvalue, the chains $x^1, x^2, \ldots, x^h$, and $y^1, y^2, \ldots, y^k$ obtained from x and y using (A.8) are linearly independent if x^1 and y^1 are such.

On the basis of the previous results, one can prove that, given an $n \times n$ matrix A, the following procedure always yields n linearly independent generalized eigenvectors (the importance of such a procedure will be clear in the next paragraph).

Procedure for the computation of n linearly independent (generalized) eigenvectors of an n×n matrix

1. Determine the distinct eigenvalues $\lambda_1, \ldots, \lambda_k$ and their algebraic multiplicities $a_1, \ldots, a_k$.

2. For each eigenvalue λ_i determine a_i linearly independent (generalized) eigenvectors as follows:

    2a. Evaluate the matrices $(A - \lambda_i I)^h$, $h = 1, 2$, and so on, until the rank of $(A - \lambda_i I)^m$ is equal to the rank of $(A - \lambda_i I)^{m+1}$.

    2b. Determine a generalized eigenvector x of order m by finding a nonzero solution of
    $$(A - \lambda_i I)^m x = 0 \qquad (A.9)$$

    2c. Evaluate the chain of eigenvectors $x^1, x^2, \ldots, x^m$ by means of Eq. (A.8) starting with $k = m$.

    2d. If $m = a_1$, the vectors $x^1, x^2, \ldots, x^m$, are the required ones. If, on the contrary, $m < a_i$, then determine a new generalized eigenvector y associated with λ_i with order m, such that the corresponding y^1, determined using Eq. (A.8), is linearly independent of x^1 [for this find a solution y of (A.9) such that y^1 is linearly independent of x^1]. If this is not possible, find an eigenvector of rank $(m-1)$ or $(m-2)$, and so forth, always by choosing y^1 linearly independent of x^1 and continue like so, until a_i eigenvectors have been found.

EXAMPLE 5
Consider the matrix
$$A = \begin{pmatrix} 1 & 2 & 0 & 1 \\ 0 & 2 & 0 & 0 \\ 0 & -1 & 1 & 0 \\ 0 & 0 & 0 & 1 \end{pmatrix}$$

and suppose we want to determine four linearly independent eigenvectors using the procedure described above.

The characteristic polynomial $\Delta_A(\cdot)$ is given by

$$\Delta_A(\lambda) = \det(\lambda I - A) = (\lambda - 1)^3(\lambda - 2)$$

Hence, there are two distinct eigenvalues ($\lambda_1 = 1, \lambda_2 = 2$) with algebraic multiplicity $a_1 = 3$ and $a_2 = 1$.

As for the eigenvalue λ_1 we have

$$(A - \lambda_1 I) = \begin{pmatrix} 0 & 2 & 0 & 1 \\ 0 & 1 & 0 & 0 \\ 0 & -1 & 0 & 0 \\ 0 & 0 & 0 & 0 \end{pmatrix}$$

$$(A - \lambda_1 I)^2 = \begin{pmatrix} 0 & 2 & 0 & 0 \\ 0 & 1 & 0 & 0 \\ 0 & -1 & 0 & 0 \\ 0 & 0 & 0 & 0 \end{pmatrix}$$

$$(A - \lambda_1 I)^3 = \begin{pmatrix} 0 & 2 & 0 & 1 \\ 0 & 1 & 0 & 0 \\ 0 & -1 & 0 & 0 \\ 0 & 0 & 0 & 0 \end{pmatrix}$$

so that, $\rho(A - \lambda_1 I) = 2$ and $\rho((A - \lambda_1 I)^2) = \rho((A - \lambda_1 I)^3) = 1$. Thus, we must determine an eigenvector x, of order 2, and, for this, x must satisfy the following relationships:

$$(A - \lambda_1 I)^2 x = 0 \quad (A - \lambda I) x \neq 0$$

A vector satisfying these relationships is

$$x = (\begin{matrix} 1 & 0 & 0 & 1 \end{matrix})^T$$

Hence, Eq. (A.8) yields two eigenvectors

$$x^2 = x = (\ 1 \quad 0 \quad 0 \quad 1\ )^T$$
$$x^1 = (A - \lambda_1 I)x^2 = (\ 1 \quad 0 \quad 0 \quad 0\ )^T$$

Since $a_1 = 3$, we must determine another linearly independent eigenvector associated with λ_1. Such an eigenvector cannot be of order 2, since, if this would be the case, there would be two other linearly independent eigenvectors associated with λ_1, which, together with the eigenvector associated with λ_2 would give five linearly independent eigenvectors, a contradiction since we are dealing with a 4×4 matrix. Therefore, the new eigenvector x^3 must be of rank 1 and linearly independent of x^1. Such an eigenvector is, for example,

$$x^3 = (\ 0 \quad 0 \quad 1 \quad 0\ )^T$$

Finally, as for the simple eigenvalue λ_2, we have

$$(A - \lambda_2 I) = \begin{pmatrix} -1 & 2 & 0 & 1 \\ 0 & 0 & 0 & 0 \\ 0 & -1 & -1 & 0 \\ 0 & 0 & 0 & -1 \end{pmatrix}$$

and an eigenvector x^4 is given by (check that $Ax^4 = \lambda_2 x^4$)

$$x^4 = (\ 2 \quad 1 \quad -1 \quad 0\ )^T$$

In conclusion, the four eigenvectors are

$$x^1 = \begin{pmatrix} 1 \\ 0 \\ 0 \\ 0 \end{pmatrix} \quad x^2 = \begin{pmatrix} 1 \\ 0 \\ 0 \\ 1 \end{pmatrix} \quad x^3 = \begin{pmatrix} 0 \\ 0 \\ 1 \\ 0 \end{pmatrix} \quad x^4 = \begin{pmatrix} 2 \\ 1 \\ -1 \\ 0 \end{pmatrix}$$

and all of them are of order 1 except x^2, which is of order 2.

♣

A.8 JORDAN CANONICAL FORM

In this section, we show how the "simplest" matrix similar to a given matrix A can be determined. Before describing the most general case, we consider the simple but frequent case of a matrix A with n distinct eigenvalues $\lambda_1, \ldots, \lambda_n$.

THEOREM 16 *(diagonalization of a matrix with distinct eigenvalues)*

> Let A be an $n \times n$ matrix with n distinct eigenvalues $\lambda_i, i = 1, \ldots, n$, and n linearly independent eigenvectors $x^i, i = 1, \ldots, n$. The diagonal matrix
>
> $$A_D = \begin{pmatrix} \lambda_1 & & & 0 \\ & \lambda_2 & & \\ & & \ddots & \\ 0 & & & \lambda_n \end{pmatrix}$$
>
> is similar to the matrix A and $A_D = TAT^{-1}$ where the ith column of the matrix T^{-1} is the ith eigenvector x^i.

The proof of this theorem is very simple. In fact,

$$Ax^i = \lambda_i x^i \tag{A.10}$$

since the eigenvectors x^i are of order 1. Consequently, the transformation A is represented in the basis $\bar{\mathcal{X}} = \{x^1, x^2, \ldots, x^n\}$ by a matrix A_D whose ith column is the representation of the vector Ax^i in the $\bar{\mathcal{X}}$ basis. From Eq. (A.10), it follows that the vector Ax^i is represented by a vector with zero entries except the ith one, which is equal to λ_i. From this, *Theorem 16* follows.

A slightly more general case is that of matrix A with multiple eigenvalues but with n linearly independent eigenvectors of order 1. If we use the set of these eigenvectors as a basis $\bar{\mathcal{X}}$, the transformation is still represented by a matrix A_D in diagonal form. More precisely, if a_i is the multiplicity of the ith eigenvalue $(i = 1, \ldots, k, k < n)$, matrix A_D will be a block diagonal matrix

$$A_D = \begin{pmatrix} D_1 & & & 0 \\ & D_2 & & \\ & & \ddots & \\ 0 & & & D_k \end{pmatrix}$$

where each block D_i is a square diagonal matrix of dimension $a_i \times a_i$, namely,

$$D_i = \begin{pmatrix} \lambda_i & & & 0 \\ & \lambda_i & & \\ & & \ddots & \\ 0 & & & \lambda_i \end{pmatrix}$$

We now consider the most general case in which some of the n eigenvectors $x^1, x^2, \ldots, x^n$ (obtained by means of the previously outlined procedure) are of order > 1.

THEOREM 17 *(Jordan form)*

Let A be an $n \times n$ matrix with k distinct eigenvalues $\lambda_1, \ldots, \lambda_k$ with multiplicities $a_1, \ldots, a_k$ and assume that a_i linearly independent eigenvectors associated with each λ_i are known. Moreover, suppose that these eigenvectors are divided into g_i groups (or chains) of the form (A.8) with dimensions $n_i^1 \geq n_i^2 \geq \cdots \geq n_i^{g_i}$ (in the following, n_i^1 will be denoted by n_i). The following matrix:

$$J_i^h = \begin{pmatrix} \lambda_i & 1 & 0 & \cdots & 0 \\ 0 & \lambda_i & 1 & \cdots & 0 \\ \vdots & \vdots & \vdots & & \vdots \\ 0 & 0 & 0 & \cdots & 1 \\ 0 & 0 & 0 & \cdots & \lambda_i \end{pmatrix}$$

of dimension $n_i^h \times n_i^h$ is associated with the hth group of eigenvectors associated with λ_i. Thus a Jordan block

$$J_i = \begin{pmatrix} J_i^1 & & & 0 \\ & J_i^2 & & \\ & & \ddots & \\ 0 & & & J_i^{g_i} \end{pmatrix}$$

of dimension $a_i \times a_i$ is associated with each eigenvalue. The block-diagonal matrix

$$A_J = \begin{pmatrix} J_1 & & & 0 \\ & J_2 & & \\ & & \ddots & \\ 0 & & & J_k \end{pmatrix}$$

is similar to the matrix A, that is, $A_J = TAT^{-1}$ and the columns of the matrix T^{-1} are the n eigenvectors $x^1, x^2, \ldots, x^n$. Matrix A_J is known as *the Jordan canonical form*.

The proof of this theorem is very similar to that of the previous one.

EXAMPLE 6
Suppose we want to determine the Jordan canonical form of the matrix

$$A = \begin{pmatrix} 1 & 2 & 0 & 1 \\ 0 & 2 & 0 & 0 \\ 0 & -1 & 1 & 0 \\ 0 & 0 & 0 & 1 \end{pmatrix}$$

which has two distinct eigenvalues $\lambda_1 = 1$ ($a_1 = 3$) and $\lambda_2 = 2$ ($a_2 = 1$). We have already seen in the examples that two chains of eigenvectors are associated with λ_1 so that $n_1 = n_1^1 = 2$ and $n_1^2 = 1$. Matrix A_J is, therefore, given by

$$A_J = \begin{pmatrix} 1 & 1 & 0 & 0 \\ 0 & 1 & 0 & 0 \\ 0 & 0 & 1 & 0 \\ 0 & 0 & 0 & 2 \end{pmatrix}$$

while matrix T^{-1} is (see the eigenvectors $x^i, i = 1, \ldots, 4$, in *Example 5*)

$$T^{-1} = \begin{pmatrix} 1 & 1 & 0 & 2 \\ 0 & 0 & 0 & 1 \\ 0 & 0 & 1 & -1 \\ 0 & 1 & 0 & 0 \end{pmatrix}$$

so that

$$T = \begin{pmatrix} 1 & -2 & 0 & -1 \\ 0 & 0 & 0 & 1 \\ 0 & 1 & 1 & 0 \\ 0 & 1 & 0 & 0 \end{pmatrix}$$

The reader can easily check that $TAT^{-1} = A_J$.

♣

The dimension n_i of the first block of each Jordan block is equal to the dimension of the longest chain of eigenvectors associated with the ith eigenvalue λ_i and is called the *index* of the ith eigenvalue. The number g_i of blocks composing each Jordan block J_i is called a *geometric multiplicity* of the eigenvalue λ_i. The reason for this terminology is that an invariant subspace X^i of dimension g_i is associated with λ_i, and each vector x belonging to X^i is an eigenvector associated with λ_i, that is,

$$Ax = \lambda_i x \quad \forall x \in X^i$$

Such a subspace X^i is the subspace generated by all the eigenvectors of order 1 associated with λ_i, which are as many as the chains of eigenvectors associated with λ_i and, consequently, as many as the blocks composing J_i.

A.9 ANNIHILATING POLYNOMIAL AND MINIMAL POLYNOMIAL

We have already seen that it is interesting to associate with a matrix A a polynomial, namely, the characteristic polynomial

$$\Delta_A(\lambda) = \det(\lambda I - A)$$

Moreover, we have verified that such a polynomial is invariant with respect to the similarity transformation $\bar{A} = TAT^{-1}$, that is, $\Delta_A(\cdot) = \Delta_{\bar{A}}(\cdot)$. This is an important property, since it allows us to associate the characteristic polynomial to the transformation $L : X \to X$. Other polynomials enjoying the same property are the *annihilating polynomials*. Before defining them, let us remark that if $p(\cdot)$ is a polynomial

$$p(\lambda) = k_0 + k_1\lambda + k_2\lambda^2 + \cdots + k_m\lambda^m$$

then, $p(A)$ is the matrix

$$p(A) = k_0 I + k_1 A + k_2 A^2 + \cdots + k_m A^m$$

Moreover, given a matrix A and one of its similar matrices $\bar{A} = TAT^{-1}$, we have

$$(\bar{A})^i = (TAT^{-1})(TAT^{-1})\ldots(TAT^{-1}) = TA^i T^{-1}$$

so that

$$p(\bar{A}) = Tp(A)T^{-1}$$

DEFINITION 23 *(annihilating polynomial and minimal polynomial)*

A polynomial $\vartheta_A(\cdot)$ is said to be an *annihilating polynomial* of the matrix A if $\vartheta_A(A) = 0$. Among all monic annihilating polynomials $\vartheta_A(\cdot)$

$$\vartheta_A(\lambda) = \lambda^m + a_1\lambda^{m-1} + \cdots + a_m$$

that with the lowest degree m is the *minimal polynomial* and is denoted by $\Psi_A(\cdot)$.

Since the minimal polynomial $\Psi_A(\cdot)$ is particularly important in applications, it is interesting to know how it can be found. *Theorem 18* links the minimal polynomial of a matrix with its eigenvalues. Since the index of an eigenvalue is the dimension of its largest Jordan block, the minimal polynomial is easily determined, once the Jordan form A_J of matrix A is known (the opposite statement is false, namely the Jordan form contains more "information" than the minimal polynomial).

THEOREM 18 *(eigenvalues and minimal polynomial)*

Let $\lambda_1, \ldots, \lambda_k$ be the distinct eigenvalues of a matrix A of dimension $n \times n$ and let $n_1, \ldots, n_k$ be the indices of such eigenvalues. Then, the minimal polynomial $\Psi_A(\cdot)$ is given by

$$\Psi_A(\cdot) = (\lambda - \lambda_1)^{n_1}(\lambda - \lambda_2)^{n_2}\ldots(\lambda - \lambda_k)^{n_k} = \prod_{i=1}^{k}(\lambda - \lambda_i)^{n_i}$$

The proof of this theorem is not very complicated and is based on the Jordan form discussed in *Section A.8*.

An important consequence of *Theorem 18* is that the minimal polynomial of an $n \times n$ matrix has a degree that is $\leq n$, since $\sum_{i=1}^{k} n_i \leq n$. Moreover, $\Psi_A(\cdot)$ divides $\Delta_A(\cdot)$ because

$$\Delta_A(\lambda) = \prod_{i=1}^{k}(\lambda - \lambda_i)^{a_i}$$

and $a_i \geq n_i$, so that $\Psi_A(A) = 0$ implies $\Delta_A(A) = 0$ (*Cayley–Hamilton's* theorem, known since 1858!)

Theorem 18 often allows one to determine the Jordan form of a matrix A without computing the eigenvectors. For example, matrix A considered in *Example 5* has two distinct eigenvalues $\lambda_1 = 1$ and $\lambda_2 = 2$, the first with algebraic multiplicity $a_1 = 3$. Consequently, the minimal polynomial $\Psi_A(\cdot)$ is one of the three following polynomials:

$$p_1(\lambda) = (\lambda - 1)(\lambda - 2)$$
$$p_2(\lambda) = (\lambda - 1)^2(\lambda - 2)$$
$$p_3(\lambda) = (\lambda - 1)^3(\lambda - 2)$$

Since $p_1(A) \neq 0$ and $p_2(A) = 0$ we have $\Psi_A(\lambda) = p_2(\lambda)$ so that the Jordan form of A is

$$A_J = \begin{pmatrix} 1 & 1 & 0 & 0 \\ 0 & 1 & 0 & 0 \\ 0 & 0 & 1 & 0 \\ 0 & 0 & 0 & 2 \end{pmatrix}$$

as already found in *Example 6*.

A.10 NORMED SPACES

Vector spaces have often more "structure" than that considered so far (sum of elements and scalar multiplication). The most important concept, which allows the study of vector spaces from a "topological" point of view, is that of *norm*, which generalizes the usual notion of distance in three-dimensional Euclidean spaces.

DEFINITION 24 *(normed space)*

A *normed* vector space is a vector space X for which there exists a transformation, denoted by $\|\cdot\|$ and called *norm*, that transforms each element $x \in X$ in a real number $\|x\|$ and has the following three properties:

$$\|x\| > 0 \quad \forall x \neq 0 \quad \text{and} \quad \|0\| = 0$$
$$\|x + y\| \leq \|x\| + \|y\|$$
$$\|ax\| = |a| \cdot \|x\|$$

where $|a|$ is the modulus of the scalar a.

The three properties of the norm have a straightforward geometric interpretation; in particular, the second one is known as the *triangular inequality*. Obviously, the concept of norm applies both to finite dimensional spaces and to functional spaces. An example of a normed functional space is the space $C[a, b]$ of continuous functions $x(\cdot)$ over the interval $[a, b]$ with the norm given by $\max_{a \leq t \leq b} |x(t)|$.

It is worth noting that a space X can have many norms. For example, the space of continuous functions over the interval $[a, b]$, which generates the normed space $C[a, b]$, may generate also other normed spaces such as those with the norm

$$\|x(\cdot)\| = \int_b^a |x(t)| dt$$

The best known normed functional space is the space L_p, which is composed of the Lebesgue p-integrable functions ($p \geq 1$), namely, the functions $x(\cdot)$ with a finite Lebesgue integral of $|x(t)|^p$. In this space, the norm of $x(\cdot)$ is

$$\|x(\cdot)\|_p = \left[\int_b^a |x(t)|^p dt \right]^{1/p}$$

Analogously, the spaces $\mathbb{R}^n$ and C^n of n-tuples of real and complex numbers $(x_1, \ldots, x_n)$ can be normed using the norm ($p \geq 1$)

$$\|x\|_p = \left[\sum_{i=1}^n |x_i|^p \right]^{1/p}$$

or, alternatively, the norm

$$\|x\|_\infty = \max_i |x_i|$$

Among these, the most significant ones are $\|\cdot\|_1, \|\cdot\|_2$ and $\|\cdot\|_\infty$, which are represented in *Fig. A.6* with their unitary contour lines (for the case $n = 2$).

The norm $\|\cdot\|_2$ is the so-called Euclidean norm and, for this reason, the space $\mathbb{R}^n$, normed with $\|\cdot\|_2$, is denoted by E^n and called Euclidean space. In the following (except when explicitly stated), we will assume that the norm adopted for $\mathbb{R}^n$ or C^n is $\|\cdot\|_2$ and, for the sake of brevity, we will simply write, $\|\cdot\|$.

In normed spaces, it is possible to define closed and open sets, bounded sets (composed of elements with bounded norm), complete sets, and compact sets (which in finite dimensional vector spaces are closed and bounded). All such notions are straightforward generalizations of the corresponding, and well-known, concepts in the plane.

The relevance of the norm becomes clear when studying transformations (linear or not) between vector spaces, because in the case where the spaces are normed, it is possible then to define continuous transformations.

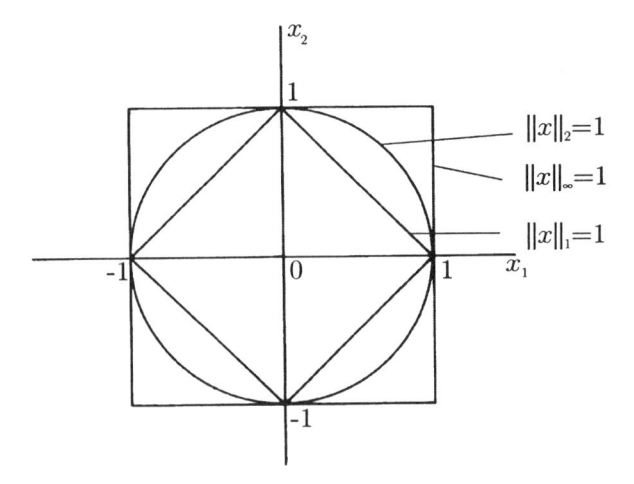

Figure A.6 Unitary norm contour lines in $\mathbb{R}^2$.

DEFINITION 25 *(continuous transformation)*

> A transformation $T(\cdot)$ (possibly nonlinear), which transforms the elements x of a normed space X into elements $T(x)$ of a normed space Y, is *continuous* at x_0 if for any $\varepsilon > 0$ there exists a $\delta(\varepsilon, x_0) > 0$ such that $\|T(x) - T(x_0)\| < \varepsilon$ for all x such that $\|x - x_0\| < \delta$. If $T(\cdot)$ is continuous at every point x_0, the transformation is said to be continuous and if $\delta(\varepsilon, x_0)$ is independent of x_0, the transformation is said to be uniformly continuous.

It is pretty clear from *Definition 25* that the continuity property is not only a characteristic of the transformation $T(\cdot)$, but also of the norms defined in the two spaces X and Y. From this point of view, the problem of an appropriate choice of the norms in the spaces X and Y appears to be extremely critical. It is important, however, to observe that in finite dimensional vector spaces, the continuity of a transformation is independent of the norms of the spaces X and Y (in other words, in finite dimensional spaces all norms are equivalent).

The linear transformations $L : X \to Y$ form a vector space: It is then natural to ask whether such a space may be normed or not. The answer is given by *Definition 26*.

DEFINITION 26 *(bounded transformations and their norm)*

> A linear transformation $L : X \to Y$ between normed spaces is said to be *bounded* if there exists a constant M such that $\|Lx\| \leq M\|x\|$ for all $x \in X$. The smallest of such constants M is the norm $\|L\|$ of the transformation L.

It easy to see that $\|L\|$ has the following properties:

$$\|L\| = \sup_{\|x\|=1} \|Lx\|$$

$$\|L\| = \sup_{\|x\|\neq 0} \frac{\|Lx\|}{\|x\|}$$

$$\|Lx\| \leq \|L\| \cdot \|x\|$$

$$\|L_1 + L_2\| \leq \|L_1\| + \|L_2\|$$

$$\|L_1 L_2\| \leq \|L_1\| \cdot \|L_2\|$$

Moreover, it can be proven that the linear transformation is continuous if and only if it is bounded.

From *Definition 26*, it follows that the norm of a linear transformation depends on the norms of the spaces X and Y. In the case of transformations $A: C^n \to C^n$, where A is a matrix with complex entries, if the norm in C^n is

$$\|x\|_1 = \sum_{i=1}^{n} |x_i|$$

it is easy to check that the transformed vertices of the hypersquare representing the unitary norm contour line are vectors whose norm is equal to the sum of the moduli of the elements of the columns of the matrix A. From the property

$$\|A\| = \sup_{\|x\|=1} \|Ax\|$$

it follows that

$$\|A\|_1 = \max_{j} \left[\sum_{i=1}^{n} |a_{ij}| \right]$$

However, if the norm in C^n is $\|\cdot\|_\infty$, then the norm of A is given by

$$\|A\|_\infty = \max \left[\sum_{j=1}^{n} |a_{ij}| \right]$$

Other possible norms for the transformations $A: C^n \to C^n$ are given by

$$\|A\| = \sum_{i=1}^{n} \sum_{j=1}^{n} |a_{ij}|$$

$$\|A\| = \left[\sum_{i=1}^{n} \sum_{j=1}^{n} a_{ij}^2 \right]^{\frac{1}{2}}$$

$$\|A\| = \max_i |\lambda_i|$$

where λ_i is the ith distinct eigenvalue of A.

A.11 SCALAR PRODUCT AND ORTHOGONALITY

In some vector spaces another operation (besides the sum) between pairs of vectors, called a *scalar product*, can be defined. Such spaces admit a norm, while normed spaces may not admit a scalar product. The scalar product allows the definition of *orthogonality* of two vectors, as a natural extension of the well-known geometrical concept of orthogonal straight lines and planes. By using this concept, it is possible to decompose each space into the direct sum of a subspace and its orthogonal subspace.

Though the scalar product can be defined in a pretty general way, we will assume in the sequel that the scalar field is the field of the reals.

DEFINITION 27 *(scalar product)*

A vector space is said to be a *pre-Hilbert* vector space if each pair of vectors (x, y) is associated with a real number, called a *scalar product* and denoted by $(x|y)$, with the following properties:

$$(x|y) = (y|x)$$
$$(x + y|z) = (x|z) + (y|z)$$
$$(ax|y) = a(x|y) \quad a \in \mathbb{R}$$
$$(x|x) > 0 \quad \text{per} \quad x \neq 0 \quad \text{and} \quad (0|0) = 0$$

The simplest example of a pre-Hilbert space is $\mathbb{R}^2$, which can be viewed as a plane in which the scalar product of two vectors x and y is given by the real number $\|x\| \cdot \|y\| \cdot \cos\vartheta$, where ϑ is the angle between the two vectors x and y and $\|x\|$ and $\|y\|$ are the "lengths" of the two vectors (see *Fig. A.7*).

A trivial generalization is the space $\mathbb{R}^n$ of the n-tuples of real numbers with

$$(x|y) = \sum_{i=1}^{n} x_i y_i = x^T y$$

where x^T denotes the row vector obtained by transposition of the column vector x. As already announced, the pre-Hilbert spaces can be normed; in fact, *Theorem 19* holds.

THEOREM 19 *(norm of pre-Hilbert spaces)*

In a pre-Hilbert space, the quantity $(x|x)^{1/2}$ is a norm.

In the space $\mathbb{R}^n$ with $(x|y) = x^T y$, one has

$$(x|x)^{1/2} = \left(\sum_{i=1}^{n} x_i^2\right)^{1/2} = \|x\|_2$$

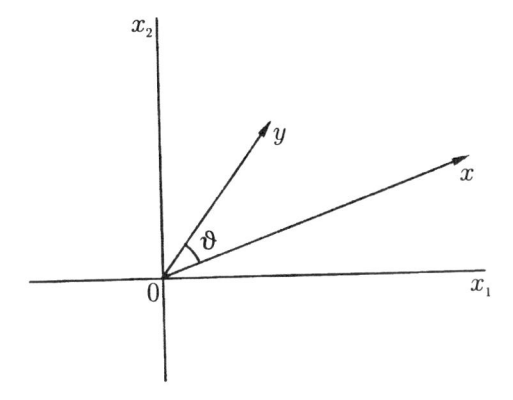

Figure A.7 The scalar product of two vectors x and y is $\|x\| \cdot \|y\| \cdot \cos \vartheta$.

that is, the norm induced by the scalar product $(x|y) = x^T y$ is $\|\cdot\|_2$. From now on, the symbol $\|\cdot\|$ will always refer to the norm induced by the scalar product, that is $\|x\| = (x|x)^{1/2}$.

It is also possible to extend to pre-Hilbert spaces the well-known "parallelogram law", which states that the sum of the squares of the diagonals of a parallelogram is equal to the sum of the squares of the sides. Indeed, it is easy to verify that $\|x+y\|^2 + \|x-y\|^2 = 2\|x\|^2 + 2\|y\|^2$.

Definition 28 of orthogonality is of paramount importance.

DEFINITION 28 *(orthogonality)*

In a pre-Hilbert space, two vectors x and y are said to be *orthogonal* (and denoted as $x \perp y$) if $(x|y) = 0$. Moreover, a vector y is said to be orthogonal to a set $\mathcal{X}$ if $(x|y) = 0 \ \forall x \in \mathcal{X}$.

For orthogonal vectors x and y, the celebrated Pythagoras' theorem holds, that is, $\|x\|^2 + \|y\|^2 = \|x+y\|^2$.

DEFINITION 29 *(orthogonal complement)*

Given a set $\mathcal{X}$ of a pre-Hilbert space, the *orthogonal complement* of $\mathcal{X}$, denoted by $\mathcal{X}^\perp$, is the set of all vectors orthogonal to $\mathcal{X}$.

Since the linear combination of orthogonal vectors to a set $\mathcal{X}$ is also orthogonal to the set $\mathcal{X}$, the orthogonal complement $\mathcal{X}^\perp$ of a set $\mathcal{X}$ is a subspace.

The results presented so far are valid both in finite dimensional spaces and in functional spaces. In contrast, the following results (*Theorem 20*) are valid only in finite-dimensional spaces:

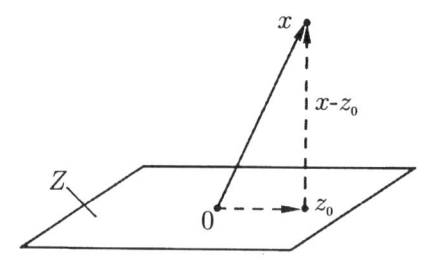

218 ELEMENTS OF LINEAR ALGEBRA AND MATRIX THEORY

Figure A.8 The point $z_0 \in Z$ closest to x is the orthogonal projection of x on Z.

THEOREM 20 *(orthogonal projection theorem)*

> Let Z be a subspace of a finite dimensional space X. Given a vector $x \in X$, there exists a unique vector $z_0 \in Z$ such that
>
> $$\|x - z_0\| \leq \|x - z\| \quad \forall z \in Z$$

Theorem 20 has straightforward geometric interpretation (see *Fig. A.8*) and is extremely important in applications. Moreover, it allows one to establish the decomposition *Theorem 21*.

THEOREM 21 *(decomposition theorem)*

> If Z is a subspace of a finite dimensional space X, then
>
> $$X = Z \oplus Z^\perp$$

Theorem 21 is illustrated in *Fig. A.9* for the case $X = \mathbb{R}^3$.

The projection of a vector on a subspace Z is, clearly, a linear transformation P_z and, as such, it admits a representation with respect to any given basis of X. Suppose $\{e^1, \ldots, e^m\}$ is a basis of Z and that $\{e^{m+1}, \ldots, e^n\}$ is a basis of $Z^\perp$. From *Theorem 21*, the vectors $\{e^1, \ldots, e^n\}$ represent a basis of X and the projection on Z leaves the vectors belonging to Z unchanged and transforms into the zero vector all the vectors belonging to $Z^\perp$. Thus, the transformation P_z is represented, in the basis $\{e^1, \ldots, e^n\}$, by the $n \times n$ matrix

$$A = \begin{pmatrix} I_m & 0 \\ 0 & 0 \end{pmatrix}$$

where I_m is the identity matrix of dimension $m \times m$.

SCALAR PRODUCT AND ORTHOGONALITY 219

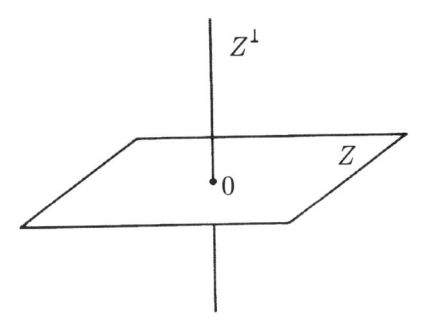

Figure A.9 Decomposition of a vector space into the direct sum of orthogonal subspaces.

DEFINITION 30 *(orthonormal vectors)*

A set of vectors $\{x^1, \ldots, x^k\}$ is called *orthonormal* (as well as the vectors) if $x^i \perp x^j$ $\forall i \neq j$ and $\|x^i\| = 1$ $\forall i$.

In an n-dimensional space, a set of n orthonormal vectors is a basis (also called orthonormal) and such a basis is particularly useful in applications. Moreover, there exists a simple procedure (known as *Gram–Schmidt procedure*) that allows one to obtain from a set $\{x^1, \ldots, x^k\}$ of linearly independent vectors an orthonormal set $\{e^1, \ldots, e^k\}$ with the property

$$[\{x^1, \ldots, x^i\}] = [\{e^1, \ldots, e^i\}] \quad i = 1, \ldots, k \tag{A.11}$$

In fact, let

$$e^1 = \frac{1}{\|x^1\|} x^1$$

Such a vector has unitary norm and $[x^1] = [e^1]$. Moreover, if (see *Fig. A.10*)

$$z^2 = x^2 - (x^2|e^1)e^1$$

the vector

$$e^2 = \frac{1}{\|z^2\|} z^2$$

has unitary norm, is orthogonal to e^1 and satisfies (A.11) for $i = 2$. By induction, one can show that the vectors

$$e^i = \frac{1}{\|z^i\|} z^i \quad i = 1, \ldots, k$$

where

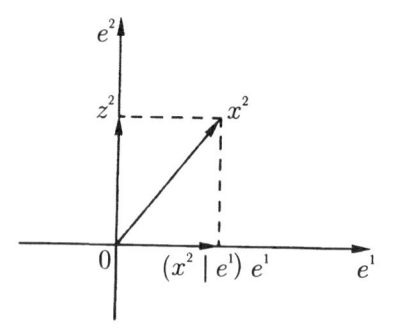

Figure A.10 The Gram–Schmidt procedure.

$$z^i = x^i - \sum_{h=1}^{i-1}(x^i|e^h)e^h$$

represent an orthonormal set satisfying (A.11).

As for vectors and transformations, the scalar product also admits, in finite dimensional spaces, a simple representation. In fact, if $\mathcal{X} = \{e^1, \ldots, e^n\}$ is a basis of the n-dimensional space X, and a and b are the representations of two vectors x and y, namely,

$$x = \sum_{i=1}^n a_i e^i \quad y = \sum_{i=1}^n b_i e^i$$

one has

$$(x|y) = \sum_{i=1}^n \sum_{j=1}^n a_i b_j (e^i|e^j) \tag{A.12}$$

Thus, setting

$$S = \begin{pmatrix} (e^1|e^1) & (e^1|e^2) & \cdots & (e^1|e^n) \\ \vdots & \vdots & & \vdots \\ (e^n|e^1) & (e^n|e^2) & \cdots & (e^n|e^n) \end{pmatrix} \quad a = \begin{pmatrix} a_1 \\ \vdots \\ a_n \end{pmatrix} \quad b = \begin{pmatrix} b_1 \\ \vdots \\ b_n \end{pmatrix}$$

Eq. (A.12) becomes

$$(x|y) = a^T S b \tag{A.13}$$

which is a formula for the computation of the scalar product of two vectors from their representations.

Clearly, Eq. (A.13) must be appropriately modified whenever a change of basis $\mathcal{X} \to \bar{\mathcal{X}}$ is performed. If T is the matrix corresponding to such a change of basis, we have

$$\bar{a} = Ta \quad \bar{b} = Tb$$

and, therefore,

$$(x|y) = \bar{a}^T \bar{S} \bar{b} = a^T T^T \bar{S} T b$$

so that, in view of (A.13),

$$S = T^T \bar{S} T$$

Finally, it is worth noting that Eq. (A.13) simplifies greatly whenever one chooses as a basis $\mathcal{X}$ of X an orthonormal basis. In fact, in this case, $(e^i|e^j) = 0$ for $i \neq j$ and $(e^i|e^i) = 1$, so that $S = I$. Thus, Eq. (A.13) becomes $(x|y) = a^T b$, which coincides with the expression already written for the scalar product in $\mathbb{R}^n$.

A.12 ADJOINT TRANSFORMATIONS

Though adjoint transformations could be presented with reference to functional spaces, we restrict ourselves to finite-dimensional spaces in which the scalar field is the field of the reals.

DEFINITION 31 *(adjoint transformation)*

Let $L : X \to Y$ be a linear transformation between two finite dimensional spaces in which a scalar product is defined. The *adjoint transformation* of L (often called adjoint operator) is the transformation $L^* : Y \to X$ with the following property:

$$(Lx|y) = (x|L^*y)$$

It is easy to see that *Definition 31* makes sense since the adjoint transformation exists and is unique and linear. Other properties of the adjoint transformation (not all easy to prove) are the following:

$$(L_1 + L_2)^* = L_1^* + L_2^*$$

$$(aL)^* = aL^*$$

$$(L_1 L_2)^* = L_2^* L_1^*$$

$$\|L\| = \|L^*\|$$

$$L^{**} = L$$

Moreover, if L^{-1}, exists then
$$(L^{-1})^* = (L^*)^{-1}$$
The adjoint transformation allows one to identify the null space of a transformation with the orthogonal complement of the image of the adjoint transformation. In fact, *Theorem 22* holds.

THEOREM 22 (*null space and adjoint transformation*)

Let $L : X \to Y$ be a linear transformation between two finite-dimensional spaces X and Y. Then,
$$\mathcal{N}(L) = \mathcal{I}(L^*)^\perp \qquad (A.14)$$

Recalling that $(X^\perp)^\perp = X$ and that $(L^*)^* = L$ Eq. (A.14) can be also written as
$$\mathcal{I}(L) = \mathcal{N}(L^*)^\perp \quad \mathcal{I}(L)^\perp = \mathcal{N}(L^*) \quad \mathcal{N}(L)^\perp = \mathcal{I}(L^*)$$
In the particular case in which $L : X \to Y$ is a matrix A, the following result (*Theorem 23*) can be established:

THEOREM 23 (*adjoint transformations and transposed matrices*)

If $A : \mathbb{R}^n \to \mathbb{R}^m$, the adjoint transformation A^* is A^T.

Such a result, together with *Theorem 22*, yields
$$\mathcal{N}(A) = \mathcal{I}(A^T)^\perp$$
that is, the null space of a matrix A is the subspace complementary to the subspace generated by the columns of A^T (viz, the rows of A).

EXAMPLE 7

Let us determine the null space of the transformation $A : \mathbb{R}^3 \to \mathbb{R}^2$ given by
$$A = \begin{pmatrix} 0 & 1 & 1 \\ 1 & 1 & 0 \end{pmatrix}$$
From the above remark, it follows that the null space $\mathcal{N}(A)$ is the subspace complementary to the space generated by the vectors
$$x^1 = (\ 0\ \ 1\ \ 1\ )^T \quad x^2 = (\ 1\ \ 1\ \ 0\ )^T$$
Since these two vectors are linearly independent, the subspace they generate is two dimensional and, therefore, $\mathcal{N}(A)$ is one dimensional.

The subspace $\mathcal{N}(A)$ is generated, for example, by the vector
$$x^3 = (\ 1\ \ -1\ \ 1\ )^T$$
which is orthogonal to x^1 and x^2, as illustrated in *Fig. A.11*.

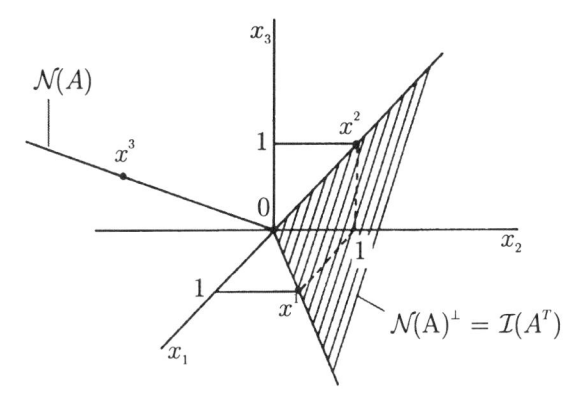

Figure A.11 Determinations of the null space $\mathcal{N}(A)$ of a matrix A.

Appendix B: Elements of Linear Systems Theory

B.1 DEFINITION OF LINEAR SYSTEMS

Linear systems are a particular, but very important, class of dynamic systems. As such, they are characterized by *input*, *state*, and *output* variables, denoted by u, x, and y, respectively. The symbol t denotes time, and can be either an integer (*discrete-time system*) or a real number (*continuous-time system*). We will consider *finite dimensional* systems with a single *input* and a single *output*, that is,

$$u(t) \in \mathbb{R} \quad x(t) \in \mathbb{R}^n \quad y(t) \in \mathbb{R}$$

where the dimension n of the state vector is called *order* of the system.

In discrete-time linear systems, the state vector is updated through a linear equation, called a *state equation*,

$$x(t+1) = Ax(t) + bu(t) \tag{B.1}$$

where A is an $n \times n$ matrix and b is an $n \times 1$ vector, while the output depends on the state and input through a linear equation, called an *output transformation*,

$$y(t) = c^T x(t) + du(t) \tag{B.2}$$

where c^T is a $1 \times n$ row vector and d is a real. Written for each component of the state vector, (B.1) corresponds to

$$\begin{aligned} x_1(t+1) &= a_{11}x_1(t) + \ldots + a_{1n}x_n(t) + b_1 u(t) \\ x_2(t+1) &= a_{21}x_1(t) + \ldots + a_{2n}x_n(t) + b_2 u(t) \\ &\vdots \\ x_n(t+1) &= a_{n1}x_1(t) + \ldots + a_{nn}x_n(t) + b_n u(t) \end{aligned}$$

while (B.2) becomes

$$y(t) = c_1 x_1(t) + \ldots + c_n x_n(t) + du(t)$$

Next, we will consider only time-invariant systems, that is, systems with A, b, c^T, and d constant over time.

Analogously, we can define continuous-time linear systems as systems with the following state equation:

$$\dot{x}(t) = Ax(t) + bu(t) \tag{B.3}$$

where $\dot{x}(t)$ is the derivative of $x(t)$ with respect to time, and

$$y(t) = c^T x(t) + du(t) \tag{B.4}$$

Thus, continuous-time and discrete-time linear systems are identified by the quadruple (A, b, c^T, d), which can be conveniently ordered as follows:

$$A = \boxed{} \qquad b = \boxed{}$$

$$c^T = \boxed{} \qquad d = \boxed{}$$

They are often graphically represented in one of the forms shown in *Fig. B.1*. The first form shows only the input and output variables, called *external* variables, since they are those through which the system interacts with the rest of the world. The second form also shows the state variables, called *internal*.

In many systems, the input does not directly influence the output, that is, $d = 0$. Such systems, called *proper*, are identified by the triple (A, b, c^T), while those with $d \neq 0$, called *improper*, are identified by the quadruple $(A, b, c^T d)$. Systems without input ($b = 0, d = 0$), are called *autonomous* and described, by the pair (A, c^T).

Next, we will discuss the main features of linear systems, starting from those depending only on the matrix A (reversibility and internal stability), and continuing with those characterized by the pair (A, b) (reachability), or by the pair (A, c^T) (observability), and ending with those depending on the triple (A, b, c^T) or on the quadruple (A, b, c^T, d) (external stability, minimum phase, minimality, etc.).

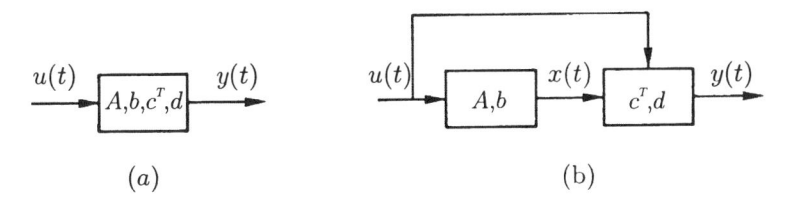

Figure B.1 Representations of a linear system: (a) compact form; (b) disaggregated form in which the first block represents the state equation and the second the output transformation.

EXAMPLE 1 (Newton's law)
Suppose that a point mass m moves without friction along a straight line and that a force $u(t)$ is applied to it along the same direction. If $y(t)$ is the position of the point mass, measured with respect to a fixed point, Newton's law states that

$$\dot{x}_1(t) = x_2(t)$$
$$\dot{x}_2(t) = \frac{1}{m}u(t)$$

while the output transformation is

$$y(t) = x_1(t)$$

In conclusion, Newton's law is described by a proper linear system identified by the triple

$$A = \begin{pmatrix} 0 & 1 \\ 0 & 0 \end{pmatrix} \quad b = \begin{pmatrix} 0 \\ 1/m \end{pmatrix}$$
$$c^T = (\ 1 \ \ 0\ )$$

♣

EXAMPLE 2 (Fibonacci's rabbits)
Maybe the oldest example of a discrete-time linear system is that concerning a rabbit population described by *Leonardo Fibonacci* from Pisa (1180–1250) in the book *Liber Abaci*. Let us denote the year by t, the number of pairs of young and adult rabbits at the beginning of year t by $x_1(t)$ and $x_2(t)$, the number of pairs of adult rabbits killed by hunters during year t by $u(t)$, and the total number of pairs of rabbits with $y(t)$. The assumptions made by Fibonacci (some of them are a bit extreme) are the following:

- Young rabbits do not reproduce.

- Young rabbits become adult after 1 year.

- Adult rabbits reproduce once a year.
- Each pair of adult rabbits generate a pair of young rabbits.
- Rabbits do not die.

Under these assumptions, a simple balance of young and adult rabbits leads to the following state equation:

$$x_1(t+1) = x_2(t)$$
$$x_2(t+1) = x_1(t) + x_2(t) - u(t)$$

while the output transformation is

$$y(t) = x_1(t) + x_2(t)$$

Thus, the system is proper and identified by the triple

$$A = \begin{pmatrix} 0 & 1 \\ 1 & 1 \end{pmatrix} \quad b = \begin{pmatrix} 0 \\ -1 \end{pmatrix}$$
$$c^T = \begin{pmatrix} 1 & 1 \end{pmatrix}$$

If we assume that at time $t = 0$ there is only one pair of young rabbits, that is,

$$x(0) = \begin{pmatrix} 1 \\ 0 \end{pmatrix}$$

and that there are no hunters [$u(t) = 0$ for all t], the state equation and the output transformation can be used recursively to determine the growth of the population, namely, the output sequence $y(0), y(1), y(2), \ldots$. The reader can easily verify that each element of the sequence is equal to the sum of the two previous elements (Fibonacci's series).

♣

B.2 ARMA MODEL AND TRANSFER FUNCTION

The definition of linear systems given in the previous paragraph is often termed *internal* since it explicitly refers to the state of the system. For the same reason, the alternative definition involving only input and output variables is called *external*. The definition is as follows: In a discrete-time system of order n, a weighted sum of $(n+1)$ subsequent input values equals, at any time t, a weighted sum of the corresponding output values, namely,

$$y(t) + \alpha_1 y(t-1) + \cdots + \alpha_n y(t-n) = \beta_0 u(t) + \beta_1 u(t-1) + \cdots + \beta_n u(t-n) \quad \text{(B.5)}$$

ARMA MODEL AND TRANSFER FUNCTION   229

If $\beta_0 \neq 0$, the input $u(t)$ directly influences the output $y(t)$ and, therefore, the system is improper. If, on the contrary, $\beta_0 = 0$ the system is proper. Equation (B.5) is often used in the form

$$y(t) = \sum_{i=1}^{n}(-\alpha_i)y(t-i) + \sum_{i=0}^{n}\beta_i u(t-i) \tag{B.6}$$

in which the first term of the right-hand side is called *autoregression* and the second *moving average*. For this reason, Eq. (B.5) is known as the autoregressive moving average model often abbreviated as the *ARMA model*. The continuous-time analog of Eq. (B.5) is the differential equation of order n

$$y^{(n)}(t) + \alpha_1 y^{(n-1)}(t) + \cdots + \alpha_n y^{(0)}(t) = \beta_0 u^{(n)}(t) + \beta_1 u^{(n-1)}(t) + \cdots + \beta_n u^{(0)}(t)$$
(B.7)

where $u^{(i)}(t)$ and $y^{(i)}(t)$ are the ith derivatives of input and output. Also, this model will be called (even if unproperly) the ARMA model.

The interpretation of Newton's law (see *Example 1*) can be completed by noting that the relationship

$$\ddot{y}(t) = \frac{1}{m}u(t)$$

is a particular case of Eq. (B.7) (MA model, *i.e.* ARMA model without the autoregressive term). As for the Fibonacci's rabbits (*Example 2*) the ARMA model is (easy to check)

$$y(t) - y(t-1) - y(t-2) = -u(t-1) - u(t-2)$$

This ARMA model can be used to recursively generate the Fibonacci's series [by annihilating the input and setting $y(0) = y(1) = 1$].

Equations (B.5) and (B.7) can be written in the general form

$$D(p)y(t) = N(p)u(t) \tag{B.8}$$

where $D(\cdot)$ and $N(\cdot)$ are two polynomials of degree n

$$D(p) = p^n + \alpha_1 p^{n-1} + \cdots + \alpha_n$$
$$N(p) = \beta_0 p^n + \beta_1 p^{n-1} + \cdots + \beta_n$$

and p is a "shift" operator for the discrete-time case (*i.e.*, $py(t) = y(t+1)$, $p^2 y(t) = y(t+2), \ldots$,) and a "derivative" operator for the continuous-time case [*i.e.*, $py(t) = \dot{y}, p^2 y(t) = \ddot{y}(t), \ldots$]. An ARMA model is therefore equivalent to two polynomials $D(\cdot)$ and $N(\cdot)$ or, alternatively, to $2n + 1$ parameters β_0 and $\{\alpha_i, \beta_i\}, i = 1, \ldots, n$. Usually, the symbol p in (B.8) is replaced by $z[s]$ when dealing with discrete [continuous] -time systems. Thus, for example, Newton's law (*Example 1*) is described by

230 ELEMENTS OF LINEAR SYSTEMS THEORY

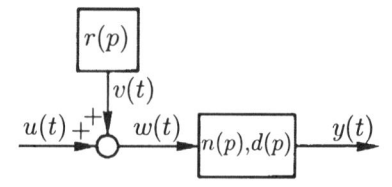

Figure B.2 Decomposition of an unreduced ARMA model $[D(p), N(p)]$ in a reduced ARMA model $[n(p), d(p)]$, and in an AR model $[r(p)]$.

$$D(s) = s^2 \quad N(s) = \frac{1}{m}$$

while Fibonacci's rabbits (*Example 2*) are described by

$$D(z) = z^2 - z - 1 \quad N(z) = -z - 1$$

If the two polynomials $D(\cdot)$ and $N(\cdot)$ are coprime (*i.e.*, they do not have common zeros), the ARMA model is said to be in *reduced form* or, simply, *reduced*. In such a case, the polynomial $D(\cdot)$ being monic, the knowledge of the pair $[D(\cdot), N(\cdot)]$ is equivalent to the knowledge of the ratio $D(\cdot)/N(\cdot)$, called the *transfer function* and denoted by $G(\cdot)$, that is,

$$G(p) = \frac{N(p)}{D(p)} \tag{B.9}$$

If the ARMA model is not in reduced form, namely, if

$$\begin{aligned} D(p) &= r(p)d(p) \\ N(p) &= r(p)n(p) \end{aligned} \tag{B.10}$$

with $n(\cdot)$ and $d(\cdot)$ coprime, the transfer function (B.9) is equal to $n(p)/d(p)$. The roots of $n(\cdot)$ and $d(\cdot)$ are called, respectively, *zeros* and *poles* of the transfer function. If we take into account Eq. (B.8) and (B.10), one can check that an unreduced ARMA model can be decomposed, as shown in *Fig. B.2*, in a reduced ARMA model identified by the pair of coprime polynomials $[d(\cdot), n(\cdot)]$.

$$d(p)y(t) = n(p)w(t) \tag{B.11}$$

and in an AR model determined by the polynomial $r(\cdot)$,

$$r(p)v(t) = 0 \tag{B.12}$$

In fact, if (B.11) is multiplied by $r(p)$ and (B.12) is taken into account together with the fact that

$$w(t) = v(t) + u(t)$$

Eq. (B.8) with $D(p)$ and $N(p)$ given by (B.10) is obtained. *Figure B.2 clearly shows that the transfer function $G(p) = n(p)/d(p)$ exclusively describes the reduced part of the ARMA model.* In other words, if only the transfer function is known, it is not possible to compute the output of the system from its input, unless the signal $v(\cdot)$ is identically zero, which occurs when the initial condition of the AR model (B.12) is zero.

B.3 COMPUTATION OF TRANSFER FUNCTIONS AND REALIZATION

Since we have given two different definitions of a dynamical system (one internal and one external) it is important to show how it is possible to move from one description to the other.

The problem of the computation of the ARMA model and of the transfer function of a system, given the quadruple (A, b, c^T, d), can be fully understood only after introducing the notions of reachability and observability. For the moment, notice that Eqs. (B.1) and (B.3), recalling the meaning of the operator p, can be written in the form

$$px(t) = Ax(t) + bu(t)$$

so that

$$x(t) = (pI - A)^{-1}bu(t)$$

From (B.2) and (B.4), it follows that

$$y(t) = [c^T(pI - A)^{-1}b + d]u(t)$$

which, compared with (B.8) and (B.9) yields

$$G(p) = c^T(pI - A)^{-1}b + d \tag{B.13}$$

The inverse of the $n \times n$ matrix $(pI - A)$ can be written in the form

$$(pI - A)^{-1} = \frac{1}{\Delta_A} P(p)$$

where $P(p)$ is an $n \times n$ matrix of polynomials with a degree $< n$ and $\Delta_A(p)$ is the characteristic polynomial of A. Then, $\Delta_A(p)$ and $P(p)$ can be computed using the following formulas (due to Souriau):

$$\Delta_A(p) = p^n + \alpha_1 p^{n-1} + \cdots + \alpha_n$$
$$P(p) = P_0 p^{n-1} + P_1 p^{n-2} + \cdots + P_{n-1}$$

where

$$P_0 = I \qquad\qquad \alpha_1 = -\mathrm{tr}(P_0 A)$$

$$P_1 = P_0 A + \alpha_1 I \qquad \alpha_2 = -\frac{1}{2}\mathrm{tr}(P_1 A)$$

$$P_2 = P_1 A + \alpha_2 I \qquad \alpha_3 = -\frac{1}{3}\mathrm{tr}(P_2 A)$$

$$\vdots \qquad\qquad \vdots$$

$$P_{n-1} = P_{n-2} A + \alpha_{n-1} I \quad \alpha_n = -\frac{1}{n}\mathrm{tr}(P_{n-1} A)$$

If the transfer function $G(p) = n(p)/d(p)$ computed by means of (B.13) has the polynomial $d(p)$ with degree n, then, from Souriau's formulas it follows that:

$$d(p) = D(p) = \Delta_A(p)$$

that is the ARMA model $[D(p), N(p)]$ of the system is in reduced form and the poles of the transfer function are n and coincide with the eigenvalues of matrix A. In contrast, if the degree of $d(\cdot)$ is $< n$, the poles of the transfer function are $< n$ but still coincide with some of the eigenvalues of matrix A.

The problem of the computation of the quadruple (A, b, c^T, d) from an ARMA model $[D(p), N(p)]$ is known as the *realization* problem [the quadruple (A, b, c^T, d), which solves the problem, is also called realization]. The solution of such a problem is not unique, so that it is particularly interesting to determine the realization with minimal dimension. In order to deal with this problem, it is necessary, however, to be aware of the notions of reachability and observability. For the moment, let us state that a particular realization, called *control canonical form*, of a reduced ARMA model

$$\begin{aligned}
D(p) &= p^n + \alpha_1 p^{n-1} + \cdots + \alpha_n \\
N(p) &= \beta_0 p^n + \beta_1 p^{n-1} + \cdots + \beta_n
\end{aligned} \qquad (\text{B}.14)$$

is the quadruple

$$A_c = \begin{pmatrix} 0 & 1 & 0 & \cdots & 0 \\ 0 & 0 & 1 & \cdots & 0 \\ \vdots & & & & \vdots \\ 0 & 0 & 0 & \cdots & 1 \\ -\alpha_n & -\alpha_{n-1} & -\alpha_{n-2} & \cdots & -\alpha_1 \end{pmatrix} \qquad b_c = \begin{pmatrix} 0 \\ 0 \\ \vdots \\ 0 \\ 1 \end{pmatrix}$$

$$c_c^T = (\;\gamma_n \quad \gamma_{n-1} \quad \gamma_{n-2} \quad \cdots \quad \gamma_1\;) \qquad d_c = \beta_0$$

with

$$\gamma_i = \beta_i - \beta_0 \alpha_i \quad i = 1, \ldots, n$$

Notice that the order n of the control canonical form is equal to the "memory" of the autoregressive component of the ARMA model.

A second realization, called *reconstruction canonical form*, is the following

$$A_r = \begin{pmatrix} 0 & 0 & \cdots & 0 & -\alpha_n \\ 1 & 0 & \cdots & 0 & -\alpha_{n-1} \\ 0 & 1 & \cdots & 0 & -\alpha_{n-2} \\ \vdots & \vdots & & \vdots & \vdots \\ 0 & 0 & \cdots & 1 & -\alpha_1 \end{pmatrix} \qquad b_r = \begin{pmatrix} \gamma_n \\ \gamma_{n-1} \\ \gamma_{n-2} \\ \vdots \\ \gamma_1 \end{pmatrix}$$

$$c_r^T = (\ 0 \quad 0 \quad \cdots \quad 0 \quad 1\ ) \qquad d_r = \beta_0$$

with

$$\gamma_i = \beta_i - \beta_0 \alpha_i \quad i = 1, \ldots, n$$

It is worth noting that

$$(A_r, b_r, c_r^T, d_r) = (A_c^T, c_c, b_c^T, d_c)$$

which is a formula that we will recall when discussing the duality principle.

EXAMPLE 3 (Fibonacci's rabbits)
Consider the ARMA model

$$D(z) = z^2 - z - 1 \qquad N(z) = -z - 1$$

which, as previously shown, is the ARMA model describing the growth of the Fibonacci's rabbits (see *Example 2*). The control and reconstruction canonical forms of this ARMA model are

$$A_c = \begin{pmatrix} 0 & 1 \\ 1 & 1 \end{pmatrix} \qquad b_c = \begin{pmatrix} 0 \\ 1 \end{pmatrix}$$
$$c_c^T = (\ -1 \quad -1\ )$$

and

$$A_r = \begin{pmatrix} 0 & 1 \\ 1 & 1 \end{pmatrix} \qquad b_r = \begin{pmatrix} -1 \\ -1 \end{pmatrix}$$
$$c_r^T = (\ 0 \quad 1\ )$$

and are, therefore, different from the triple (A, b, c^T) used in *Example 2*.

♣

234   ELEMENTS OF LINEAR SYSTEMS THEORY

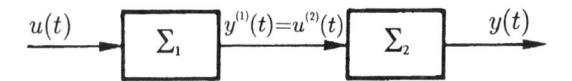

Figure B.3 Two systems connected in series.

B.4 INTERCONNECTED SUBSYSTEMS AND MASON'S FORMULA

Very often a dynamic system Σ is composed of interconnected subsystems Σ_i. Two dynamic systems, Σ_1 and Σ_2, can be interconnected in three ways: in *series*, in *parallel*, and in *feedback*. If $x^{(1)}$ and $x^{(2)}$ are the state vectors of Σ_1 and Σ_2, the state vector x of Σ is $x = [x^{(1)^T} \; x^{(2)^T}]^T$. Thus, if $\Sigma_i = (A_i, b_i, c_i^T, d_i)$, $i = 1, 2$, are the two subsystems, we are interested in the determination of the interconnected system $\Sigma = (A, b, c^T, d)$.

Series

Two systems are connected in series (*Fig. B.3*) when the output of the first system is the input of the second system.

The state equations of Σ are therefore

$$\dot{x}^{(1)}(t) = A_1 x^{(1)}(t) + b_1 u(t)$$

$$\dot{x}^{(2)}(t) = A_2 x^{(2)}(t) + b_2(c_1^T x^{(1)}(t) + d_1 u(t))$$

while the output transformation is

$$y(t) = c_2^T x^{(2)}(t) + d_2(c_1^T x^{(1)}(t) + d_1 u(t))$$

In conclusion, Σ is identified by the following quadruple:

$$A = \begin{pmatrix} A_1 & 0 \\ b_2 c_1^T & A_2 \end{pmatrix} \quad b = \begin{pmatrix} b_1 \\ b_2 d_1 \end{pmatrix}$$
$$c^T = (\; d_2 c_1^T \quad c_2^T \;) \quad d = (d_1 d_2)$$

Observe that matrix A is in block triangular form, so that its eigenvalues are those of matrices A_1 and A_2.

Parallel

Two systems are connected in parallel (*Fig. B.4*) when they have the same input and the output of the overall system is the sum of their outputs.

It is straightforward to check that Σ is identified by the following four matrices

$$A = \begin{pmatrix} A_1 & 0 \\ 0 & A_2 \end{pmatrix} \quad b = \begin{pmatrix} b_1 \\ b_2 \end{pmatrix}$$
$$c^T = (\; c_1^T \quad c_2^T \;) \quad d = (d_1 + d_2)$$

Also, in this case matrix A is block triangular (actually, diagonal) so that its eigenvalues are those of matrices A_1 and A_2.

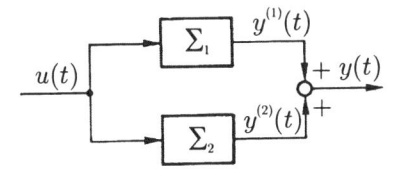

Figure B.4 Two systems connected in parallel.

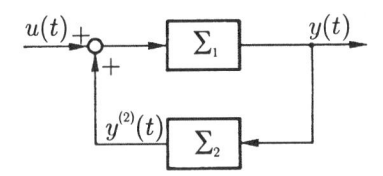

Figure B.5 Two systems connected in feedback (Σ_1 is the forward path and Σ_2 the feedback path).

Feedback

Two systems are connected in feedback (*Fig. B.5*) when the input of the first system is the sum of an external input u and of the output of the second system and the input of the second system is the output of the first one.

Obviously, interconnected subsystems can also be studied from the point of view of their external behavior. Actually, the ARMA model and the transfer function of a system Σ can be easily determinated from the ARMA models and the transfer functions of all its subsystems Σ_i. To verify this statement, we first analyze the cases of series, parallel, and feedback connections of two subsystems.

Series

With reference to *Fig. B.3*, let $\Sigma_1 = [D_1(p), N_1(p)]$ and $\Sigma_2 = [D_2(p), N_2(p)]$.

This means that the ARMA model of the first subsystem is

$$D_1(p)y^{(1)}(t) = N_1(p)u(t)$$

By applying to both sides of this equation the operator $N_2(p)$ and by noting that $y^{(1)} = u^{(2)}$, we obtain

$$N_2(p)D_1(p)u^{(2)}(t) = N_2(p)N_1(p)u(t)$$

But $N_2 D_1 = D_1 N_2$ and $N_2 N_1 = N_1 N_2$ since deriving (or shifting) a function first r times and then s times is equivalent to deriving (or shifting) it first s times and then r times. Thus, we can write

$$D_1(p)N_2(p)u^{(2)}(t) = N_1(p)N_2(p)u(t)$$

On the other hand, the ARMA equation of the second subsystem is

$$D_2(p)y(t) = N_2(p)u^{(2)}(t)$$

so that, finally, we obtain

$$D_1(p)D_2(p)y(t) = N_1(p)N_2(p)u(t)$$

In other words, if two systems Σ_1 and Σ_2 are connected in series, the resulting system Σ is characterized by an ARMA model identified by the following two polynomials:

$$D(p) = D_1(p)D_2(p) \quad N(p) = N_1(p)N_2(p)$$

This means that the transfer function $G(p) = N(p)/D(p)$ of Σ can be obtained by multiplying the two transfer functions $G_1(p)$ and $G_2(p)$ of the two subsystems, that is,

$$G(p) = G_1(p)G_2(p)$$

This result allows one to conclude that the order in which the two systems are connected is not relevant when computing the transfer function of the resulting system.

Parallel

With reference to *Fig. B.4*, proceeding as in the case of the series connection, it is easy to show that the transfer function of Σ is

$$G(p) = G_1(p) + G_2(p)$$

In other words, the transfer function of a system composed of two systems connected in parallel is the sum of their transfer functions.

Feedback

In the case of two systems Σ_1 and Σ_2 connected in feedback, as shown in *Fig. B.5*, one obtains

$$G(p) = \frac{G_1(p)}{1 - G_1(p)G_2(p)}$$

This formula is very useful in the analysis of feedback systems. It holds for the connection shown in *Fig. B.5* where the *feedback* is called *positive* since the signal $y^{(2)}$ coming from the feedback path is summed to the external signal u. In contrast, if one considers the *negative feedback*

$$u^{(1)} = u - y^{(2)}$$

the formula to be used is the following:

$$G(p) = \frac{G_1(p)}{1 + G_1(p)G_2(p)}$$

This formula is often described by saying that the transfer function of a system with a negative feedback is the ratio between the transfer function of the direct path (G_1) and the loop transfer function ($G_1 G_2$) plus one ($G_1 G_2$ is the loop transfer function because, in the loop, the two systems Σ_1 and Σ_2 are connected in series).

Mason's formula

Mason's formula allows one to compute the transfer function $G(p)$ of any system composed of interconnected subsystems. Under the nonlimiting assumption that signals are only summed (and not subtracted) the formula is

$$G(p) = \frac{\sum_k C_k(p)\Delta_k(p)}{\Delta(p)}$$

where $C_k(p)$, $\Delta(p)$, and $\Delta_k(p)$ are called, respectively, *transfer function of the kth direct input–output path*, *determinant* of the system, and *reduced determinant with respect to the kth direct path*. The transfer function $C_k(p)$ is simply the product of the transfer functions of all the systems composing the kth direct (*i.e.*, not containing cycles) path from input to output. The determinant $\Delta(p)$ is given by

$$\Delta(p) = 1 - \sum_i L_i(p) + \sum_i \sum_j L_i(p) L_j(p) - \sum_i \sum_j \sum_k L_i(p) L_j(p) L_k(p) + \cdots$$

where $L_i(p)$ is the transfer function of the ith closed path (loop), that is, the product of the transfer functions of all the subsystems composing the ith closed path exiting in the system. The first sum in the formula concerns all the loops, the second all the disjoint pairs of loops (*i.e.*, loops that do not touch each other) and so on. Finally, the reduced determinant Δ_k is the determinant Δ without all the terms corresponding to loops that are touched by the kth direct path. On occasions, it may not be easy to find all the direct paths and all the loops by inspection of the graph representing the interconnected system. However, in many cases of practical interest, Mason's formula is straightforward to apply, particularly when there are no disjoint loops.

B.5 CHANGE OF COORDINATES AND EQUIVALENT SYSTEMS

The quadruple (A, b, c^T, d) describing a linear system depends on the units chosen for time, input, state and output variables and on the order in which the state variables are listed. But, the choice of the variables to be considered as state variables is also not unique and has an impact on the quadruple (A, b, c^T, d) that identifies the system. For example, in a chemical reactor characterized by two species, one can consider as state variables the concentrations x_1 and x_2 of such species or, alternatively, their sum z_1 and their difference z_2. Obviously, the quadruple (A, b, c^T, d) corresponding to the state variables (x_1, x_2) is different from that corresponding to the state variables (z_1, z_2) while the system, from a physical point of view, is the same. For this reason, the two quadruples are called *equivalent*. In order to find the relationship among equivalent quadruples, it is necessary to determine the effect of a change of coordinates

$$z = Tx$$

In the case of the chemical reactor, for example, the change of coordinates $z = Tx$ is given by

$$\begin{pmatrix} z_1 \\ z_2 \end{pmatrix} = \begin{pmatrix} x_1 + x_2 \\ x_1 - x_2 \end{pmatrix} = \begin{pmatrix} 1 & 1 \\ 1 & -1 \end{pmatrix} \begin{pmatrix} x_1 \\ x_2 \end{pmatrix}$$

It is straightforward to check that a change of coordinates $z = Tx$ transforms the discrete-time system (B.1), (B.2) into the equivalent system

$$z(t+1) = TAT^{-1}z(t) + Tbu(t)$$
$$y(t) = c^T T^{-1} x(t) + du(t)$$

Analogously, the continuous-time system (B.3), (B.4) is transformed into the system

$$\dot{z}(t) = TAT^{-1}z(t) + Tbu(t)$$
$$y(t) = c^T T^{-1} x(t) + du(t)$$

In conclusion, a change of coordinates $z = Tx$ transforms the quadruple (A, b, c^T, d) into the quadruple $(TAT^{-1}, Tb, c^T T^{-1}, d)$.

B.6 MOTION, TRAJECTORY, AND EQUILIBRIUM

Once the initial state $x(0)$ and the input $u(t)$ for $t \geq 0$ are fixed, the state equations (B.1) and (B.3) admit a unique solution $x(t)$ for $t \geq 0$ (this is pretty obvious for discrete-time systems while, for continuous-time systems, it follows from results of existence and uniqueness of ordinary differential equations). The function $x(\cdot)$ thus obtained, is called *motion*, while the set $\{x(t), t \geq 0\}$ in the space $\mathbb{R}^n$ is called *trajectory*. The trajectory of a continuous-time system is a line originating from point $x(0)$ and with a specified direction [see *Fig. B.6(a)*]. In the case of discrete-time systems, the trajectory is a sequence of points $\{x(0), x(1), \ldots\}$ that, for the sake of clarity, are often linked one to the next with a segmented straight line, as shown in *Fig. B.6(b)*.

As shown in *Fig. B.6(a)*, it may happen that the trajectory passes through the same point x at different instants of time t_1, t_2, and so on, and that the vectors tangent to the trajectory at that point are different. In fact, the tangent vector is $\dot{x}$ and therefore, the case shown in *Fig. B.6(a)* can occurs if

$$Ax + bu(t_1) \neq Ax + bu(t_2)$$

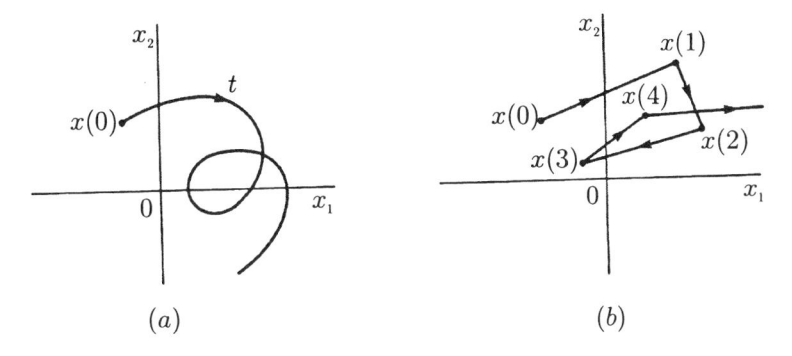

Figure B.6 Trajectories of second-order systems: (a) continuous-time systems; (b) discrete-time systems.

that is,

$$bu(t_1) \neq bu(t_2)$$

Obviously, such conditions cannot hold if the input u is constant over time.

It may occur that the motion $x(\cdot)$ corresponding to a particular initial state $x(0)$ and to a particular input function is periodic with period T, that is,

$$x(t) = x(t+T) \quad \forall t$$

In this case, the trajectory is a closed line (cycle) repeatedly visited every T time units. If x is periodic, then so is $\dot{x}$, so that

$$Ax + bu(t) = Ax(t+T) + bu(t+T) \quad \forall t$$

and therefore

$$bu(t) = bu(t+T) \quad \forall t \tag{B.15}$$

This means that a cycle can possibly be obtained only if the input function satisfies condition (B.15). It is important to note that condition (B.15) holds for every periodic input function with period T and, consequently, for any constant input.

A degenerate case occurs when the state of the system does not change over time, so that the cycle is represented by a point $\bar{x}$ called an *equilibrium state*. To this purpose, we give *Definition 1*.

DEFINITION 1 *(equilibrium)*

A system is said to be at *equilibrium* if input and state (and, therefore, also output) are constant, that is, if

$$u(t) = \bar{u} \quad x(t) = \bar{x} \quad y(t) = \bar{y} \quad \forall t$$

Vector $\bar{x}$ is called an equilibrium state.

Since for continuous-time systems, $x(t) = \bar{x}$ $\forall t$ implies $\dot{x}(t) = 0$, it follows that in such systems

$$A\bar{x} + b\bar{u} = 0 \qquad (B.16)$$

$$\bar{y} = c^T\bar{x} + d\bar{u} \qquad (B.17)$$

If A is nonsingular (i.e., if $\det A \neq 0$ or, equivalently, if A has no zero eigenvalues), then there exists a unique solution $\bar{x}$ to Eq. (B.16) for each $\bar{u}$, and, therefore, there is also only one solution $\bar{y}$ of (B.17), which are formally given by

$$\bar{x} = -A^{-1}b\bar{u} \qquad \bar{y} = (d - c^T A^{-1} b)\bar{u} \qquad (B.18)$$

In the case A is singular [$\det A = 0$] and $\bar{u}$ is fixed, either no solutions $\bar{x}, \bar{y}$ of (B.16), (B.17) exist or they are infinite.

For discrete-time systems, Eq. (B.16) and (B.17) must be replaced by the relations

$$(I - A)\bar{x} = b\bar{u}$$

$$\bar{y} = c^T\bar{x} + d\bar{u}$$

so that uniqueness of the equilibrium state (and output) for any fixed $\bar{u}$ is guaranteed by nonsingularity of the matrix $(I - A)$, that is,

$$\det(I - A) \neq 0$$

or, equivalently, from the fact that A has no unitary eigenvalues. In such cases, one has

$$\bar{x} = (I - A)^{-1} b\bar{u} \qquad \bar{y} = (d + c^T(I - A)^{-1} b)\bar{u} \qquad (B.19)$$

Equations (B.18) and (B.19) show that in nonsingular cases the relationship between input and output at the equilibrium is linear. Since for single input and single output systems it is usual to define the *gain* of the system as the ratio μ between output and input at the equilibrium

$$\mu = \frac{\bar{y}}{\bar{u}}$$

then, for continuous-time systems the following formula holds:

$$\mu = d - c^T A^{-1} b$$

while for discrete-time systems one has

$$\mu = d + c^T (I - A)^{-1} b$$

Obviously, the same formulas show that it is meaningless to define a gain in singular cases.

It is important to note that the calculation of the gain is straightforward whenever the input–output relation (B.7) of a continuous-time system is known, since the equilibrium condition implies $y^{(i)} = u^{(i)} = 0, i = 1, \ldots, n, y^{(0)} = \bar{y}$ and $u^{(0)} = \bar{u}$, so that

$$\mu = \frac{\beta_n}{\alpha_n} \tag{B.20}$$

For discrete-time systems, one has [see (B.6)]

$$\mu = \frac{\sum_{i=0}^{n} \beta_i}{1 + \sum_{i=1}^{n} \alpha_i} \tag{B.21}$$

Note that, denoting with $G(s)$ the transfer function of a continuous-time system, Eq. (B.20) is equivalent to $\mu = G(0)$, so that the gain μ is equal to the value of the transfer function for $s = 0$. For discrete-time systems with transfer function $G(z)$, from (B.21) it follows that $\mu = G(1)$, that is, the gain is equal to the value of the transfer function for $z = 1$.

The gain is also easy to compute in the case of systems composed of interconnected subsystems. It is in fact immediate to verify that the gain μ of a system composed of two subsystems connected in series is the product of the gains of the two subsystems, that is,

$$\mu = \mu_1 \mu_2$$

while for parallel connections, the following formula holds

$$\mu = \mu_1 + \mu_2$$

and for feedback connections we have

$$\mu = \frac{\mu_1}{1 + \mu_1 \mu_2}$$

B.7 LAGRANGE'S FORMULA AND TRANSITION MATRIX

From state equations of a linear system, it follows that the state at time t is a function of the initial state at time $t = 0$, of the input during the interval of time $[0, t)$ and, obviously, of the considered interval of time t. Obtaining an explicit solution, in the usual sense, of the state equations is possible only in particularly simple cases (typically, for first- and second-order systems). The solution can be, however, specified and written in a particularly useful form for the understanding of many problems and for the proof of several properties. In the case of continuous-time systems, the formula is credited to Lagrange; for the sake of simplicity, we will give the same name to the corresponding formula holding for discrete-time systems.

THEOREM 1 (Lagrange formula)

In a continuous-time linear system

$$\dot{x}(t) = Ax(t) + bu(t)$$

the state $x(t)$ for $t \geq 0$ is given by (*Lagrange formula*)

$$x(t) = e^{At}x(0) + \int_0^t e^{A(t-\xi)} bu(\xi) d\xi \qquad (B.22)$$

where

$$e^{At} = I + At + A^2 \frac{t^2}{2!} + A^3 \frac{t^3}{3!} + \cdots$$

Analogously, in a discrete-time linear system

$$x(t+1) = Ax(t) + bu(t)$$

for $t > 0$, the following holds:

$$x(t) = A^t x(0) + \sum_{i=0}^{t-1} A^{t-i-1} bu(i) \qquad (B.23)$$

Equation (B.23) is also called the Lagrange formula.

The Lagrange formulas (B.22) and (B.23) can be rewritten in a more compact form as follows:

$$x(t) = \Phi(t)x(0) + \Psi(t)u_{[0,t)}(\cdot) \qquad (B.24)$$

where $\Phi(t)$ and $\Psi(t)$ are linear transformations acting, respectively, on the initial state $x(0)$ and on the segment $u_{[0,t)}(\cdot)$ of the input function $u(\cdot)$. By comparing Eq. (B.24) with (B.22) and (B.23), it follows that the matrix $\Phi(t)$, called the *transition matrix*, is given by

$$\Phi(t) = \begin{cases} e^{At} & \text{for continuous-time systems} \\ A^t & \text{for discrete-time systems} \end{cases}$$

Equation (B.24) states that the state of the system is at any time given by the sum of two terms, the first linearly depending on the initial state and the second linearly depending on the input. These two contributions to the motion of a dynamic system are called, respectively, *free motion* and *forced motion*. The reason for such names is obvious: $\Phi(t)x(0)$ represents the evolution of the "free" system, that is, of the system with a null input (or without input, as usually said), while $\Psi(t)u_{[0,t)}(\cdot)$ represents the evolution of the system initially at rest $[x(0) = 0]$ but forced by the input $u(\cdot)$. By applying the output transformation given by (B.24), one obtains

$$y(t) = c^T \Phi(t)x(0) + c^T \Psi(t)u_{[0,t)}(\cdot) + du(t) \qquad (B.25)$$

which clearly shows that the output is the sum of free and forced evolutions.

Equation (B.25), if appropriately interpreted, allows us to formulate the so-called *superposition principle*, often recalled when talking about dynamic linear systems.

THEOREM 2 *(superposition principle)*

> If the pair $(x'(0), u'(\cdot))$ gives rise to the output $y'(\cdot)$ and the pair $(x''(0), u''(\cdot))$ to the output $y''(\cdot)$, then the pair $[\alpha x'(0) + \beta x''(0), \alpha u'(\cdot) + \beta u''(\cdot)]$ gives rise to the same linear combination $\alpha y'(\cdot) + \beta y''(\cdot)$ of the outputs.

Among the most frequently used formulas of any discipline, it is almost invariably possible to find some that are nothing but the Lagrange formula applied to simple first- or second-order systems. The law of the falling of bodies, the law of charge and discharge of a capacitor, the law governing the rise of temperature in a thermometer, and the one describing the release from a reservoir, are just examples of the application of the Lagrange formula to continuous-time systems. But the same holds for laws regarding discrete-time systems, as shown in *Example 4*.

EXAMPLE 4 (amortization)
If a debt D is amortized by returning for N consequent years an amount A, the debt x varies during the years according to the equation

$$x(t+1) = (1+\rho)x(t) - A$$

where ρ is the annual interest rate. One can then apply the Lagrange formula (B.23) with $t = N$ and $u(i) = A$ to such a system thus obtaining

$$x(N) = (1+\rho)^N D - A \sum_{i=0}^{N-1} (1+\rho)^{N-i-1}$$

By imposing the final condition $x(N) = 0$ and by solving with respect to A, one obtains the famous amortization formula

$$A = \frac{\rho}{1 - (1+\rho)^{-N}} D$$

♣

The Lagrange formula should *not* be regarded as a formula useful for the calculation of the state evolution of a linear system. Such a statement is particularly simple to illustrate in the case of discrete-time systems. In fact, in such systems, the state evolution can be computed by a repeated use for $t = 0, 1, 2$, and so on, of the equation

$$x(t+1) = Ax(t) + bu(t)$$

By doing so, at each step $n^2 + n$ multiplications are needed and about the same number of sums, so that the computation of $x(1), x(2), \ldots, x(N)$ requires $Nn(n+1)$

elementary operations. The computation of the same vectors by means of (B.23) is indeed much more onerous since the evaluation the matrix powers $A^2, A^3, \ldots, A^N$ is an operation requiring Nn^3 elementary operations. The importance of the Lagrange formula is then mostly related to conceptual and formal aspects of the theory of linear systems.

B.8 REVERSIBILITY

In a dynamic system, the input in an interval of time $[0, t]$ and the initial state $x(0)$ uniquely determine the state $x(t)$ and the output $y(t)$ at the final time t. In other words, the future evolution of the system is always guaranteed and uniquely determined. In the case of a linear system, this is clear from the Lagrange formulas (B.22) and (B.23) holding for $t \geq 0$. In some systems, the existence of the evolution is guaranteed and uniquely determined also in the past. Such systems are called *reversible*. For a linear system, *Theorem 3* holds.

THEOREM 3 *(reversibility condition)*

> Continuous-time linear systems are reversible, while discrete-time systems are such if and only if their matrix A is nonsingular.

The proof of *Theorem 3* follows from the fact that in continuous-time systems the transition matrix $\Phi(t) = e^{At}$ is invertible [its inverse is, in fact, $\Phi^{-1}(t) = e^{-At}$]. On the contrary, in the case of discrete-time systems $\Phi(t) = A^t$, so that $\Phi(t)$ is invertible if and only if A^t and, therefore, A is invertible.

Theorem 3 let us envisage a strong analogy between reversible continuous and discrete -time systems. On the other hand, it is clear that discrete-time systems need more attention. The peculiarity of irreversibility is often not emphasized as it should be, mainly because the discrete-time systems that are more frequently studied are the *sampled-data systems* that, as it will be shown in Section B.9, are reversible. However, there are important classes of discrete-time systems that are irreversible, such as the *finite memory systems*. Such systems have the property that the initial state influences the systems evolution only for a finite period of time. Since

$$x(t) = \Phi(t)x(0) + \Psi(t)u_{[0,t)}(\cdot)$$

the free motion is zero from a certain time, for any initial state $x(0)$. This implies that $\det \Phi(t) = \det(A^t) = (\det A)^t = 0$, that is, the system is irreversible.

B.9 SAMPLED-DATA SYSTEMS

The input of many continuous-time systems is often changed at precise instants of time, and then kept constant for an interval of time. This may occur in a production system, in which the production rate is fixed each week, in the reactions induced by a drug treatment delivered by perfusion each day, in the exploitation of water supplies for the production of electric power in which the power of an hydraulic turbine is scheduled to vary each hour, and in many other systems characterized by the presence of a supervisor who, for different reasons, considers it not appropriate to control the system continuously. Once a decision is taken at each instant, the input of the system (production rate, drug delivery rate, turbine power) is kept constant for a certain interval of time, to the end of which a new decision is taken. Unlike inputs, the state variables characterizing the system (stocks, concentrations, water supplies) vary (sometimes quite heavily) during such an interval of time. An analogous situation can often be found in industrial automation, where computers are used for controlling various processes: During a certain interval of time, the computer processes the information received and determines the value $\tilde{u}$ of the input to be applied to the system during the subsequent interval of time. As shown in *Fig. B.7*, an interface is needed between the computer and the system, called *hold circuit*, able to transform the digital input of the computer into a constant analog signal (input of the system).

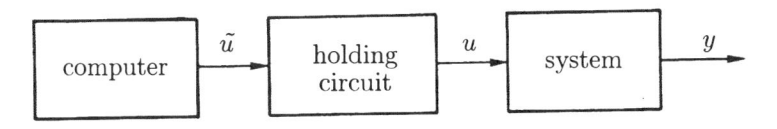

Figure B.7 Sampled-data system.

Also, for state (x) and output (y) variables, the assumption considered in order to define sampled-data systems is consistent with the modern measurement techniques that repeatedly "read" the values $\tilde{x}$ and $\tilde{y}$ at specific times, called *sampling times*. The interval of times occurring between successive sampling times is called a *sampling interval*.

The simplest sampling scheme, depicted in *Fig. B.8* is characterized by having the same sampling time for all variables (state and output) and sampling interval T constant and equal to the interval in which the input is kept constant.

More complex sampling schemes are obtained when T is not constant over time (random sampling and adaptive sampling), when the sampling interval is not the same for all variables (multirate sampling), when, though T is the same for all variables, the sampling instants are shifted over time (asynchronous sampling); or when the holding circuit, instead of keeping constant the input to the system, allows it to vary following a fixed law (*e.g.*, linearly). If we consider the simplest case, we are then dealing with a continuous-time system

$$\dot{x}(t) = Ax(t) + bu(t)$$
$$y(t) = c^T x(t) + du(t)$$

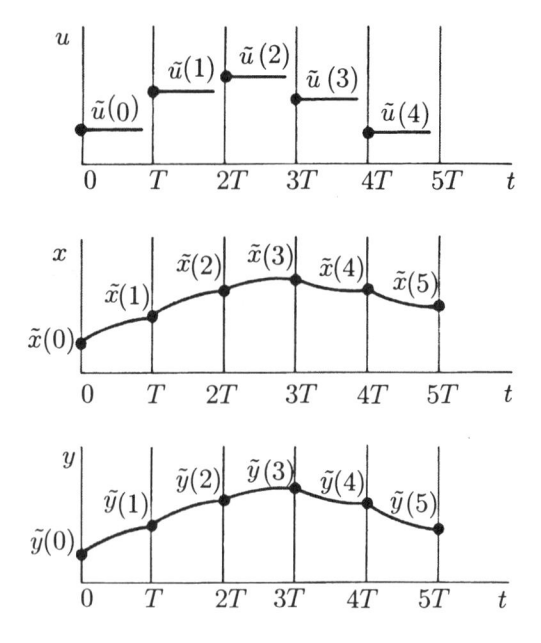

Figure B.8 Input, state, and output of the simplest sampled-data system.

with piecewise constant input. After having numbered the sampling instants with the index $k = 0, 1, 2$, and so on, we can then define input $\tilde{u}$, state $\tilde{x}$, and output $\tilde{y}$ variables of the sampled-data system, in the following way (see *Fig. B.8*):

$$\tilde{u}(k) = u(t) \quad kT \le t < (k+1)T$$
$$\tilde{x}(k) = x(kT)$$
$$\tilde{y}(k) = y(kT)$$

By applying the Lagrange formula (B.22) to the continuous-time system with initial time kT and final time $(k+1)T$, and by taking into account that between those two instants the input is constant and equal to $\tilde{u}(k)$, one gets

$$x((k+1)T) = e^{AT}x(kT) + \int_0^T e^{A(T-\xi)}d\xi\, b\tilde{u}(k)$$

By substituting in this equation $x(kT)$ and $x((k+1)T)$ with $\tilde{x}(k)$ and $\tilde{x}(k+1)$ and taking into account that

$$\int_0^T e^{A(T-\xi)}d\xi = \int_0^T e^{A\xi}d\xi$$

one obtains

$$\tilde{x}(k+1) = e^{AT}\tilde{x}(k) + \left(\int_0^T e^{A\xi}d\xi\right)b\tilde{u}(k)$$

which is the state equation of the sampled-data system, interpreted as a discrete-time system. Since, clearly, the output transformation holds also for sampled variables, we can conclude that a sampled-data system is a discrete-time system

$$\tilde{x}(k+1) = \tilde{A}\tilde{x}(k) + \tilde{b}\tilde{u}(k)$$
$$\tilde{y}(k) = \tilde{c}^T\tilde{x}(k) + \tilde{d}\tilde{u}(k)$$

with

$$\tilde{A} = e^{At} \quad \tilde{b} = \left(\int_0^T e^{A\xi}d\xi\right)b \quad \tilde{c}^T = c^T \quad \tilde{d} = d$$

EXAMPLE 5 (mechanical system)
Let us consider again the mechanical system described in *Example 1*, composed of a point mass m moving along a straight line and to which a force $u(t)$ is applied. If $x_1 = y$ and x_2 are the position and velocity of the mass and there is no friction, the system is described by the triple

$$A = \begin{pmatrix} 0 & 1 \\ 0 & 0 \end{pmatrix} \quad b = \begin{pmatrix} 0 \\ 1/m \end{pmatrix}$$
$$c^T = (1 \quad 0)$$

Since $A^2 = 0$ (and, therefore, $A^i = 0, i \geq 2$) the transition matrix results to be

$$e^{AT} = I + At = \begin{pmatrix} 1 & t \\ 0 & 1 \end{pmatrix}$$

so that the sampled-data system is described by the triple

$$\tilde{A} = e^{AT} = \begin{pmatrix} 1 & T \\ 0 & 1 \end{pmatrix} \quad \tilde{b} = \left(\int_0^T e^{A\xi}d\xi\right)b = \begin{pmatrix} T & T^2/2 \\ 0 & T \end{pmatrix}\begin{pmatrix} 0 \\ 1/m \end{pmatrix} = \begin{pmatrix} T^2/2m \\ T/m \end{pmatrix}$$
$$\tilde{c}^T = c^T = (1 \quad 0)$$

♣

It is important to note that an entire family of sampled-data systems $\tilde{\Sigma}$ is associated to each continuous-time system Σ, since, even if the notation adopted does not show this clearly, the matrix $\tilde{A}$ and the vector $\tilde{b}$ depend on the sampling interval T. It is therefore interesting to know whether a property holding for the continuous-time system Σ is maintained for the family $\tilde{\Sigma}$ or whether a property that does not hold for Σ can be "gained" by sampling the system using an appropriate sampling interval. As stated in Section B.8, we can find reversibility among the properties that are maintained under sampling. All sampled-data systems $\tilde{\Sigma}$ are in fact reversible (just as continuous-time systems), since the matrix $\tilde{A} = e^{AT}$ is nonsingular for any sampling interval (actually, e^{AT} admits as inverse the matrix e^{-AT}).

B.10 INTERNAL STABILITY: DEFINITIONS

Stability is certainly the most studied property of a dynamical system. As we will see, it allows us to characterize the asymptotic behavior ($t \to \infty$) of the system, which is a very important feature in applications.

DEFINITION 2 *(asymptotic stability, simple stability, and instability)*

> A linear system is *asymptotically* stable if and only if its free motion tends to zero as ($t \to \infty$) for any initial state. If, instead, the free motion is bounded but does not tend to zero for some initial state, the system is said to be *simply stable*. Finally, if the free motion is unbounded for some initial state, the system is said to be *unstable*.

On the basis of this definition, it is immediate to see that the two systems discussed in the first two Examples (Newton's law and Fibonacci's rabbits) are both unstable. The first, however, is a *weakly unstable* system since the free motion, though unbounded, grows with time following a polynomial law, which is linear in this case. On the other hand, the second is *strongly unstable*, since the free motion grows exponentially.

From *Definition 2* it follows that a system is asymptotically stable if and only if

$$\lim_{t \to \infty} \Phi(t) = 0$$

that is, if and only if all the entries of the transition matrix tend to zero as $t \to \infty$. Finite memory systems are then asymptotically stable.

The most important property (easy to prove) of asymptotically stable systems, sometimes used as an alternative definition of asymptotic stability, is the following:

THEOREM 4 *(asymptotic stability and convergence to equilibrium)*

> A system is asymptotically stable if and only if for any input $\bar{u}$ there exists a single equilibrium state $\bar{x}$ and $x(t)$ tends to $\bar{x}$ for $t \to \infty$ for any $x(0)$ when $u(t) = \bar{u}$.

It is worth noting that in the unstable system corresponding to Newton's law, we have for $\bar{u} = 0$ an infinite number of equilibria $\bar{x}^T = |\bar{x}_1 \; 0|^T$, while in the system describing Fibonacci's rabbits, the equilibrium state $\bar{x}$ is unique but $x(t)$ does not tend to $\bar{x}$ as $t \to \infty$.

B.11 EIGENVALUES AND STABILITY

Stability of linear systems can be fully understood by making reference to the *Jordan canonical form* A_J of the matrix A. More precisely, by means of an appropriate change of the state variables

$$z = T_J x$$

it is possible to transform the given system (A, b, c^T, d) into an equivalent one $(T_J, AT_J^{-1}, T_J b, c^T T_J^{-1}, d)$ in which the matrix T_J, AT_J^{-1} is the Jordan matrix A_J. The free motion of the system is then described by the equations

$$\dot{z}(t) = A_J z(t)$$

in the case of a continuous-time system and by the equations

$$z(t+1) = A_J z(t)$$

in the case of a discrete-time system. The advantage of such a transformation is that, due to the structure of the matrix A_J (see *Appendix A*), the system is decomposed in a number of noninteracting subsystems, one for each Jordan block

$$J_i^h = \begin{pmatrix} \lambda_i & 1 & 0 & \cdots & 0 \\ 0 & \lambda_i & 1 & \cdots & 0 \\ \vdots & \vdots & \vdots & & \vdots \\ 0 & 0 & 0 & \cdots & 1 \\ 0 & 0 & 0 & \cdots & \lambda_i \end{pmatrix}$$

where λ_i is the ith distinct eigenvalue of A and J_i^h has dimension $n_i^h \times n_i^h$ with n_i^h smaller than or equal to the multiplicity of the eigenvalue λ_i in the minimal polynomial $\Psi_A(\lambda)$. In the case of continuous-time systems, the transition matrix of each of these subsystems is

$$e^{J_i^h t} = e^{\lambda_i t} \begin{pmatrix} 1 & t & \frac{t^2}{2} & \frac{t^3}{3!} & \cdots \\ 0 & 1 & t & \frac{t^2}{2} & \cdots \\ 0 & 0 & 1 & t & \cdots \\ \vdots & \vdots & \vdots & \vdots & \end{pmatrix}$$

and contains terms of the form $t^k e^{\lambda_i t}$ with k smaller than the multiplicity of the eigenvalue λ_i in the minimal polynomial $\Psi_A(\lambda)$. Since $t^k e^{\lambda_i t}$ tends to zero, as $t \to \infty$, if and only if the real part of λ_i is negative, it follows that a continuous-time system is asymptotically stable if and only if all the eigenvalues of A have a negative real part. In contrast, if eigenvalues with a positive real part exist, some of the terms of the transition matrix are unbounded and grow exponentially ($k = 0$) or more than exponentially ($k \geq 1$). In both cases, the system is strongly unstable. In the remaining cases, that is, when there are zero or purely imaginary eigenvalues λ_{i^*} but there are no eigenvalues with a positive real part, one has simple stability if the term $t^k e^{\lambda_{i^*} t}$ is bounded ($k = 0$) and weak instability in the opposite case ($k \geq 1$). By taking into account that k is necessarily zero only in the case in which the eigenvalue λ_{i^*} with zero real part is a simple root of the minimal polynomial $\Psi_A(\lambda)$ and noting that in the case of discrete-time systems the exponential term $e^{\lambda_i t}$ is replaced by the power λ_i^t, which tends to zero if and only if $|\lambda_i| < 1$, we can summarize the previous discussion with *Theorem 5*.

THEOREM 5 *(stability conditions)*

A continuous-time [discrete-time] linear system (A, b, c^T, d) is		
a. Asymptotically stable if and only if	$\text{Re}(\lambda_i) < 0$ $[|\lambda_i| < 1]$ $\forall i$	
b. Simply stable if and only if	$\text{Re}(\lambda_i) \leq 0$ $[|\lambda_i| < 1]$ $\forall i$	
	$\exists i^* : \text{Re}(\lambda_{i^*}) = 0$ $[|\lambda_{i^*}| = 1]$	
	all λ_{i^*} are simple roots of Ψ_A	
c. Weakly unstable if and only if	$\text{Re}(\lambda_i) \leq 0$ $[|\lambda_i| \leq 1]$ $\forall i$	
	$\exists i^* : \text{Re}(\lambda_{i^*}) = 0$ $[|\lambda_{i^*}| = 1]$	
	at least one λ_{i^*} is not a simple root of Ψ_A	
d. Strongly unstable if and only if	$\exists i : \text{Re}(\lambda_i) > 0$ $[|\lambda_i| > 1]$	

The system representing Newton's law (*Example 1*), which has the matrix A in Jordan form, has a zero eigenvalue that is a double root of the minimal polynomial. As previously stated, it is then weakly unstable. The Fibonacci's system (*Example 2*) has, instead, two eigenvalue $\lambda_{1,2} = (1 \pm \sqrt{5})/2$ so that one of them is > 1 and the system is therefore strongly unstable.

The n eigenvalues of matrix A of a continuous-time linear system can be divided into three classes, depending on the sign of their real part: n^- eigenvalues, called *stable*, have a negative real part, n^0 have a zero real part and are called *critical*, and n^+ have a positive real part and are called unstable. Obviously, $n = n^- + n^0 + n^+$. The corresponding eigenvectors define three disjoint invariant subspaces X^-, X^0, and X^+ with dimension n^-, n^0, and n^+ respectively. Initial states in the subspace X^- give rise to free motions that tend to zero, while initial states in the subspace X^+ give rise to free motions that tend to infinity at least at an exponential rate. For this reason, these two subspaces are called, respectively, *stable manifold* and *unstable manifold*. The subspace X^0 is called *center manifold*: The free motions corresponding to initial states in X^0 remain in X^0, do not tend to zero and eventually tend to infinity at a polynomial rate. Systems without center manifold (*i.e.*, without critical eigenvalues) are called *hyperbolic* and are divided into *attractors* $(X^- = \mathbb{R}^n)$, *saddles* $(X^- \oplus X^+ = \mathbb{R}^n)$, and *repellors* $(X^+ = \mathbb{R}^n)$. Systems possessing a center manifold are called *nonhyperbolic*. Figure B.9 shows the trajectories corresponding to the free motion of eight different second-order continuous-time systems. Each figure also shows the two eigenvalues of the system. The first five systems (stable focus, stable node, unstable focus, unstable node, saddle) are hyperbolic and the last three are non-hyperbolic. The last system (pure imaginary eigenvalues) is called *center* and this explains the choice of the term "center manifold".

The advantage of the decomposition of the state space $\mathbb{R}^n$ into the direct sum of three subspaces X^-, X^0, and X^+ is particularly clear when visualizing the

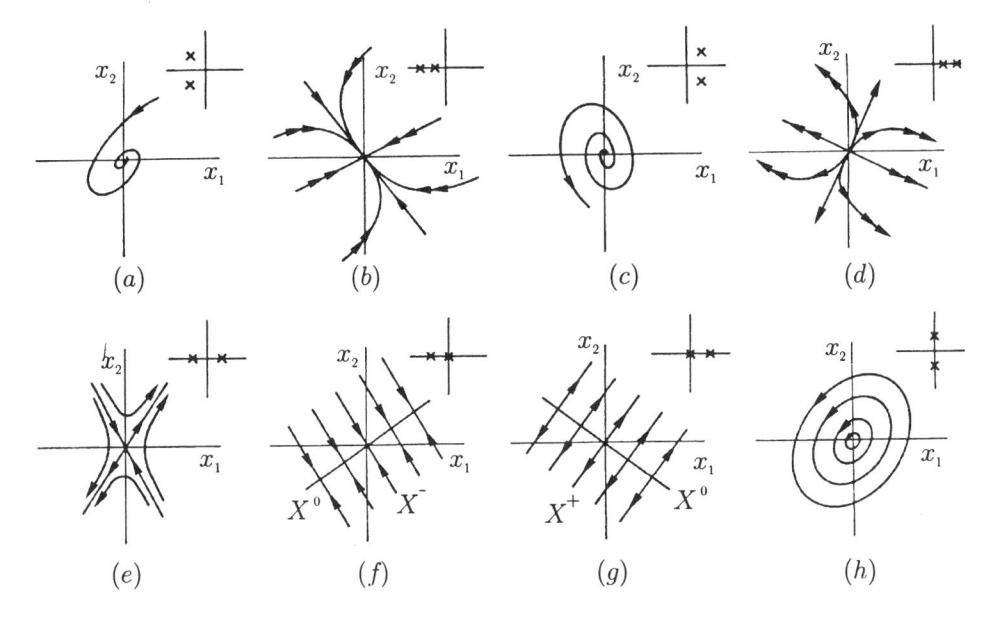

Figure B.9 Trajectories corresponding to the free motion of second-order continuous-time systems: (a) (stable focus) and (b) (stable node) are attractors; (c) (unstable focus) and (d) (unstable node) are repellors; (e) is a saddle; (f), (g), and (h) (center) are systems with center manifold X^0. The straight trajectories correspond to eigenvectors associated to real eigenvalues. The double arrow indicates parts of the trajectories where the state of the system moves more rapidly.

geometry of the free motion, in particular for third-order systems, as the two saddles shown in *Fig. B.10*.

Obviously, what has been said for continuous-time systems also holds for discrete-time systems when separately considering the cases in which we have stable eigenvalues ($|\lambda_i| < 1$), critical ($|\lambda_i| = 1$), and unstable ($|\lambda_i| > 1$).

B.12 TESTS OF ASYMPTOTIC STABILITY

In Section B.11, we showed that knowing the eigenvalues of matrix A of a linear system one can establish whether such a system is asymptotically stable (or not). Unfortunately, the computation of the eigenvalues of a matrix can be very onerous if the matrix is large, as it is often the case in real applications. For this reason, it is convenient to use some tests or methods that, avoiding the computation of the eigenvalues, allow us to infer the asymptotic stability or instability of a system.

One of the most popular of such tests, which is a sufficient condition for instability, is the *trace criterion*, which states that a continuous-time [discrete-time]

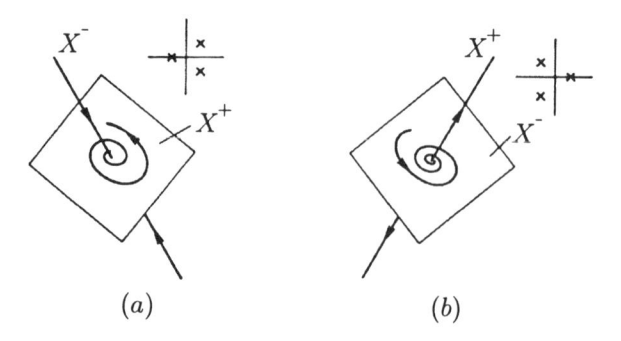

Figure B.10 Two third-order saddles: (a) $n^- = 1, n^+ = 2$; (b) $n^- = 2, n^+ = 1$.

system of dimension n, has the trace of its matrix A positive [$> n$ in modulus], then the system is unstable. For proving this theorem it suffices to remember that the trace of a matrix equals the sum of its eigenvalues.

A condition that requires a much greater computational burden (but is still more effective than the computation of the eigenvalues) is known as the *Hurwitz criterion*. Such criterion (whose proof is not reported here) is a necessary and sufficient condition for the n roots of a polynomial equation with real coefficients

$$\alpha_0 \lambda^n + \alpha_1 \lambda^{n-1} + \cdots + \alpha_n = 0$$

to have a negative real part. When applied to the characteristic equation $\Delta_A(\lambda) = 0$, which can be determined using the Souriau method cited in section B.3, the criterion allows one to establish whether a continuous-time system is asymptotically stable or not.

THEOREM 6 *(Hurwitz criterion)*

Let
$$\Delta_A(\lambda) = \lambda^n + \alpha_1 \lambda^{n-1} + \cdots + \alpha_n$$
be the characteristic polynomial of a continuous-time linear system $\dot{x}(t) = Ax(t)$. Consider the following $n \times n$ matrix (called the *Hurwitz matrix*)

$$H = \begin{pmatrix} \alpha_1 & 1 & 0 & 0 & \cdots \\ \alpha_3 & \alpha_2 & \alpha_1 & 1 & \cdots \\ \alpha_5 & \alpha_4 & \alpha_3 & \alpha_2 & \cdots \\ \alpha_7 & \alpha_6 & \alpha_5 & \alpha_4 & \cdots \\ \vdots & \vdots & \vdots & \vdots & \end{pmatrix}$$

in which $\alpha_i = 0$ for $i > n$. Then, a necessary and sufficient condition for asymptotic stability of the system is that all the principal minors of the Hurwitz matrix be positive. That is, setting

$$D_1 = \alpha_1 \quad D_2 = \det \begin{pmatrix} \alpha_1 & 1 \\ \alpha_3 & \alpha_2 \end{pmatrix} \quad D_3 = \det \begin{pmatrix} \alpha_1 & 1 & 0 \\ \alpha_3 & \alpha_2 & \alpha_1 \\ \alpha_5 & \alpha_4 & \alpha_3 \end{pmatrix} \ldots D_n = \det H$$

a necessary and sufficient condition for asymptotic stability of the system is that $D_i > 0, i = 1, \ldots, n$.

Another important criterion for asymptotic stability, equivalent to the Hurwitz criterion, is the following:

THEOREM 7 *(Routh criterion)*

Let
$$\Delta_A(\lambda) = \lambda^n + \alpha_1 \lambda^{n-1} + \cdots + \alpha_n$$
be the characteristic polynomial of a continuous-time linear system $\dot{x}(t) = Ax(t)$. Consider the $(n+1) \times (n+1)$ matrix (called the *Routh matrix*)

$$R = \begin{pmatrix} 1 & \alpha_2 & \alpha_4 & \cdots \\ \alpha_1 & \alpha_3 & \alpha_5 & \cdots \\ r_{21} & r_{22} & r_{23} & \cdots \\ r_{31} & r_{32} & r_{33} & \cdots \\ \vdots & \vdots & \vdots & \\ r_{n1} & r_{n2} & r_{n3} & \cdots \end{pmatrix}$$

> in which
> $$\alpha_{n+i} = 0 \quad \text{for} \quad i = 1, 2, \ldots$$
> $$r_{2j} = \alpha_{2j} - \frac{\alpha_{2j+1}}{\alpha_1}$$
> $$r_{ij} = r_{i-2,j+1} - \frac{r_{i-2,1} r_{i-1,j+1}}{r_{i-1,1}} \quad \text{for} \quad i = 3, 4, \ldots$$
>
> (note that r_{ij} can be computed only if $r_{i-1,1} \neq 0$). Then, a necessary and sufficient condition for asymptotic stability of the system is that all the entries of the first column of R be positive.

Clearly, there exist criteria for asymptotic stability for discrete-time systems analogous to the Hurwitz and Routh ones. It is worth noting, however, that by means of the change of variables

$$z = \frac{s+1}{s-1}$$

one can transform the problem of checking whether the zeros of a polynomial in z are within the unitary circle to that of checking whether the zeros of a polynomial in s have a negative real part. In other words, if

$$\Delta_A(z) = \det(zI - A) = z^n + \alpha_1 z^{n-1} + \cdots + \alpha_n$$

is the characteristic polynomial of a discrete-time system $x(t+1) = Ax(t)$, one can write the characteristic equation in the form

$$\left(\frac{s+1}{s-1}\right)^n + \alpha_1 \left(\frac{s+1}{s-1}\right)^{n-1} + \cdots + \alpha_n = 0$$

which yields

$$s^n + \alpha'_1 s^{n-1} + \cdots + \alpha'_n = 0$$

By applying the Hurwtitz or Routh criterion to this equation, one can determine whether the discrete-time system is asymptotically stable or not.

However, we report next one of the most popular criteria for asymptotic stability of discrete-time systems.

THEOREM 8 *(Jury criterion)*

Let
$$\Delta_A(\lambda) = \lambda^n + \alpha_1 \lambda^{n-1} + \cdots + \alpha_n$$
be the characteristic polynomial of a discrete-time system $x(t+1) = Ax(t)$. Consider the following table of dimension $2n \times (n+1)$ composed of n pairs of rows:

$$\begin{pmatrix} p_{11} & p_{12} & \cdots & p_{1n} & p_{1,n+1} \\ q_{11} & q_{12} & \cdots & q_{1n} & q_{1,n+1} \\ \cdots & \cdots & \cdots & \cdots & \cdots \\ p_{21} & p_{22} & \cdots & p_{2n} & \\ q_{21} & q_{22} & \cdots & q_{2n} & \\ \cdots & \cdots & & & \\ \cdots & \cdots & & & \\ \cdots & \cdots & & & \\ p_{n1} & p_{n2} & & & \\ q_{n1} & q_{n2} & & & \end{pmatrix}$$

where

a. The elements of the first row are $\alpha_n, \alpha_{n-1}, \ldots, \alpha_1, 1$, that is, the coefficients of the characteristic polynomial are in reversed order.

b. Each even row coincides with the preceding row in reversed order.

c. The elements p_{ji} can be calculated as follows:

$$p_{j+1,i} = \det \begin{pmatrix} p_{j1} & q_{ji} \\ q_{j1} & p_{ji} \end{pmatrix} \quad j = 1, 2, \ldots, n-1$$

Then, a necessary and sufficient condition for asymptotic stability of the system is that the following conditions hold:

$$\Delta_A(1) > 0 \qquad (-1)^n \Delta_A(-1) > 0$$
$$p_{21} < 0 \qquad p_{j1} > 0, j = 3, 4, \ldots, n$$

In the particular case of second-order systems (matrix A of dimension 2×2), special conditions hold that often allow us to check asymptotic stability by simple inspection of the matrix A.

THEOREM 9 *(trace and determinant criterion)*

A second-order continuous-time system $\dot{x}(t) = Ax(t)$ is asymptotically stable if and only if

$$\operatorname{tr} A < 0 \quad \det A > 0$$

Analogously, a second-order discrete-time system $x(t+1) = Ax(t)$ is asymptotically stable if and only if

$$|\operatorname{tr} A| < 1 + \det A \quad \det A < 1$$

The reader may check the efficacy of this criterion by applying it to the systems described in the *Examples 1* and *2*.

B.13 ENERGY AND STABILITY

It is known that in some systems (*e.g.*, electrical and mechanical ones) it is possible to define a function of the state $V(x(t))$, called energy, which has the property to be quadratic and nonnegative and decreases with time (and tends to zero) whenever the system is asymptotically stable and evolves freely. This property, studied by the Russian mathematician *Alexander Liapunov* (more than a century ago), allows us to analyze the stability of any linear system in a very synthetic and elegant way.

In order to introduce this topic, we need to define positive definite matrices. First of all, we say that a function $V(x(t))$ with $x \in \mathbb{R}^n$ is *quadratic* if

$$V(x) = x^T P x$$

where P is an $n \times n$ matrix. This means that $V(x)$ is a weighted sum of all the products $x_i x_j$. Since the weight of the term $x_i x_j$ is $(p_{ij} + p_{ji})$, there is no loss of generality in assuming that the matrix P is symmetric. Moreover, a matrix P is said to be *positive definite* if the associated quadratic form $x^T P x$ is positive for all the vectors $x \neq 0$. To know whether a given matrix P is positive definite, one can apply the following criterion:

THEOREM 10 *(Sylvester criterion)*

A symmetric matrix P is positive definite if and only if all the principal minors $D_1, D_2, \ldots, D_n$ are positive, that is,

$$D_1 = p_{11} > 0 \quad D_2 = \det \begin{pmatrix} p_{11} & p_{12} \\ p_{21} & p_{22} \end{pmatrix} > 0 \quad \ldots \quad D_n = \det P > 0$$

It is clear from the previous discussion that any positive definite matrix P induces a metric in the space $\mathbb{R}^n$: in other words, $V(x) = x^T P x$ can be interpreted as the distance of point x from the origin. In this metric, the points x at the same distance from the origin lie on the manifold $x^T P x =$ constant, which in $\mathbb{R}^2$ is an ellipse.

Suppose now that a positive definite matrix P is associated to an autonomous continuous-time linear system $\dot{x}(t) = Ax(t)$. The "distance" of the point $x(t)$ from the origin is $V(x(t)) = x^T(t) P x(t)$ and such a distance varies in time since x depends on t. More precisely

$$\begin{aligned} \dot{V} &= \dot{x} P x + x^T P \dot{x} = x^T A^T P x + x^T P A x \\ &= x^T (A^T P + PA) x \end{aligned}$$

so that the distance of $x(t)$ from the origin decreases continuously with time ($\dot{V} < 0$) if the matrix $-(A^T P + PA)$ is positive definite, that is, if the so-called *Liapunov equation*

$$A^T P + PA = -Q \tag{B.26}$$

is satisfied with Q positive definite. Obviously, if this equation holds, then the system is asymptotically stable since the free motion of the system asymptotically tends to zero, since $\dot{V} < 0$ for $\neq 0$.

In the case of discrete-time systems $x(t+1) = Ax(t)$, one has to determine the quantity ΔV defined as

$$\begin{aligned} \Delta V &= V(x(t+1)) - V(x(t)) = (Ax(t))^T P A x(t) - x^T(t) P x(t) \\ &= x^T(t)(A^T P A - P) x(t) \end{aligned}$$

so that the discrete-time Liapunov equation is the following:

$$A^T P A - P = -Q \tag{B.27}$$

In conclusion, if a continuous-time [discrete-time] system satisfies the Liapunov equation (B.26) [(B.27)] with P and Q positive definite matrices, the system is asymptotically stable and the function $V(x) = x^T P x$, called the *Liapunov function*, has the properties of any energy function, that is, it is positive and decreasing with time for $x \neq 0$. Clearly, this does not imply that, given an asymptotically stable autonomous system in which $x(t)$ tends toward the origin as $t \to \infty$, any quadratic function $V(x) = x^T P x$ with P positive definite systematically decreases with time (even though V tends to zero as $t \to \infty$). This fact is illustrated in *Fig. B.11* for a second-order continuous-time system (stable focus). In such a system, $V(x) = x^T P x$ is positive definite, but $\dot{V} = x^T(A^T P + PA)x$ is positive at some points and negative at others, so that V tends to zero as $t \to \infty$ but not monotonically.

The above discussion is further specified in the following theorem.

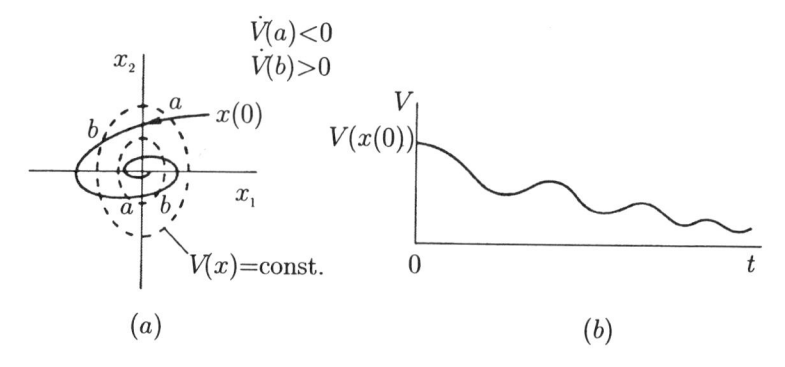

Figure B.11 Second-order asymptotically stable autonomous system (stable focus): (a) trajectory (−) and contour lines (...) of the function $V(x) = x^T P x$; (b) time evolution of V.

THEOREM 11 *(Liapunov theorem)*

> A continuous-time [discrete-time] linear system $(A, -, -)$ is asymptotically stable if and only if there exists a quadratic function $V(x) = x^T P x$ with P positive definite, which is strictly decreasing in time (i.e., such that $\dot{V} < 0$ [$\Delta V < 0$]) along the free motion of the system with $x \neq 0$. Moreover, such a function exists if and only if the Liapunov equation (B.26) [(B.27)] admits a solution (P, Q) with P and Q positive definite. Finally, if such a pair (P, Q) exists, then one can find an infinite number of them, one for each positive definite matrix Q.

The last statement of the Liapunov theorem allows one to derive a practical criterion for testing if a linear system is asymptotically stable. If we chose any symmetric and positive definite matrix Q in the Liapunov equation [e.g., $Q = I$ (identity matrix)], one can solve the equation in the unknown P [note that if P is symmetric there are $n(n+1)/2$ linear equations in the same number of unknowns]; if the solution P exists and is positive definite (this can be checked using Sylvester criterion) the system is asymptotically stable.

It is worth noting that the Liapunov theorem does not require any particular structure for the matrix P. Often, however, the matrix P is diagonal, that is, the Liapunov function $V(x)$ does not contain mixed terms $x_i x_j$ with $i \neq j$ (as an example, consider the energy function of an electrical network). If this is the case, that is, when the Liapunov equation admits a solution (P, Q) with P and Q positive definite and P diagonal, the system, besides being asymptotically stable, remains such for any structural perturbation (*Arrow–McManus* theorem). This means that, if there exists a diagonal matrix P with $p_{ii} > 0$, $i = 1, \ldots, n$ such that the matrix $-(A^T P + P A)$ is positive definite, we can assert that the system $\dot{x} = Ax$ is asymptotically stable as well as all other systems of the kind $\dot{x} = DAx$ with D diagonal and $d_{ii} > 0$. In other words, the system $\dot{x} = Ax$ is asymptotically stable and remains such under the effect of any perturbation that results in the multiplication of any one of its state equations by a positive arbitrary constant.

B.14 DOMINANT EIGENVALUE AND EIGENVECTOR

The free motion $x(t) = \Phi(t)x(0)$ of a linear system is completely identified by the transition matrix

$$\Phi(t) = \begin{cases} e^{At} & \text{for continuous-time systems} \\ A^t & \text{for discrete-time systems} \end{cases}$$

Recalling what has been previously stated about eigenvectors, eigenvalues, and the Jordan canonical form of a matrix A, we can conclude that in the case of real and distinct eigenvalues the free motion is

$$x(t) = \begin{cases} \sum_{i=1}^{n} c_i x^{(i)} e^{\lambda_i t} & \text{for continuous-time systems} \\ \sum_{i=1}^{n} c_i x^{(i)} \lambda_i^t & \text{for discrete-time systems} \end{cases}$$

where $x^{(i)}$, $i = 1, \ldots, n$ are the eigenvectors of A [satisfying the equation $Ax^{(i)} = \lambda_i x^{(i)}$] and c_i are the components of the initial state $x(0)$ in the basis composed of the n eigenvectors $x^{(i)}$ [i.e., $\sum_{i=1}^{n} c_i x^{(i)} = x(0)$]. Clearly, for particular initial conditions, some c_i may be zero, but for generic initial conditions, all the c_i's are different from zero and the free motion is the weighted sum of n exponential terms. As time passes, one of these exponential terms necessarily dominates the other since, for large values of t, $e^{\lambda_i t} \gg e^{\lambda_j t}$ [$|\lambda_i^t| \gg |\lambda_j^t|$] if $\lambda_i > \lambda_j$ [$|\lambda_i| > |\lambda_j|$] (notice that this occurs also in the case of simple stability or instability). The eigenvalue and eigenvector associated with this exponential term are called *dominant*. The dominant eigenvalue λ_{dom} is clearly the highest eigenvalue in continuous-time systems and the eigenvalue with maximal modulus in discrete-time systems. In Fig. B.9, five cases [$(b), (d), (e), (f), (g)$] deal with continuous-time systems with real eigenvalues. They show that all generic trajectories tend to align with the dominant eigenvector as time passes. This property is very important and must be accounted for when characterizing the asymptotic behavior of linear systems. Therefore, the free motion can be approximated in the long run with a single exponential term

$$x(t) \cong \begin{cases} cx_{\text{dom}} e^{\lambda_{\text{dom}} t} & \text{for continuous-time systems} \\ cx_{\text{dom}} \lambda_{\text{dom}}^t & \text{for discrete-time systems} \end{cases}$$

and if the system is asymptotically stable, this exponential term tends to zero. In these cases, the exponential term is often written in the form $e^{-t/T_{\text{dom}}}$, where $T_{\text{dom}} (> 0)$ is the so-called *dominant time constant*, linked to the dominant eigenvalue by the relationship

$$T_{\text{dom}} = \begin{cases} -\dfrac{1}{\lambda_{\text{dom}}} & \text{for continuous-time systems} \\ -\dfrac{1}{\log |\lambda_{\text{dom}}|} & \text{for continuous-time systems} \end{cases}$$

Moreover, in applications it is rather common to say that the exponential term is "practically" over after a period of time equal to five times the time constant T_{dom}. If the eigenvalues of A, though distinct, are not all real, the free motion is also characterized by terms of the form $e^{\lambda t}$ and λ^t with λ complex. In the continuous-time case, these terms correspond to an exponentially dumped sinusoidal function $e^{at}\sin(bt + \varphi)$, where a is the real part of the eigenvalue and b is the imaginary part. It follows that the dominant term of the free motion is the one associated with the (real or complex) eigenvalue with the maximal real part. In conclusion, in the long run, the free motion can be approximated by an exponential if the dominant eigenvalue is real, or by a sinusoid with an exponentially varying amplitude if the dominant eigenvalue is complex. Obviously, in the case of asymptotically stable continuous-time systems the dominant time constant is given by

$$T_{\text{dom}} = -\frac{1}{\text{Re}(\lambda_{\text{dom}})}$$

Finally, if the eigenvalues of A are not all distinct, the free motion of a continuous-time [discrete-time] system may contain terms of the form $t^k e^{\lambda t}$ $[t^k \lambda^t]$; this fact, however, does not change the previous conclusions, since the free motion is still dominated by the term associated with the (real or complex) eigenvalue with maximal real part [maximal modulus].

B.15 REACHABILITY AND CONTROL LAW

The motion of a linear system is given by

$$x(t) = \Phi(t)x(0) + \Psi(t)u_{[0,t)}(\cdot)$$

that is, the sum of the free and forced motion. The forced motion

$$\Psi(t)u_{[0,t)}(\cdot) = \begin{cases} \int_0^t e^{A(t-\xi)}bu(\xi)d\xi & \text{for continuous-time systems} \\ \sum_{i=0}^{t-1} A^{t-i-1}bu(i) & \text{for discrete-time systems} \end{cases}$$

describes the set $X_r(t)$ of all states reachable from the origin of the state space at time t. Obviously, such a set is a subspace for which the following property holds

$$X_r(t_1) \subset X_r(t_2) \quad t_1 \leq t_2$$

Since $X_r(t)$ cannot grow indefinitely, there exists a time t^* such that $X_r(t) = X_r$ for $t > t^*$. Finally, if $X_r = \mathbb{R}^n$, the system is said to be *completely reachable*. Theorem 12, known as the *Kalman theorem*, holds.

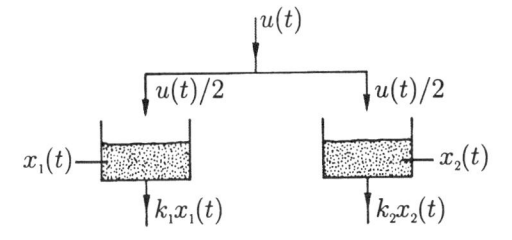

Figure B.12 Two reservoirs fed in parallel.

THEOREM 12 *(complete reachability)*

> In a linear system (A, b) of order n, the reachability subspace X is spanned by the n vectors $b, Ab, \ldots, A^{n-1}b$, called reachability vectors. Thus the system is completely reachable if and only if these n vectors are linearly independent. Moreover, each state belonging to X_r is reachable in any time if the system is continuous-time and in at most n transitions if the system is discrete-time.

This theorem is often formulated by making reference to the reachability matrix (also called the *Kalman matrix*)

$$R = (\, b \quad Ab \quad \ldots \quad A^{n-1}b \,)$$

This matrix is $n \times n$ and its image is the reachability subspace, that is,

$$X_r = \mathcal{I}[R]$$

so that the complete reachability of the system is equivalent to the nonsingularity of the matrix R (*i.e.*, to the existence of R^{-1}).

EXAMPLE 6

Consider the system shown in *Fig. B.12* composed of two reservoirs $i = 1, 2$ feed in parallel with a flow $u(t)/2$ and with an output flow-rate proportional (through a coefficient k_i) to the storage $x_i(t)$.

The conservation of mass gives

$$\dot{x}_1 = -k_1 x_1 + \frac{u}{2}$$

$$\dot{x}_2 = -k_2 x_2 + \frac{u}{2}$$

which are the state equations of a linear system with

$$A = \begin{pmatrix} -k_1 & 0 \\ 0 & -k_2 \end{pmatrix} \qquad b = \begin{pmatrix} 1/2 \\ 1/2 \end{pmatrix}$$

Thus, the reachability matrix is

$$R = \begin{pmatrix} 1/2 & -k_1/2 \\ 1/2 & -k_2/2 \end{pmatrix}$$

so that the system is complete reachable if and only if $k_1 \neq k_2$. The reason for this is that in the case of identical reservoirs ($k_1 = k_2$) it is impossible to unbalance the system $[x_1(t) \neq x_2(t)]$ since, by assumption, the initial condition is balanced $[x_1(0) = x_2(0) = 0]$.

♣

It is worth noting that all systems with $A = A_c$, $b = b_c$, where

$$A_c = \begin{pmatrix} 0 & 1 & 0 & \cdots & 0 \\ 0 & 0 & 1 & \cdots & 0 \\ \vdots & \vdots & \vdots & & \vdots \\ 0 & 0 & 0 & \cdots & 1 \\ -\alpha_n & -\alpha_{n-1} & -\alpha_{n-2} & \cdots & -\alpha_1 \end{pmatrix} \quad b_c = \begin{pmatrix} 0 \\ 0 \\ \vdots \\ 0 \\ 1 \end{pmatrix}$$

that is, all systems in control canonical form (see Section B.3) are completely reachable. In fact, it is immediate to check that the Kalman matrix

$$R_c = \begin{pmatrix} b_c & A_c b_c & \cdots & A_c^{n-1} b_c \end{pmatrix}$$

is nonsingular for all values of the coefficients α_i, $i = 1, 2, \ldots, n$ and that

$$R_c^{-1} = \begin{pmatrix} \alpha_{n-1} & \alpha_{n-2} & \cdots & \alpha_1 & 1 \\ \alpha_{n-2} & \alpha_{n-3} & \cdots & 1 & 0 \\ \vdots & \vdots & & \vdots & \vdots \\ 1 & 0 & \cdots & 0 & 0 \end{pmatrix}$$

The reason for the interest in this canonical form is justified by *Theorem 13*.

THEOREM 13 *(control canonical form)*

> A completely reachable system (A, b) can be put in control canonical form by means of the coordinate transformation $z = R_c R^{-1} x$, where R and R_c are the reachability matrices of (A, b) and (A_c, b_c).

This means that any system (A, b) can be assumed to be in control canonical form provided it is completely reachable. Moreover, the control canonical form (A_c, b_c) can be easily determined by evaluating (*e.g.*, with the Souriau formula) the coefficients α_i.

The importance of complete reachability emerges whenever one tries to modify the dynamics of a given system by linking its input $u(t)$ to the state $x(t)$ by means of a linear feedback rule

$$u(t) = k^T x(t) + v(t)$$

REACHABILITY AND CONTROL LAW 263

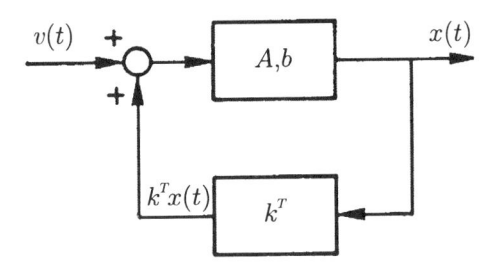

Figure B.13 Controlled system composed of a system (A, b) and its controller k^T.

known as the *control law* (algebraic and linear). The resulting system, called the *controlled system*, has $v(t)$ as input, and is shown in *Fig. B.13* through a block diagram. The feedback block, called *controller*, performs a simple weighted sum $k_1 x_1 + \ldots + k_n x_n \ (= k^T x(t))$ of the state variables.

If (A, b) is a continuous-time system, then the controlled system is described by the state equations

$$\dot{x} = Ax + b(k^T x + v) = (A + bk^T)x + bv$$

In other words, the system (A, b) is transformed, by means of the feedback controller k^T, into the controlled system $(A + bk^T, b)$. Thus, the dynamics of the system are now different since the characteristic polynomial has been modified from $\Delta_A(\lambda)$ into $\Delta_{A+bk^T}(\lambda)$. Obviously, the same holds for discrete-time systems. We can now present the main theorem of this section. It states that complete reachability is a necessary and sufficient condition for the free assignment of the eigenvalues of the controlled system.

THEOREM 14 *(eigenvalue assignment)*

> The eigenvalues of the controlled system $(A + bk^T)$ can be arbitrarily assigned by means of a controller k^T, if and only if the system (A, b) is completely reachable. If α_i are the coefficients of the characteristic polynomial of A, and α_i^* those of the characteristic polynomial of $A + bk^T$, then
>
> $$k^T = ((\alpha_n - \alpha_n^*) \ \ldots \ (\alpha_1 - \alpha_1^*)) R_c R^{-1}$$
>
> where R and R_c are the reachability matrices of (A, b) and (A_c, b_c).

Theorem 14 implies that the dynamics of a completely reachable system can be modified at will through a linear feedback. The most spectacular consequence of this theorem is the possibility of stabilizing unstable systems.

Since complete reachability is a generic property of linear systems [$\det R \neq 0$ for a generic pair (A, b)], it can be understood that the control scheme shown in *Fig. B.13* is of great practical interest.

B.16 OBSERVABILITY AND STATE RECONSTRUCTION

The observability of a dynamical system refers to the possibility of computing its initial state $x(0)$ once input and output have been recorded during time interval $[0,t)$. Analogously, reconstructability refers to the possibility of computing the final state $x(t)$. Thus, observability implies reconstructability, since once $x(0)$ and $u_{[0,t)}(\cdot)$ are known, it is possible to compute $x(t)$ (Lagrange's formula), while the inverse is possible only if the system is reversible.

In order to study observability, it is worth considering the output free motion of the system

$$c^T \Phi(t) x(0) = \begin{cases} c^T e^{At} x(0) & \text{for continuous-time systems} \\ c^T A^t x(0) & \text{for discrete-time systems} \end{cases}$$

and define the set $X_{no}(t)$ of the states indistinguishable from the origin as the set of the initial states $x(0)$ for which the output free motion is identically zero in the interval $[0,t)$. Obviously, such a set is a subspace enjoying the property

$$X_{no}(t_1) \supset X_{no}(t_2) \quad t_1 \leq t_2$$

But $X_{no}(t)$ cannot decrease indefinitely, so that a time t^* exists such that $X_{no}(t) = X_{no}$ for $t > t^*$. Finally, if $X_{no} = \{0\}$, all states $x(0) \neq 0$ are distinguishable from the zero state and can be computed from $u_{[0,t)}(\cdot)$ and $y_{[0,t)}(\cdot)$. This is the reason why a system with $X_{no} = \{0\}$ is called completely observable. It is customary to consider also the subspace $X_{no}^\perp$, that is, the subspace orthogonal to X_{no}. This subspace, called the *observability subspace* and denoted by X_o, is useful for explicitly formulating the condition of complete observability.

THEOREM 15 *(complete observability)*

> In a linear system of order n, the observability subspace X_o is spanned by the n vectors $c, A^T c, \ldots, (A^T)^{n-1} c$ called observability vectors. Thus, the system is completely observable if and only if these vectors are linearly independent. Moreover, in a completely observable system the initial state can be computed if input and output have been recorded for a time period of any length in the case of continuous-time systems and of length n in the case of discrete-time systems.

The above condition is often formulated with reference to the *observability matrix* (also called the Kalman matrix)

$$O = \begin{pmatrix} c^T \\ c^T A \\ \vdots \\ c^T A^{n-1} \end{pmatrix}$$

Then, we have

$$X_{no} = \mathcal{N}[O] \quad X_o = \mathcal{I}[O^T]$$

and a system is completely observable if its observability matrix is nonsingular (*i.e.*, if the matrix O^{-1} exists). The fact that the initial state of a completely observable system can be computed from input and output records can be verified by explicitly writing the first n output values as a function of the initial state and of the input value, that is,

$$y(0) = c^T x(0) + du(0)$$
$$y(1) = c^T A x(0) + c^T b u(0) + du(1)$$
$$y(2) = c^T A^2 x(0) + c^T A b u(0) + c^T b u(1) + du(2)$$
$$\vdots$$
$$y(n-1) = c^T A^{n-1} x(0) + c^T A^{n-2} b u(0) + \cdots + c^T b u(n-2) + du(n-1)$$

This is a system of n linear equations with n unknowns [the components of the vector $x(0)$] which admits a unique solution if and only if the observability matrix O is nonsingular.

EXAMPLE 7
Suppose that 10 pairs of adult rabbits have been captured at the beginning of the season in a population described by the Fibonacci model (see *Example 2*) and assume that at the beginning and at the end of the same year 50 and 60 pairs of rabbits (young and adult) where present. This means that

$$u(0) = 10 \quad y(0) = 50 \quad y(1) = 60$$

Since the system is described by the triple

$$A = \begin{pmatrix} 0 & 1 \\ 1 & 1 \end{pmatrix} \quad b = \begin{pmatrix} 0 \\ -1 \end{pmatrix}$$
$$c^T = \begin{pmatrix} 1 & 1 \end{pmatrix}$$

we have

$$O = \begin{pmatrix} c^T \\ c^T A \end{pmatrix} = \begin{pmatrix} 1 & 1 \\ 1 & 2 \end{pmatrix}$$

so that the system is completely observable and

$$O^{-1} = \begin{pmatrix} 2 & -1 \\ -1 & 1 \end{pmatrix}$$

Thus, the system of 2 equations and 2 unknowns

$$y(0) = c^T x(0)$$
$$y(1) = c^T A x(0) + c^T b u(0)$$

can be solved with respect to $x(0)$

$$x(0) = O^{-1} \begin{pmatrix} y(0) \\ y(1) - c^T b u(0) \end{pmatrix} = \begin{pmatrix} 2 & -1 \\ -1 & 1 \end{pmatrix} \begin{pmatrix} 50 \\ 70 \end{pmatrix} = \begin{pmatrix} 30 \\ 20 \end{pmatrix}$$

We can, therefore, conclude that at the beginning of the year the population was composed of 30 pairs of young rabbits and 20 pairs of adult rabbits.

♣

The comparison of *Theorems 12* and *15* allows one to note a strong analogy between reachability and observability, which can be formalized in the following duality principle:

THEOREM 16 *(duality principle)*

> A system Σ is completely reachable [observable] if and only if its dual $\Sigma^* = (A^T, c, b^T, d)$ is completely observable [reachable]. Moreover, the reachability matrix of the system is the transposed of the observability matrix of the dual system.

The duality principle enables us to obtain from *Theorem 13* and from the properties of the control canonical form, the following result:

THEOREM 17 *(reconstruction canonical form)*

> A system in reconstruction canonical form
>
> $$A_r = \begin{pmatrix} 0 & 0 & \cdots & 0 & -\alpha_n \\ 1 & 0 & \cdots & 0 & -\alpha_{n-1} \\ 0 & 1 & \cdots & 0 & -\alpha_{n-2} \\ \vdots & \vdots & & \vdots & \vdots \\ 0 & 0 & \cdots & 1 & -\alpha_1 \end{pmatrix}$$
>
> $$c_r^T = (\,0 \quad 0 \quad \cdots \quad 0 \quad 1\,)$$
>
> is completely observable. Conversely, a completely observable system (A, c^T) can be put into reconstruction canonical form by means of a suitable coordinate transformation.

Also, *Theorem 14* on eigenvalues assignment can be dualized. For this, we first introduce the notion of state *reconstructor*, illustrated in *Fig. B.14*.

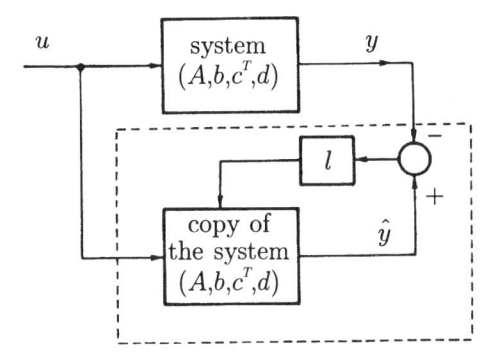

Figure B.14 A system and its state reconstructor.

The reconstructor is a copy of the system [with state $\hat{x}(t)$] with two inputs: the input $u(t)$ of the system and the difference $[\hat{y}(t) - y(t)]$ between the reconstructed output $\hat{y}(t)$ and the output of the system. The vector l identifies the reconstructor uniquely, and will be assumed to be constant in time. Thus, if the system is continuous-time, that is,

$$\dot{x}(t) = Ax(t) + bu(t)$$
$$y(t) = c^T x(t) + du(t)$$

the time invariant reconstructor is described by

$$\dot{\hat{x}}(t) = A\hat{x}(t) + bu(t) + l(\hat{y}(t) - y(t))$$
$$\hat{y}(t) = c^T \hat{x}(t) + du(t)$$

If the reconstruction error is the difference between the state $\hat{x}(t)$ of the reconstructor and the state $x(t)$ of the system

$$e(t) = \hat{x}(t) - x(t)$$

it is straightforward to check that

$$\dot{e}(t) = (A + lc^T)e(t) \tag{B.28}$$

and this means that the dynamics of the reconstruction error [Eq. (B.28) with initial state $e(0) = \hat{x}(0) - x(0)$], are independent upon the input applied to the system. In the case of discrete-time systems, Eq. (B.28) simply becomes

$$e(t+1) = (A + lc^T)e(t) \tag{B.29}$$

so that, we can conclude that the reconstructed state $\hat{x}(t)$ tends toward the state of the system $x(t)$, for any initial error $\hat{x}(0) - x(0)$, if and only if the system (B.28)

268 ELEMENTS OF LINEAR SYSTEMS THEORY

or (B.29) is asymptotically stable. If this is the case, we say that the reconstructor l is an asymptotic state reconstructor. Clearly, the rate of convergence of the reconstructed state $\hat{x}(t)$ toward $x(t)$ is determinated by the dominant eigenvalue of the matrix $A + lc^T$. In this context, the following result, which is the dual of those concerning the assignment of the eigenvalues of the controlled system, if of interest.

THEOREM 18 *(eigenvalues of the reconstructor)*

> The eigenvalues of the matrix $A+lc^T$, which describes the reconstruction error dynamics (B.28) or (B.29), can be arbitrarily fixed by means of a suitable choice of the vector l, if and only if the system (A, c^T) is completely observable.

This result states that it is possible to rapidly and precisely reconstruct the state of a completely observable linear system by elaborating its inputs and outputs in real time. Since complete observability is a property that holds generically for a linear system, one can argue that the scheme of *Fig. B.14* is of great interest in applications.

B.17 DECOMPOSITION THEOREM

The notions of reachability and observability enable us to interpret any linear system as the interconnection of four subsystems called, respectively,

  a. Reachable and unobservable part (r,no).

  b. Reachable and observable part (r,o).

  c. Unreachable and unobservable part (nr,no).

  d. Unreachable and observable part (nr,o).

If the dimension of the system is n and $n_a, n_b, n_c,$ and n_d are the dimensions of the four parts, obviously

$$n = n_a + n_b + n_c + n_d$$

A system is rarely composed of the four parts. In contrast, very often a system is composed only of part b: This happens when the system is completely reachable and completely observable. The interactions among the four subsystems $\Sigma_a, \Sigma_b, \Sigma_c,$ and Σ_d are pointed out in *Fig. B.15* where, for simplicity, we have assumed that the system is proper $(d = 0)$.

The figure shows that the input u directly influences parts a and b but does not influence even indirectly, the parts c and d. This means that if the subsystems (c) and (d) are initially at rest $[x_c(0) = 0, x_d(0) = 0]$ they will remain at rest forever. On the contrary, the state vectors z_a and z_b of the first two parts, may

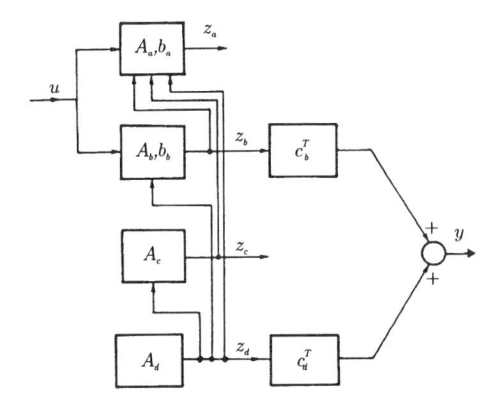

Figure B.15 A proper system decomposed in parts.

vary since they are influenced by the input. Moreover, the decomposition theorem (due to Kalman) states that the system composed of the first two parts is completely reachable. The figure shows also that the output is influenced by the input through part b, which is, therefore, the only channel through which the information flows from the input to the output of a dynamical system. The output is also influenced by part d but is completely insensitive to what is going on in parts a and c. This means that it will not be possible to compute the initial state of parts a and c, from the knowledge of the input and output functions. However, this will be possible for parts b and d, which compose a completely observable system.

THEOREM 19 (*decomposition theorem*)

Given a linear system (A, b, c^T), it is possible to perform a change of state variables $z = Tx$ in such a way that the equivalent system $(TAT^{-1}, Tb, c^T T^{-1})$ is decomposed into four parts (as shown in *Fig. B.15*), that is,

$$TAT^{-1} = \begin{pmatrix} A_a & A_{ab} & A_{ac} & A_{ad} \\ 0 & A_b & 0 & A_{bd} \\ 0 & 0 & A_c & A_{cd} \\ 0 & 0 & 0 & A_d \end{pmatrix} \quad Tb = \begin{pmatrix} b_a \\ b_b \\ 0 \\ 0 \end{pmatrix}$$

$$c^T T^{-1} = \begin{pmatrix} 0 & c_b^T & 0 & c_d^T \end{pmatrix}$$

with the following properties. System (A_b, b_b, c_b^T) is completely reachable and observable, while system (A_r, b_r, c_r^T) given by

$$A_r = \begin{pmatrix} A_a & A_{ab} \\ 0 & A_b \end{pmatrix} \quad b_r = \begin{pmatrix} b_a \\ b_b \end{pmatrix}$$

$$c_r^T = \begin{pmatrix} 0 & c_b^T \end{pmatrix}$$

is completely reachable and system (A_o, b_o, c_o^T) given by

$$A_o = \begin{pmatrix} A_b & A_{bd} \\ 0 & A_d \end{pmatrix} \quad b_o = \begin{pmatrix} b_b \\ 0 \end{pmatrix}$$

$$c_o^T = \begin{pmatrix} c_b^T & c_d^T \end{pmatrix}$$

is completely observable

The proof of *Theorem 19* is constructive. It suggest the following procedure for the computation of the coordinate transformation T.

Decomposition procedure

1. Compute the reachability and observability matrices R and O of system (A, b, c^T).

2. Determine the four subspaces

$$X_r = \mathcal{I}[R] \quad X_o = \mathcal{I}[O^T]$$

$$X_{nr} = \mathcal{N}[R^T] \quad X_{no} = \mathcal{N}[O]$$

3. Determine the four subspaces (*i.e.*, their basis)

$$X_a = X_r \cap X_{no}$$

$$X_b = X_r \cap (X_{nr} + X_o)$$

$$X_c = X_{no} \cap (X_{nr} + X_o)$$

$$X_d = X_{nr} \cap X_o$$

4. The columns of the matrix T^{-1} are the vectors of the basis of the four subspaces determined at step (3).

5. Determine $T = (T^{-1})^{-1}$ and $(TAT^{-1}, Tb, c^T T^{-1})$.

It is important to note that the four parts composing a linear system, though interconnected one to the other, do not form cycles, as pointed out by *Fig. B.15* and by the block triangular structure of the matrix TAT^{-1}. This implies that the eigenvalues of the system are the union of the eigenvalues of the four parts or, equivalently, that the characteristic polynomial of the system is the product of the characteristic polynomials of the four parts. For this reason, many properties of linear systems are linked to the stability of one or more of its parts. We now confirm this by discussing three properties of linear systems: stabilizability, detectability, and external stability.

The first property, *stabilizability*, concerns the possibility of transforming a given system (A, b, c^T, d) into an asymptotically stable system, using a linear control law. Assuming that the system is decomposed into its four parts with state vectors $z_a, z_b, z_c,$ and z_d, the linear control law becomes

$$u(t) = k_a^T z_a(t) + k_b^T z_b(t) + k_c^T z_c(t) + k_d^T z_d(t)$$

By looking at *Fig. B.15* it is clear that such a control law does not modify the dynamics of parts c and d, whose eigenvalues remain eigenvalues of the controlled system. For the system (A, b, c^T, d) to be stabilizable, it is then necessary that its parts c and d be asymptotically stable. But, this condition is also sufficient since, parts a and b compose a completely reachable system so that from *Theorem 14* it follows that their eigenvalues can be modified at will. In conclusion, the following result (*Theorem 20*) holds:

THEOREM 20 *(stabilizability condition)*

> A system is stabilizable if and only if its unreachable parts (c and d) are asymptotically stable.

A dual result holds for the so-called *detectability*, namely, for the possibility of reconstructing, at least asymptotically, the state of a system by means of a linear time-invariant reconstructor.

THEOREM 21 *(stabilizability condition)*

> A system is detectable if and only if its unobservable parts (a and c) are asymptotically stable.

This result can be understood by a simple inspection of *Fig. B.15*. In fact, since parts b and d compose a completely observable system, their state variables $z_b(t)$ and $z_d(t)$ can be reconstructed from the input and output functions (see *Theorem 18*). But then, the inputs $u(t), z_b(t)$ and $z_d(t)$ of the system composed of parts a and c are known so that the forced motion of such a system can be computed. But, if the parts a and c are asymptotically stable, the forced motion tends, as time goes on, toward the state vectors $z_a(t)$ and $z_c(t)$, so that, in conclusion, the system is detectable.

We now give the definition of *external stability*, also known as *bounded input bounded output* (BIBO) stability.

DEFINITION 3 *(external stability)*

> A linear system is externally stable if its forced output is bounded for any bounded input.

From *Fig. B.15*, one can readily see that external stability is a property of part b of the system, since the forced motion is characterized by $z_c(t) = 0$ and $z_d(t) = 0$, while part a gives no contribution to the output. It is not surprising, then, that the external stability of a system is equivalent to the asymptotic stability of its part b as stated in *Theorem 22*.

272 ELEMENTS OF LINEAR SYSTEMS THEORY

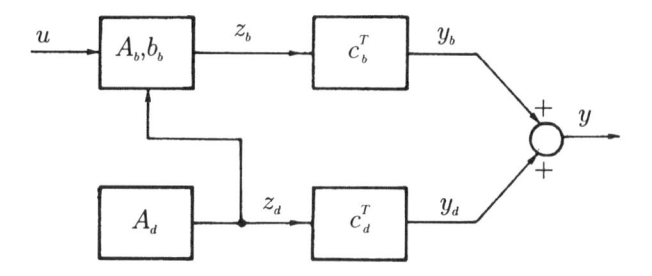

Figure B.16 Observable parts b and d of a system.

THEOREM 22 *(external stability condition)*

> A system is externally stable if and only if its reachable and observable part is asymptotically stable.

This result is due to the fact that only part b is responsible for the relationship between input and output whenever the system is initially at rest. If the initial state is nonzero, the output depends also on part d, which gives, however, a bounded contribution if it is simply or asymptotically stable. In conclusion, the output of a linear system is bounded for any initial state if and only if its reachable and observable part b is asymptotically stable and its observable and unreachable part d is stable.

B.18 DETERMINATION OF THE ARMA MODELS

We can now be more precise on the problem dealt with in *Section B.3*, namely, the determination of the ARMA model of a given system (A, b, c^T, d). For this, recall that an ARMA model is the pair of polynomials $[N(p), D(p)]$ identifying the input–output equation (B.8), which is the difference equation (B.6) in the discrete-time case and the differential equation (B.7) in the continuous-time case. Recall also that the ARMA model is said to be *reduced* if the polynomials $N(p)$ and $D(p)$ are coprime. Moreover, the ratio between $N(p)$ and $D(p)$ is the *transfer function* of the system denoted by $G(p)$, that is,

$$G(p) = \frac{N(p)}{D(p)}$$

Obviously, the ARMA model $[N(p), D(p)]$ and the transfer function $G(p)$ are not equivalent unless the ARMA model is reduced. Since the ARMA model represents the relation between input and output in the general case of a nonzero initial state, it must be associated with the observable parts b and d of the system, which, for the sake of clarity, are depicted in *Fig. B.16*.

The first subsystem with output y_b is described by the (multiple inputs) ARMA model

$$\Delta_b(p)y_b(t) = N_b(p)u(t) + N_b^1(p)z_{d1}(t) + N_b^2(p)z_{d2}(t) + \cdots \tag{B.30}$$

where $\Delta_b(p)$ is the characteristic polynomial of A_b and $z_{d1}(t), z_{d2}(t)$, and so on, are the components of the state vector $z_d(t)$. The second subsystem with output y_d has no input and is therefore described by the AR model

$$\Delta_d(p)y_b(t) = 0 \tag{B.31}$$

where $\Delta_d(p)$ is the characteristic polynomial of A_d. Moreover, since each component of the state vector $z_d(t)$ can be interpreted as an output of the second subsystem, we have

$$\Delta_d(p)z_{di}(t) = 0 \quad i = 1, 2, \ldots, n_d \tag{B.32}$$

If Eqs. (B.30) and (B.31) are multiplied by $\Delta_d(p)$ and $\Delta_b(p)$, respectively, and then summed up, the resulting equation in view of Eq. (B.32) becomes

$$\Delta_b(p)\Delta_d(p)y(t) = N_b(p)\Delta_d(p)u(t)$$

where $y(t) = y_b(t) + y_d(t)$. Thus, the ARMA model of the system is not reduced, since

$$D(p) = \Delta_b(p)\Delta_d(p)$$

and

$$N(p) = N_b(p)\Delta_d(p)$$

are not coprime.

If part d is missing, that is, if the system (A, b, c^T, d) does not have the unreachable and observable part, then $\Delta_d(p) = 1$ and the ARMA model

$$\Delta_b(p)y(t) = N_b(p)u(t)$$

is reduced, since $\Delta_b(p)$ and $N_b(p)$ are coprime (part b being completely reachable and observable). We can then conclude this section stating *Theorem 23*.

274   ELEMENTS OF LINEAR SYSTEMS THEORY

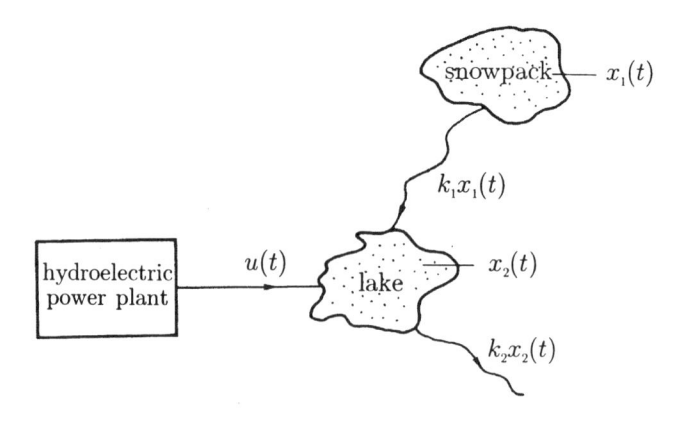

Figure B.17 A simple hydraulic system.

THEOREM 23 *(characterization of the ARMA model)*

> The ARMA model $[N(p), D(p)]$ of a system (A, b, c^T, d) is the ARMA model of the system composed of its observable parts (b and d) and
>
> $$N(p) = N_b(p)\Delta_d(p) \quad D(p) = \Delta_b(p)\Delta_d(p)$$
>
> where $[\Delta_b(p), N_b(p)]$ is the ARMA model of the reachable and observable part and $\Delta_d(p)$ is the characteristic polynomial of the unreachable and observable part (equal to 1 if such a part is missing). Then, the ARMA model of a system is in reduced form if and only if the system does not have the unreachable and observable part. Moreover, the transfer function $G(p) = N(p)/D(p)$ of the system is the transfer function $G_b(p) = N_b(p)/\Delta_b(p)$ of the reachable and observable part.

EXAMPLE 8

Consider the hydraulic system represented in *Fig. B.17* composed of a lake with two inflows, one with flow rate $u(t)$ (discharge of a plant) and the other $k_1 x_1(t)$ [melting of a snow-pack of volume $x_1(t)$].

If the flow rate of the effluent is assumed to be proportional through a coefficient k_2 to the water storage $x_2(t)$ of the lake, the mass conservation law gives

$$\begin{aligned} \dot{x}_1 &= -k_1 x_1 \\ \dot{x}_2 &= k_1 x_1 - k_2 x_2 + u \end{aligned}$$

Thus, if one considers the flow rate of the effluent as output variable $y(t)$, the system is described by the triple

$$A = \begin{pmatrix} -k_1 & 0 \\ k_1 & -k_2 \end{pmatrix} \quad b = \begin{pmatrix} 0 \\ 1 \end{pmatrix}$$

$$c^T = \begin{pmatrix} 0 & k_2 \end{pmatrix}$$

Such a system is completely observable but not completely reachable since the variable x_1 cannot be influenced. The system is therefore composed of part b (lake) and part d (snow-pack) as depicted in *Fig. B.16*. Since the eigenvalues of A are $-k_1$ (snow-pack) and $-k_2$ (lake) we have

$$\Delta_b(s) = (s + k_2) \quad N_b(s) = k_2 \quad \Delta_d(s) = (s + k_1)$$

so that

$$D(s) = (s + k_1)(s + k_2) \quad N(s) = k_2(s + k_1)$$

In conclusion, the flow rates $u(t)$ and $y(t)$ are linked by the second-order differential equation

$$\ddot{y}(t) + (k_1 + k_2)\dot{y}(t) + k_1 k_2 y(t) = k_2 \dot{u}(t) + k_1 k_2 u(t)$$

In contrast, if the snow-pack is missing, the model becomes

$$\dot{y}(t) + k_2 y(t) = k_2 u(t)$$

which is a reduced ARMA model.

♣

In many cases of practical interest, the model of the system (A, b, c^T, d) is not known, but a pair $u(\cdot), y(\cdot)$ of input and output records of length T is available. This is, for example, the case of a river basin in which rainfall and outflow have been recorded for a period of a few months. Another example is the case of an electrical amplifier in which the input and output signals have been measured for some seconds. In these cases, it is interesting to know if it is possible to determine the model (A, b, c^T, d) of the system by processing the recorded input and output data. This problem is known as *identification* of the model and is of paramount importance in applications. Very often a solution is obtained by assuming that input and output measurements are affected by noise, and by using suitable notions of the theory of stochastic processes. However, the problem is theoretically and practically interesting even in the absence of noise. For this assume, that the system is proper and discrete-time and that the dimension $n^\circ = n_b + n_d$ of its observable part is known. Under these assumptions, the ARMA model of the system is

$$y(t) = -\alpha_1 y(t-1) - \cdots - \alpha_{n^\circ} y(t-n^\circ) + \beta_1 u(t-1) + \cdots + \beta_{n^\circ} u(t-n^\circ)$$

which can be written in the more compact form

$$y(t) = -(y_{t-n^\circ}^{t-1})^T \alpha + (u_{t-n^\circ}^{t-1})^T \beta \qquad (B.33)$$

where

$$\alpha = \begin{pmatrix} \alpha_1 \\ \vdots \\ \alpha_{n^\circ} \end{pmatrix} \quad \beta = \begin{pmatrix} \beta_1 \\ \vdots \\ \beta_{n^\circ} \end{pmatrix} \quad y_{t-n^\circ}^{t-1} = \begin{pmatrix} y(t-1) \\ \vdots \\ y(t-n^\circ) \end{pmatrix} \quad u_{t-n^\circ}^{t-1} = \begin{pmatrix} u(t-1) \\ \vdots \\ u(t-n^\circ) \end{pmatrix}$$

Suppose, now, that a recorded time series composed of N input and output values

$$u(0), u(1), \ldots, u(N-2), u(N-1)$$
$$y(0), y(1), \ldots, y(N-2), y(N-1)$$

is known and write Eq. (B.33) in the $2n^\circ$ unknowns α_i and $\beta_i, i = 1, \ldots, n^\circ$, for $N - n^\circ$ successive values of t, that is, for $t = n^\circ, n^\circ + 1, \ldots, N - 1$. This leads to the following system of $N - n^\circ$ linear equations in $2n^\circ$ unknowns:

$$\begin{pmatrix} -(y_0^{n^\circ-1})^T & (u_0^{n^\circ-1})^T \\ -(y_1^{n^\circ})^T & (u_1^{n^\circ})^T \\ \vdots & \vdots \\ -(y_{N-n^\circ-1}^{N-2})^T & (u_{N-n^\circ-1}^{N-2})^T \end{pmatrix} \begin{pmatrix} \alpha \\ \beta \end{pmatrix} = \begin{pmatrix} y(n^\circ) \\ y(n^\circ+1) \\ \vdots \\ y(N-1) \end{pmatrix}$$

This is an algebraic linear system of the kind

$$Fp = y_{N-1}^{n^\circ} \tag{B.34}$$

where p is the unknown vector of parameters α_i and β_i identifying the ARMA model and F is an $(N - n^\circ) \times (2n^\circ)$ matrix depending on the input and output data. By excluding special critical cases, this algebraic system can be solved if

$$N \geq 3n^\circ$$

In the case $N = 3n^\circ$, the solution is

$$\hat{p} = F^{-1} y_{N-1}^{n^\circ}$$

while in the case $N > 3n^\circ$ the solution can be given in the form

$$\hat{p} = (F^T F)^{-1} F^T y_{N-1}^{n^\circ} \tag{B.35}$$

The critical cases, of nonidentifiability are those in which the matrix F is not full rank, so that the matrix $F^T F$ is not invertible. These cases occur, for example, when the input and output data are collected during a period of time in which the system is at equilibrium (steady state). In fact, in such conditions the first [second] n° columns of the matrix F are identical because the output [input] does not vary in time. Another case of nonidentifiability occurs when the initial state of the unreachable and observable part is zero. In fact, under this circumstance, the output is not influenced by part d of the system, see *Fig. B.16*, so that the coefficients of the characteristic polynomial $\Delta_d(p)$ are not identifiable. This means that the ARMA model of the system is not identifiable since $N(p) = N_b(p)\Delta_d(p)$

and $D(p) = \Delta_b(p)\Delta_d(p)$. The previous discussion can be summarized in *Theorem 24*.

THEOREM 24 (*identifiability of ARMA models*)
> The ARMA model of a proper discrete-time system with known dimension $n° = n_b + n_d$ of the observable parts cannot be identified from a series of N input and output values if $N < 3n°$. On the contrary, if $N \geq 3n°$ the ARMA model is uniquely identified, apart from some critical cases (of nonidentifiability).

Once the ARMA model has been identified, it is possible to construct a triple (A, b, c^T, d), which realizes it. Such a triple is the reconstruction canonical form

$$A_r = \begin{pmatrix} 0 & 0 & \cdots & 0 & -\alpha_{n°} \\ 1 & 0 & \cdots & 0 & -\alpha_{n°-1} \\ 0 & 1 & \cdots & 0 & -\alpha_{n°-2} \\ \vdots & \vdots & & \vdots & \vdots \\ 0 & 0 & \cdots & 1 & -\alpha_1 \end{pmatrix} \quad b_r = \begin{pmatrix} \beta_{n°} \\ \beta_{n°-1} \\ \beta_{n°-2} \\ \vdots \\ \beta_1 \end{pmatrix}$$

$$c_r^T = (\,0 \quad 0 \quad \cdots \quad 0 \quad 1\,)$$

In fact, from *Theorem 17*, such a system is completely observable, so that it is composed of parts b and d, which characterize the ARMA model. If the ARMA model is in reduced form, that is, if the polynomials

$$D(p) = p^{n°} + \alpha_1 p^{n°-1} + \cdots + \alpha_{n°}$$
$$N(p) = \beta_1 p^{n°-1} + \cdots + \beta_{n°}$$

are coprime, the triple (A_r, b_r, c_r^T) is also reachable, namely it is composed of part (b) only. In contrast, if the ARMA model is not in reduced form, namely, if

$$D(p) = r(p)d(p)$$
$$N(p) = r(p)n(p)$$

then, the system is composed of a reachable and observable part described by the ARMA model $[n(p), d(p)]$ and of an unreachable and observable part, described by an AR model $r(p)$, which coincides with the characteristic polynomial $\Delta_d(p)$ of such a part.

The control canonical form

$$A_c = A_r^T \quad b_c = c_r \quad c_c^T = b_r^T$$

obtained by duality from the realization in reconstruction canonical form, is not a realization of the ARMA model $[N(p), D(p)]$, if such a model is not in reduced

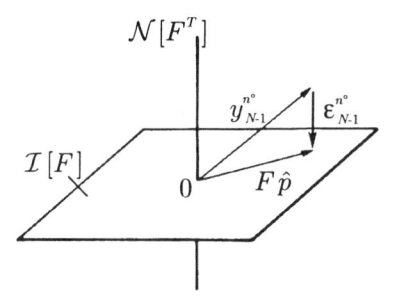

Figure B.18 Illustration of the least-square estimation principle.

form. In fact, the system (A_c, b_c, c_c^T) is completely reachable (see *Theorem 13*) so that it cannot be composed of parts b and d. Therefore, the ARMA model of the control canonical form (A_c, b_c, c_c^T) is $[n(p), d(p)]$ instead of $[N(p), D(p)]$. In other words, the control canonical form realizes the transfer function $G(p) = N(p)/D(p)$ of the system but not the ARMA model. Obviously, if the system is completely reachable and observable, the ARMA model is in reduced form and the control canonical form is one of its realizations.

If the input and output values contain errors, Eq. (B.34) must be replaced by

$$Fp - y_{N-1}^{n^o} = \varepsilon_{N-1}^{n^o}$$

where the vector ε represents the difference between the output values predicted by the ARMA model and the measured output values. In *Fig. B.18*, the measured output vector and the subspace $\mathcal{I}[F]$ of the output predicted by the ARMA model are shown in the space of dimension $N - n^o$. It is then natural to choose the ARMA model that is, the value $\hat{p}$ of p, that minimizes the distance between the measured and the predicted output vectors.

As illustrated in *Figure B.18*, this amounts to choosing $\hat{p}$ in such a way that the vector $\varepsilon_{N-1}^{n^o}$ is orthogonal to $\mathcal{I}[F]$. But, since $\mathcal{I}[F]^\perp = \mathcal{N}[F^T]$, this is equivalent to

$$F^T \varepsilon_{N-1}^{n^o} = 0$$

namely,

$$F^T (F\hat{p} - y_{N-1}^{n^o}) = 0$$

from which, assuming that $F^T F$ is invertible, it follows that:

$$\hat{p} = (F^T F)^{-1} F^T y_{N-1}^{n^o} \qquad (B.36)$$

which coincides with Eq. (B.35).

The estimation $\hat{p}$ given by (B.36) is known as the *least-square estimation* because it minimizes the sum of the squares of the differences between predictions and actual measurements. This estimation, here interpreted in geometrical terms, possesses a number of peculiar properties for specific statistical characteristics of the input and output measurement errors. Moreover, formula (B.35) can be fruitfully given in a recursive form, in such a way that the computation of $\hat{p}$ can be updated in real time, without the need of computing the inverse of a $2n° \times 2n°$ matrix each time a new pair of input–output data is available.

B.19 POLES AND ZEROS OF THE TRANSFER FUNCTION

We have already mentioned that the transfer function $G(p)$ of a linear system is, by definition, the ratio of the two polynomials $N(p)$ and $D(p)$ that identify the ARMA model of the system

$$D(p)y(t) = N(p)u(t)$$

From *Theorem 23*, we can immediately conclude that $G(p)$ coincides with the transfer function of the reachable and observable part of the system, that is,

$$G(p) = \frac{N_b(p)}{\Delta_b(p)}$$

where $N_b(p)$ and $\Delta_b(p)$ are the two coprime polynomials characterizing the ARMA model of the reachable and observable part. The zeros of the polynomials $N_b(\cdot)$ and $\Delta_b(\cdot)$ are called, respectively, *zeros* and *poles of the transfer function* (or of the system) and are denoted by z_i and p_i. The transfer function of a proper system with a reachable and observable part of dimension n can be written in the following form:

$$G(p) = \frac{\beta_r p^{n-r} + \beta_{r+1} p^{n-r-1} + \cdots + \beta_n}{p^n + \alpha_1 p^{n-1} + \cdots + \alpha_n}$$

where $r \geq 1$ is the so-called *relative degree*, or in the form

$$G(p) = \rho \frac{(p - z_1)(p - z_2) \ldots (p - z_{n-r})}{(p - p_1)(p - p_2) \ldots (p - p_n)}$$

where ρ is called *transfer constant*. Poles and zeros are of paramount importance in a number of problems in systems and control theory. We shall see in a moment that it is particularly interesting to know if a continuous [discrete] -time system has all its poles and zeros with negative real part [modulus < 1]. In other words, we are interested to know whether the poles and zeros are "stable" or not. In view of the decomposition theorem, it follows that the poles of the transfer function are the eigenvalues of the reachable and observable part, so that (see *Theorem 22*) a system is externally stable if and only if its poles are stable. Moreover, the output of a system is bounded for any bounded input if and only if its poles are stable and the unreachable and observable part d does not exist or is asymptotically or simply stable. Under these conditions, the solutions of the ARMA model

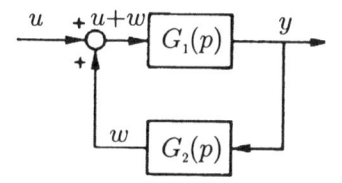

Figure B.19 Structure of a system equivalent to a completely reachable and observable system [$G_1(p)$ has no zeros].

$$\Delta_b(p)\Delta_d(p)y(t) = N_b(p)\Delta_d(p)u(t)$$

for different initial conditions and for the same bounded input $\hat{u}(t)$, are different bounded outputs $\hat{y}(t)$ which do not diverge unlimitatedly one from each other. Moreover, if the unreachable and observable part is missing or asymptotically stable, the outputs $\hat{y}(t)$ of the system tend toward the same function $y(t)$ which can be computed with great accuracy for t sufficiently large using the reduced ARMA model

$$\Delta_b(p)y(t) = N_b(p)u(t)$$

To fully understand the role of the zeros in the dynamics of a linear system, it is necessary to refer to the particular canonical form shown in *Fig. B.19*. This is always possible since a completely reachable and observable system (A, b, c^T) with n poles and $(n - r)$ zeros, can always be put, by means of an appropriate change of coordinates $z = Tx$, in the form reported in *Fig. B.19*, where the subsystem in the forward path, has dimension r and has no zeros.

If $G_1(p)$ and $G_2(p)$ are the transfer functions of the two subsystems, from the formula

$$G(p) = \frac{G_1(p)}{1 - G_1(p)G_2(p)}$$

we can conclude that the poles of $G_2(p)$ are the zeros of $G(p)$. Therefore, if (A_2, b_2, c_2^T) is the triple that defines the subsystem in the feedback path, the eigenvalues of A_2 are the zeros of the system and the free output of the feedback subsystem, obtained with $y(t)$ identically zero, is

$$w(t) = c_2^T e^{A_2 t} z_2(0)$$

If the system in the forward path is initially at rest [$z_1(0) = 0$] and the signal $w(t)$ is compensated by the input $u(t) = -w(t)$, the system in the forward path is not excited from the outside and remains, consequently, at rest [*i.e.*, $y(t) \equiv 0$). This means that the output of the system can be identically zero even if its input is not. This happens when the initial state is appropriately chosen [$z_1(0) = 0$] and the input $u(t)$ is the output of an autonomous system $(A_2, -, c_2^T)$ with eigenvalues equal to the zeros of the system. In other words, the zeros of a system completely determine the dynamics of its "hidden" inputs.

Systems with no zeros or with strictly stable zeros are called *minimum phase systems*. They have no hidden inputs or hidden inputs that asymptotically tend to zero at a speed dictated by the "dominant" zero. In contrast, continuous-time [discrete-time] nonminimum phase systems have zeros with nonnegative real part [modulus not < 1] and therefore have hidden inputs not tending to zero. The knowledge of a record of the output of a completely reachable and observable linear system allows one to reconstruct an input $\hat{u}(t)$ of the system, which is the sum of the true input $u(t)$ and of a hidden input. But if the system is minimum phase, the hidden input tends to zero as $t \to \infty$ so that $\hat{u}(t)$ tends toward the true input $u(t)$. The reconstruction algorithm is still an ARMA model

$$\Delta_b(p)y(t) = N_b(p)u(t)$$

which must be solved with respect to $u(t)$. In the case of a discrete-time system, this implies the recursive solution of the equation

$$y(t) + \alpha_1 y(t-1) + \cdots + \alpha_n y(t-n) = \beta_r \hat{u}(t-r) + \beta_{r+1}\hat{u}(t-r-1) + \cdots + \beta_n \hat{u}(t-n)$$

with respect to $\hat{u}(t-r)$. Note that this operation cannot be performed in real time since the evaluation of $\hat{u}(t-r)$ requires the knowledge of $y(t)$. At best, the input can be reconstructed after r transitions.

Clearly, the problem of the hidden inputs and of the reconstruction of the inputs from the outputs is well posed also when the system has an unreachable and observable part. In this case, taking into account *Fig. B.2*, one obtains the block diagram shown in *Fig. B.20*.

Such a diagram shows that the hidden inputs can be divided into two groups: those "generated" by the zeros of the system and those "generated" by the eigenvalues of the unreachable and observable part [zeros of the polynomial $\Delta_d(p)$]. Therefore, the hidden inputs tend to zero, if the system is minimum phase and its unreachable and observable part is asymptotically stable. In this case, the input can be reconstructed by solving the nonreduced ARMA model

$$\Delta_b(p)\Delta_d(p)y(t) = N_b(p)\Delta_d(p)u(t)$$

with respect to $u(t)$ or, alternatively the reduced ARMA model

$$\Delta_b(p)y(t) = N_b(p)u(t)$$

This section can be summarized by noting that poles and zeros play roles that are "dual" in some way. In fact, in the long run, the output of a completely reachable and observable system also can be computed, given its input, with no information on the initial state provided that the poles of the system are stable (external stability).

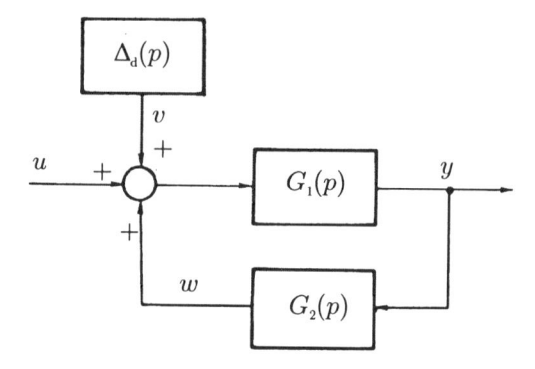

Figure B.20 Block diagram of a system with a nonreduced ARMA model: $\Delta_d(p)$ is the characteristic polynomial of the unreachable and observable part and $G_1(p)$ has no zeros.

Dually, in the long run, the input of a completely reachable and observable system can be computed, given its output, with no information on the initial state, provided that the zeros of the system are stable (minimum phase). In other words, the stability of the poles allows one to neglect the free motion in the long run, while the stability of the zeros allows one to neglect the hidden inputs. The reader can easily formulate these properties for systems with an unreachable and observable part.

B.20 POLES AND ZEROS OF INTERCONNECTED SYSTEMS

As shown in Section B4, the transfer function of two interconnected systems (*Figs. B.3, B.4,* and *B.5*) is given by

$$G(p) = G_1(p)G_2(p) \qquad \text{series connection}$$
$$G(p) = G_1(p)G_2(p) \qquad \text{parallel connection}$$
$$G(p) = \frac{G_1(p)}{1 + G_1(p)G_2(p)} \qquad \text{(negative) feedback connection}$$

where $G_1(p)$ and $G_2(p)$ are the transfer functions of the two subsystems. Apart from critical cases (related to the nonreachability and nonobservability of the resulting system), in which the denominator of the transfer function $G(p)$ turns out to be a polynomial of degree smaller than the sum of the degrees of the denominators of the two transfer functions $G_1(p)$ and $G_2(p)$, we can immediately conclude the following:

Series. The poles and the zeros of $G(p)$ are the union of those of $G_1(p)$ and $G_2(p)$.

Parallel. The poles of $G(p)$ are the union of those of $G_1(p)$ and $G_2(p)$.

Feedback. The zeros of $G(p)$ are the union of the zeros of $G_1(p)$ and of the poles of $G_2(p)$.

The computation of poles and zeros of interconnected systems is, therefore, immediate, except for the computation of the zeros of systems connected in parallel and of the poles of systems connected in feedback. These two cases however, are equivalent since the zeros of the transfer function

$$G(p) = G_1(p) + G_2(p)$$

of the system in *Fig. B.21(a)* coincide with the poles of the transfer function

$$G(p) = \frac{\frac{G_1(p)}{G_2(p)}}{1 + \frac{G_1(p)}{G_2(p)}} = \frac{G_1(p)}{G_1(p) + G_2(p)}$$

of the system in *Fig. B.21(b)*.

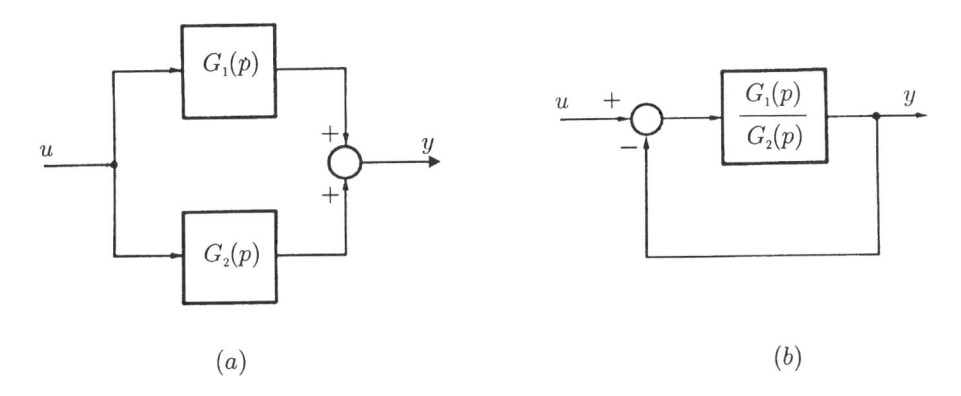

Figure B.21 The zeros of system (a) coincide with the poles of system (b).

We can then conclude that there is only one significant problem, namely, the determination of the poles of a system composed of two subsystems connected in feedback. This is the central problem of classical control theory, mainly focused on the determination of feedback systems with appropriate dynamic properties as, for example, external stability.

In applications, it is often important to determine poles and zeros when some parameter is varied (typically, a design parameter). Though today this can be simply done by means of specific software, we briefly describe a method called *root locus*, which has been often used in the past for the design of control systems, and is of great value still today for the discussion of the stability of feedback systems.

284   ELEMENTS OF LINEAR SYSTEMS THEORY

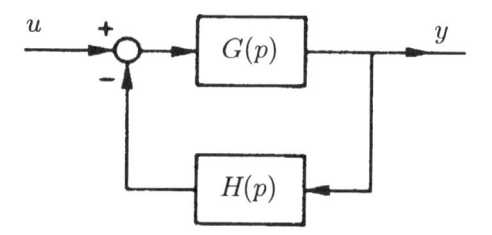

Figure B.22 Feedback system.

The root locus is by definition, the locus described in the complex plane by the poles of the feedback system shown in *Fig. B.22* when the transfer constants of the two subsystems are allowed to vary. It is, therefore, composed of n curves, one for each pole, called "branches" of the locus.

If the product of the two transfer constants is positive [negative] and assumes all values, from 0 to $\infty[-\infty]$ we obtain the direct [inverse] locus. When the transfer constants are related to a design parameter (as it is usually the case) and one wants to obtain an externally stable system, one must check whether the n segments of the branches of the locus are "stable" for the feasible values of the design parameters. This amounts to checking whether the n segments of the "locus" are in the left half-plane [unitary circle] of the complex plane if the system is continuous-time [discrete-time]. In *Fig. B.23*, six examples of direct root loci are depicted: poles and zeros of the two transfer functions $G(p)$ and $H(p)$ are represented by crosses ($\times$) and circles ($\circ$), respectively.

Obviously, there is no reason to distinguish the poles [zeros] of $G(p)$ from those of $H(p)$. In fact, if

$$G(p) = \rho_G \frac{\Pi(p - z_i^G)}{\Pi(p - p_i^G)} \qquad H(p) = \rho_H \frac{\Pi(p - z_i^H)}{\Pi(p - p_i^H)}$$

then the transfer function $F(p)$ of the system is

$$F(p) = \frac{G(p)}{1 + G(p)H(p)}$$

and its poles are the roots of the equation

$$1 + G(p)H(p) = 0$$

in which only the product $G(p)H(p)$ appears. Such equation can be fruitfully written in the form

$$k\Pi(p - z_i^G)\Pi(p - z_i^H) = -\Pi(p - p_i^G)\Pi(p - p_i^H) \tag{B.37}$$

where the parameter k, which is positive in the direct locus and negative in the inverse one, is the product of the two transfer constants, namely, $k = \rho_G \rho_H$. The

POLES AND ZEROS OF INTERCONNECTED SYSTEMS   285

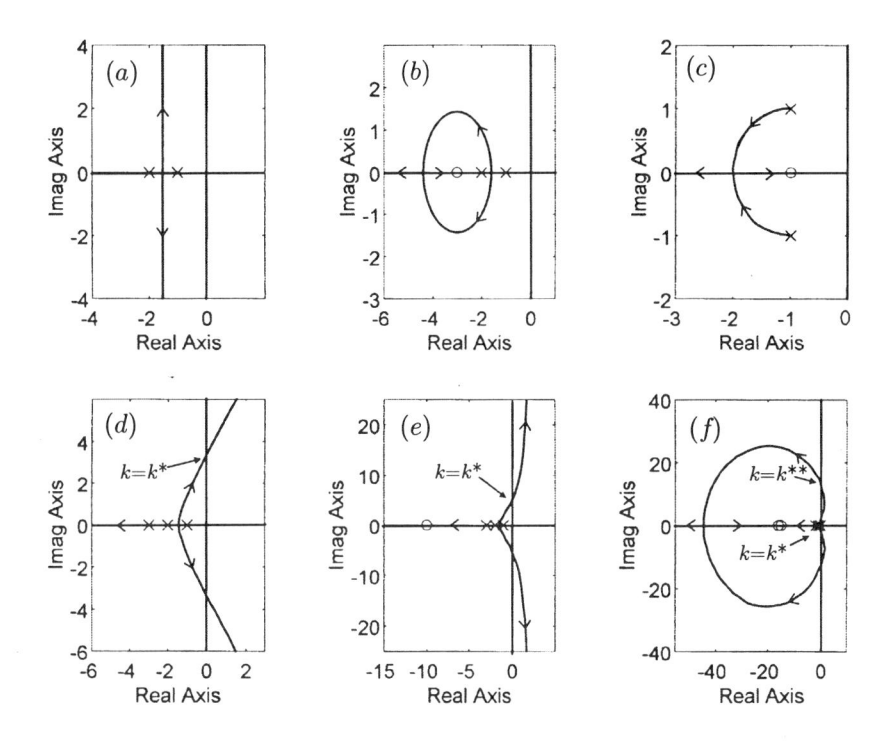

Figure B.23 Root loci of second- and third-order systems: (a) two poles; (b) and (c) two poles and one zero; (d) three poles; (e) three poles and one zero; (f) three poles and two zeros.

loci (a), (b), and (c) of *Fig. B.23* refer to a feedback system composed of two subsystems, which have only two poles, while in cases (d), (e), and (f) there are three poles. Consistently, the first three loci are composed of two branches, while the last three are composed of three branches. If we imagine that we are dealing with continuous-time systems, we can infer that the feedback system is externally stable for all values of k in cases a, b and c, only for $k < k^*$ in cases d and e and for $k < k^*$ and $k > k^{**}$ in case f. The values k^* and k^{**} are very important, since they mark the transition from stability to instability.

The six loci depicted in *Fig. B.23* show some general properties of the root locus that are worth noting. First of all, the locus is symmetrical with respect to the real axis. Moreover, each branch starts from a pole [of $F(p)$ or $G(p)$] since for $k \to 0$ the roots of (B.37) tend to p_i^G and p_i^H. On the other hand, for $k \to \infty$, $(n-r)$ branches tend to the zeros z_i^G and z_i^H [see (B.37)] while the remaining r tend to infinity forming an angle of $2\pi/r$. Finally, all the points of the real axis that have on their right an odd [even] number of singularities (poles and zeros) belong to the direct [inverse] locus. All such properties can be easily proved. In contrast, other properties as the "rule of the center of mass" are less immediate. Such a rule states that if the relative degree r is ≥ 2, the sum of the n poles is independent of k. This can be checked by writing Eq. (B.37) in the form

$$p^n + \gamma_1 p^{n-1} + \gamma_2 p^{n-2} + \cdots = 0$$

and by noting that γ_1, which is equal to the opposite of the sum of the poles, is independent of k if $r \geq 2$. The consequences of this rule are evident in the loci a and e in *Fig. B.23*. In case a, the point in which the two branches collide when k increases is the central point of the segment connecting the two poles. In case e, since for $k = 0$ the sum of the three poles is equal to -6 and for $k \to \infty$ one of the three poles tends to the zero located at -8, the other two poles must have, for $k \to \infty$, real part equal to 1.

All such rules often allows one to discuss qualitatively, but effectively, the external stability of a feedback system when a design parameter is varied. From *Fig. B.21*, it is clear that the same rules also allow the discussion of the minimum phase of systems composed by subsystems connected in parallel.

B.21 IMPULSE RESPONSE

The impulse response of a continuous-time linear system is, as the term itself suggests, the output of the system corresponding to an impulsive input. In order to uniquely define the impulse response, one must specify the initial state which, for simplicity, is assumed to be zero. The impulse response, denoted in the following by $g(t)$, is then the output of the system

$$\dot{x} = Ax + bu$$
$$y = c^T x$$

with $x(0) = 0$ and $u(t) = \text{imp } t$.

The impulse response, can often be measured directly in the field or in the laboratory. For example, *Fig. B.24* reports the impulse responses of four systems. The first concerns the position of a point mass moving along a straight line after it has been hit by another point mass (impulsive force), the second is the voltage of an $R - C$ circuit fed by an impulse of current, the third is the flow of a river after a short but intensive storm in the river basin (impulsive rainfall), and the fourth is the behavior of the wings of an airplane after an air pocket (impulsive force).

From the Lagrange formula (*Theorem 1*) it follows that:

$$g(t) = c^T \int_0^t e^{A(t-\xi)} b \,\text{imp}\xi \, d\xi = c^T e^{At} \int_0^{0+} \text{imp}\xi \, d\xi \, b$$

that is,

$$g(t) = c^T e^{At} b$$

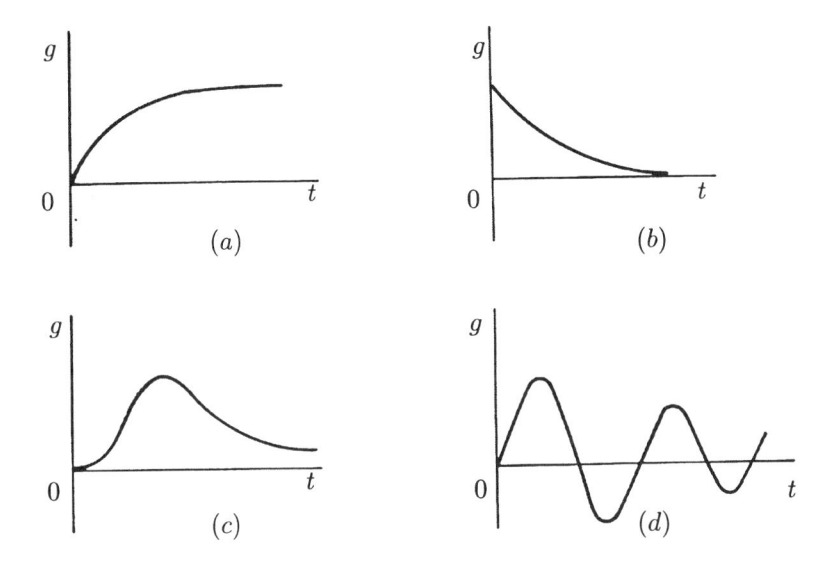

Figure B.24 Four impulse responses: (a) point mass; $R - C$ (b)electrical circuit; (c) river basin; (d) wings of an airplane.

This means that the impulse response is the free output of the system $g(t) = c^T e^{At} x(0)$ with $x(0) = b$ and this is consistent with the fact that the impulse steers the state ot the system form 0 to b in a time interval of zero measure.

By recalling that the reachable and observable part of a system (part b) is the only part determining the output in the case of a zero initial state, one can also write

$$g(t) = c_b^T e^{A_b t} b_b$$

From the Lagrange formula, it follows also that the forced evolution of a continuous-time linear system is

$$g(t) = c^T \int_0^t e^{A(t-\xi)} bu(\xi) d\xi = \int_0^t g(t - \xi) u(\xi) d\xi$$

namely, the forced output is the *convolution integral* of the impulse response and of the input.

Moreover, it is worth noting that

$$g(t) = c^T e^{At} b = c^T \left(I + At + A^2 \frac{t^2}{2!} + \cdots \right) b$$
$$= c^T b + c^T Abt + c^T A^2 b \frac{t^2}{2!} + \cdots$$

so that, recalling the formula for the Taylor expansion of a function $g(t)$ in a neighborhood of the origin, we can conclude that

$$\left.\frac{d^i g(t)}{dt^i}\right|_{t=0} = c^T A^i b \quad i = 0, 1, 2, \ldots$$

The coefficients

$$g_1 = c^T b \quad g_2 = c^T A b \quad g_3 = c^T A^2 b \ldots$$

are known as *Markov coefficients*.

Other *canonical responses* of continuous-time linear systems are the *step response* and the *ramp response*, which are the output of the system with $x(0) = 0$ and

$$u(t) = \begin{cases} 1 & t \geq 0 \quad \text{step} \\ t & t \geq 0 \quad \text{ramp} \end{cases}$$

Since the step function is the integral of the impulse function and the ramp is the integral of the step, we can conclude that the step response is the integral of the impulse response and that the ramp response is the integral of the step response. By using analogous arguments one can define the impulse response of discrete-time systems, which turn out to be given by

$$g(t) = \begin{cases} 0 & t = 0 \\ c^T A^{t-1} b & t > 0 \end{cases}$$

From the above definition of Markov coefficients, one can conclude that the impulse response $g(t)$ of a discrete-time system is zero at time zero and equal to the corresponding Markov coefficients for $t > 0$, that is,

$$g(t) = g_t$$

As a useful exercise, the reader is invited to compute the impulse response of the systems considered in *Example 1* (Newton's law) and in *Example 2* (Fibonacci's rabbits).

B.22 FREQUENCY RESPONSE

For a large class of continuous-time linear systems, the periodic output corresponding to a sinusoidal input is unique and is actually a sinusoid with the same frequency as the input. This property directly leads to the definition of *frequency response* and to the possibility of determining the transfer function by means of simple experiments.

As for the existence and uniqueness of the periodic output, *Theorem 25* holds.

THEOREM 25 *(existence and uniqueness of the periodic regime)*

> In a continuous-time linear system (A, b, c^T) with no eigenvalues with zero real part, one and only one periodic output of period T, say $y_T(\cdot)$, is associated to each periodic input function $u_T(\cdot)$ of period T. Moreover, if the observable part is asymptotically stable, the output $y(t)$ corresponding to the input $u_T(t)$ tends asymptotically to $y_T(t)$, for any initial state $x(0)$ of the system.

To verify that the theorem cannot be extended to systems with eigenvalues with zero real part (nonhyperbolic systems) it is sufficient to consider the case of an integrator $\dot{x} = u, y = x$. In fact, in such a system, the input

$$u_T(t) = U \cos(2\pi/T)t$$

gives rise to an infinite number of periodic outputs

$$y_T(t) = x(0) + (UT/2\pi)\sin(2\pi/T)t$$

parameterized in the initial state $x(0)$ of the system.

Among all periodic input functions $u_T(\cdot)$, of particular interest is the sinusoidal function

$$U \sin(2\pi/T)t = U \sin \omega t$$

since it is known that, under very general conditions (see Section B.23), any periodic function $u_T(\cdot)$ of period T can be expanded in Fourier series and expressed as an infinite linear combination of sinusoids and cosinusoids of angular frequency $n(2\pi/T)$, n being any nonnegative integer. In fact, a periodic function of period T can be written as

$$u_T(t) = a_0 + \sum_{n=1}^{\infty}\left[a_n \cos\left(n\frac{2\pi t}{T}\right) + b_n \sin\left(n\frac{2\pi t}{T}\right)\right]$$

where

$$a_0 = \frac{1}{T}\int_{-T/2}^{T/2} u_T(t)dt$$
$$a_n = \frac{2}{T}\int_{-T/2}^{T/2} u_T(t)\cos\left(n\frac{2\pi t}{T}\right)dt$$
$$b_n = \frac{2}{T}\int_{-T/2}^{T/2} u_T(t)\sin\left(n\frac{2\pi t}{T}\right)dt$$

Therefore, the periodic function $y_T(\cdot)$ corresponding to the periodic input $u_T(\cdot)$ can be computed by first determining the components of the Fourier series of $u_T(\cdot)$ and then by summing up all the corresponding output periodic functions (superposition principle). Moreover, the relevance of the sinusoidal regime is motivated by the following result:

THEOREM 26 (*frequency response*)

Continuous-time linear systems (A, b, c^T) with no eigenvalues with zero real part have one and only one sinusoidal output

$$y_T(t) = Y \sin\left(\frac{2\pi}{T}t + \varphi\right)$$

for any sinusoidal input

$$u_T(t) = U \sin\frac{2\pi}{T}t$$

Moreover, Y is linear in U and φ depends only on $\omega = \frac{2\pi}{T}$, that is,

$$Y = R(\omega)U \quad \varphi = \varphi(\omega)$$

Theorem 26 states that a system in sinusoidal regime has an output sinusoid of amplitude $R(\omega)U$ shifted with respect to the input sinusoid of an angle $\varphi(\omega)$. Therefore, the computation of the periodic function $y_T(\cdot)$ corresponding to a periodic input $u_T(\cdot)$ is straightforward once the two functions $R(\cdot)$ and $\varphi(\cdot)$ are known. This is why this pair of functions is called *frequency response*.

The frequency response of a system can be measured by means of simple experiments if the observable part of the system is asymptotically stable. In fact, if one applies to a system of this kind a sinusoidal input of amplitude U and angular frequency ω, after a sufficiently long time interval the output of the system is in practice a sinusoid of amplitude $R(\omega)U$ and phase $\varphi(\omega)$, no matter what the initial conditions of the system are. Thus, $R(\omega)$ is simply the ratio of the amplitudes of the output and input sinusoids while $\varphi(\omega)$ is the phase shift between the two sinusoids.

In contrast, if the observable part of the system is not asymptotically stable, it is not possible to experimentally determine the frequency response of the system since the output does not tend toward a sinusoid for a generic initial state. Nevertheless, this fact does not imply that the frequency response $[R(\omega), \varphi(\omega)]$ cannot be defined; indeed, the problem of the definition of a quantity is different from the problem of its determination.

The frequency response of a system can be graphically represented in two ways: by means of the Cartesian plots of the functions $R(\cdot)$ and $\varphi(\cdot)$ or by means of the polar plot (called *Nyquist plot*) representing the function $R(\cdot)e^{i\varphi(\cdot)}$.

Typical Cartesian plots of the function $R(\cdot)$ are depicted in *Fig. B.25*. In certain ranges of the angular frequency ω the function $R(\omega)$ is almost zero, which means that the input sinusoid is very strongly attenuated. Thus, for example, the system in *Fig. B.25(a)* attenuates all the sinusoids with angular frequency $> \omega_0$ while the sinusoids with $\omega < \omega_0$ are not attenuated. A system of this kind is called a *low-pass* filter and the interval $[0, \omega_0]$ is called a *bandwidth* [since it is not possible

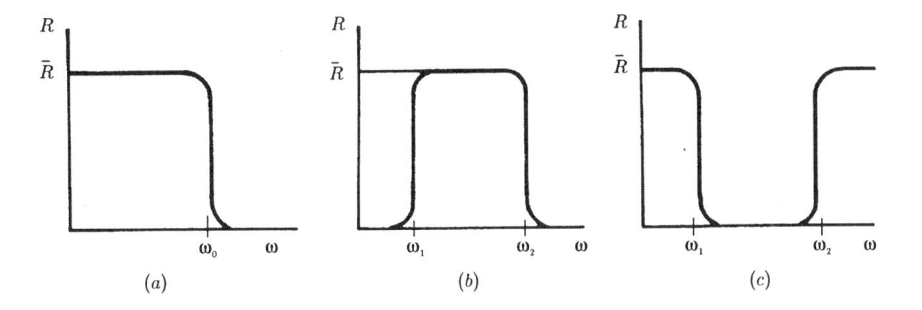

Figure B.25 Typical examples of frequency responses: (*a*) low-pass; (*b*) band-pass; (*c*) stop band.

that $R(\omega) = 1$ for $\omega \leq \omega_0$ and $R(\omega) = 0$ for $\omega > \omega_0$, the bandwidth of the system has to be defined appropriately].

On the other hand, the system with $R(\cdot)$ as in *Fig. B.25(b)*, attenuates the sinusoids with angular frequency $\omega < \omega_1$ and those with $\omega < \omega_2$, and is therefore called a *band-pass filter*. For the opposite reason, the system with $R(\cdot)$ as in *Fig. B.25(c)* is called *stop band*.

Limiting cases of particular relevance can be obtained by letting the bandwidth of a band-pass or stop-band system go to zero. By doing so one obtains systems called, respectively, *resonant filters* and *notch filters*, which are sensitive or insensitive only to a very specific angular frequency.

Obviously, not only the function $R(\cdot)$ is of interest since the function $\varphi(\cdot)$ also contributes to define the sinusoidal regime. In fact, the frequency response $[R(\omega), \varphi(\omega)]$ matters. As an example, consider a communication system which, ideally, should be able to reproduce at the output a perfect copy of the input, obviously with a certain delay τ needed to transfer the information from the input to the output. Thus, a sinusoidal input $U \sin \omega t$ must produce a sinusoid $U \sin \omega(t - \tau)$ at the output, and this must hold for any angular frequency ω. In other words, the ideal communication system is a pure delay system characterized by $R(\omega) = 1$ and $\varphi(\omega) = -\omega \tau$, as shown in *Fig. B.26*.

A pure delay system cannot be realized by a finite-dimensional linear system (A, b, c^T), so that communication systems are often designed by allowing a certain degree of distortion between input and output, that is, by approximating the shape of the functions $R(\cdot)$ and $\varphi(\cdot)$ in *Fig. B.26* in a suitable range of the angular frequency.

So far, we have shown that the frequency response $[R(\cdot), \varphi(\cdot)]$ enables one to rapidly compute the periodic regime of a linear system and its filtering properties. We have also shown how the frequency response can be experimentally measured if the observable part of the system is asymptotically stable. A third very important property is the following connection (*Theorem 27*) between frequency response and transfer function.

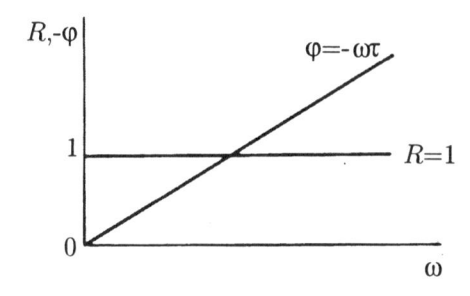

Figure B.26 Frequency response of an ideal communication system.

THEOREM 27 (*frequency response and transfer function*)

The frequency response $[R(\omega), \varphi(\omega)]$ of a continuous-time linear system is uniquely determined by its transfer function $G(s)$. More precisely, $R(\omega)$ and $\varphi(\omega)$ are, respectively, the modulus and the phase of the complex number $G(i\omega)$, that is,
$$G(i\omega) = R(\omega)e^{i\varphi(\omega)} \tag{B.38}$$

The proof of *Theorem 27* can be obtained by noting that the sinusoids $u_T(t) = U \sin \omega t$ and $y_T(t) = R(\omega)U \sin(\omega t + \varphi(\omega))$ with $R(\omega)$ and $\varphi(\omega)$ given by (B.38), satisfy the differential equation (ARMA model)

$$d(s)y(t) = n(s)u(t)$$

where $G(s) = n(s)/d(s)$. For example, if the system is the first-order system $\dot{x} = ax + bu$ with $y = x$, the transfer function is

$$G(s) = \frac{b}{s-a}$$

and the ARMA model is

$$\dot{y} - ay = bu$$

while the frequency response is

$$R(\omega) = \sqrt{\frac{b^2}{a^2 + \omega^2}} \quad \varphi(\omega) = \operatorname{arctg}\frac{\omega}{a}$$

It is therefore immediate to check (using a bit of trigonometry) that the two sinusoids

$$u_T(t) = U \sin \omega t$$
$$y_T = R(\omega)U \sin(\omega t + \varphi(\omega))$$

satisfy (B.22).

Theorem 27 is very important, since it allows one to compute the frequency response of a linear system from the triple (A, b, c^T), from the ARMA model, or from the transfer function $G(s)$. Moreover, it is the basis for a relatively simple solution of the identification problem (see *Section B.18*). In fact, if we want to model a physical system we must perform some tests on the system and, on the basis of the results, determine, for example, the triple (A, b, c^T). Such tests, must be measures of pairs of input and output functions, for example, the impulse response or the frequency response. From these functions, we must compute the transfer function of the system and, then, realize the triple (A, b, c^T) (see *Section B.3*). Among the tests that can be performed on the system, the frequency response is perhaps the most convenient. In fact, in order to measure the frequency response of an asymptotically stable system, it is not necessary that the initial state be zero, while this is necessary when dealing with the impulse or the step response of the system. Moreover, the frequency response can be measured by applying at the input of the system sinusoids of relatively small amplitude in such a way that nonlinear effects are negligible. Obviously, this is not the case when one wants to measure the impulse response. Finally, if the long-term response of an asymptotically stable system to a sinusoid is not sinusoidal, it is possible to conclude that the system is nonlinear, a conclusion, that can be hardly obtained by examining the impulse response since it is extremely difficult to check whether a function is a linear combination of exponentials or not.

B.23 FOURIER TRANSFORM

Before introducing the notions of Fourier series and Fourier transform, we give some definitions that will be used in the sequel.

DEFINITION 4 *(functions with bounded variation)*

A real function $f(\cdot)$ has a *bounded variation* in the closed interval $[a, b]$ if there exists a constant K such that for any finite set of points $t_0, t_1, t_2, \ldots, t_n$ partitioning the interval $[a, b](a = t_0 < t_1 < t_2 < \ldots < t_n = b)$ one has

$$\sum_{k=0}^{n-1} |f(t_{k+1}) - f(t_k)| \leq K$$

If a real function $f(\cdot)$ has a bounded variation in any closed interval, we say it is of bounded variation. Moreover, a complex function $f(\cdot)$ is of bounded variation if its real and imaginary parts have bounded variations.

The functions with bounded variation enjoy a number of properties that are now reported without proof.

THEOREM 28 *(properties of functions with bounded variation)*

> A real function $f(\cdot)$ has a bounded variation in the interval $[a, b]$ if and only if it is the difference between two nondecreasing functions. A function $f(\cdot)$ with bounded variation in the interval $[a, b]$ is bounded in the same interval. If a function $f(\cdot)$ has bounded variation in an interval $[a, b]$, the discontinuities of the function in such interval are numerable. If a function $f(\cdot)$ has bounded variation in an interval $[a, b]$ then, for any $t \in (a, b)$ there exist the right and left limits, that is,
>
> $$f(t^-) = \lim_{\varepsilon \to 0} f(t - \varepsilon) \quad f(t^+) = \lim_{\varepsilon \to 0} f(t + \varepsilon) \quad \varepsilon > 0$$
>
> Moreover, for $t = a$ there exists the right limit and for $t = b$ the left one.

We can now state the first important result concerning the *Fourier series*. From an intuitive point of view, the result says that, under very general assumptions, a periodic function of period T can be represented as the linear combination of sinusoids of angular frequency equal to multiples of the angular frequency $2\pi/T$. The proof of this result is not reported because it is not easy.

THEOREM 29 *(Fourier series)*

> If $f(\cdot)$ is a periodic function of period T with bounded variation, then for all t one has
>
> $$\lim_{N \to \infty} \sum_{k=-N}^{N} f_k e^{i\frac{2\pi k}{T}t} = \frac{1}{2}(f(t^+) + f(t^-)) \tag{B.39}$$
>
> where
>
> $$f_k = \frac{1}{T} \int_{-T/2}^{T/2} f(t) e^{-i\frac{2\pi k}{T}t} dt \quad k = 0, \mp 1, \mp 2, \ldots \tag{B.40}$$

Obviously, if the function $f(\cdot)$ is continuous at time t, Eq. (B.39) simplifies and becomes

$$f(t) = \lim_{N \to \infty} \sum_{k=-N}^{N} f_k e^{i\frac{2\pi k}{T}t} \tag{B.41}$$

By recalling that

$$e^{i\theta} = \cos\theta + i\sin\theta$$

From (B.40) and (B.41) one easily obtains

$$f(t) = \frac{1}{2}a_0 + \lim_{N \to \infty} \sum_{k=-N}^{N} \left\{ a_k \cos\left(\frac{2\pi k}{T}t\right) + b_k \sin\left(\frac{2\pi k}{T}t\right) \right\} \tag{B.42}$$

where

$$a_k = \frac{2}{T} \int_{-T/2}^{T/2} f(t) \cos\left(\frac{2\pi k}{T}t\right) dt$$
$$b_k = \frac{2}{T} \int_{-T/2}^{T/2} f(t) \sin\left(\frac{2\pi k}{T}t\right) dt$$

Expression (B.42) is the most popular formulation of the Fourier series, since it shows explicitly that a periodic function $f(\cdot)$ is the linear combination of sinusoids and cosinusoids. Moreover, if

$$\int_{-T/2}^{T/2} |f(t)|^2 dt < \infty$$

it follows that

$$\lim_{N\to\infty} \int_{-T/2}^{T/2} \left| \sum_{k=-N}^{N} f_k e^{i\frac{2\pi k}{T}t} - f(t) \right|^2 dt = 0$$

where f_k is given by (B.40).

Since any function $f(\cdot)$ can be considered as a periodic function of a infinite period, from the previous results it follows that any function $f(\cdot)$ is the sum of a continuum of sinusoids and cosinusoids, since the difference $2\pi/T$ between two different angular velocities tends to zero whenever T tends to infinity. This is the basic idea of the so-called Fourier transform, which is specified below.

Let $f(\cdot)$ be a function with bounded variation over R and suppose that such a function satisfies the inequality

$$\int_{-\infty}^{\infty} |f(t)| dt < \infty$$

Denote with $f_T(\cdot)$ the periodic function of period T that coincides with $f(\cdot)$ in the interval $[-T/2, T/2]$. From the previous results, it follows that $f_T(\cdot)$ can be expanded in Fourier series, namely,

$$\lim_{N\to\infty} \sum_{k=-N}^{N} \left[\frac{1}{T} \int_{-\infty}^{\infty} f_T(t) e^{-i\frac{2\pi k}{T}t} dt \right] e^{i\frac{2\pi k}{T}t} = \frac{1}{2}(f_T(t^+) + f_T(t^-)) \quad \text{(B.43)}$$

Since, by definition, $f_T(\cdot)$ and $f(\cdot)$ coincide in the interval $[-T/2, T/2]$, the relationship (B.43) can also be written with $f(t)$ instead of $f_T(t)$ provided t belongs to the interval $[-T/2, T/2]$. By setting

$$F_T(i\omega) = \int_{-T/2}^{T/2} f(t) e^{i\omega t} dt$$

from (B.40) with $T \to \infty$ one obtains

$$\frac{1}{2\pi} \int_{-\infty}^{\infty} F(i\omega)e^{i\omega t} d\omega = \frac{1}{2}(f(t^+) + f(t^-)) \tag{B.44}$$

and

$$F(i\omega) = \int_{-\infty}^{\infty} f(t)e^{i\omega t} dt = \lim_{T \to \infty} F_T(i\omega)$$

The function $F(\cdot)$ is called a *Fourier transform* or Fourier integral of the function $f(\cdot)$.

B.24 LAPLACE TRANSFORM

Suppose that a function $f(\cdot)$ has bounded variation in any closed interval contained in $[0, \infty)$ and that there exists a constant $\sigma < \infty$ such that

$$\int_{-\infty}^{\infty} |f(t)|e^{-\sigma t} dt < \infty$$

Then, consider, the following function $F(\cdot)$

$$F(\sigma + i\omega) = \int_{0}^{\infty} f(t)e^{-i\omega t}e^{-\sigma t} dt \tag{B.45}$$

and note that

$$F(\sigma + i\omega) = \int_{0}^{\infty} e^{-i\omega t}(\text{step}(t)e^{-\sigma t} f(t)) dt$$

which means that the function $F(\cdot)$ is the Fourier transform of the function

$$\text{step}(t)e^{-\sigma t} f(t)$$

Therefore, from (B.44) it follows that

$$\frac{1}{2} f(0^+) = \frac{1}{2\pi} \int_{-\infty}^{\infty} F(\sigma + i\omega) d\omega$$

and

$$\frac{1}{2} e^{-\sigma t}(f(t^+) + f(t^-)) = \frac{1}{2\pi} \int_{-\infty}^{\infty} F(\sigma + i\omega)e^{-i\omega t} d\omega$$

for $t > 0$. If we denote the complex variable by s (i.e., $s = \sigma + i\omega$) Eq. (B.45) becomes

$$F(s) = \int_{0}^{\infty} e^{-st} f(t) dt \tag{B.46}$$

The function $F(\cdot)$, often denoted by $L[f(\cdot)]$, is called a *Laplace transform* of the function $f(\cdot)$. It is a complex valued function defined on the domain $\text{Re}(s) > \sigma_0$, where σ_0 is the smallest real number such that $\sigma < \sigma_0$ implies

$$\int_{-\infty}^{\infty} |f(t)| e^{-\sigma t} dt < \infty$$

The Laplace transformation $f(\cdot) \mapsto F(\cdot)$, defined by (B.46), has a number of properties. First of all, it is a linear transformation since

$$L[\alpha f_1(\cdot) + \beta f_2(\cdot)] = \alpha L[f_1(\cdot)] + \beta L[f_2(\cdot)]$$

Moreover, the Laplace transform $F(\cdot)$ of any function $f(\cdot)$ is an analytical function in the domain $\text{Re}(s) > \sigma_0$. This implies that the function $F(\cdot)$ can often be extended to the whole complex plane [*i.e.*, there exists a unique function coinciding with $F(\cdot)$ for $\text{Re}(s) > \sigma_0$ but defined over the whole complex plane and analytical anywhere, apart from a certain number of isolated singularity points]. For example, if $f(t) = e^t$, $0 \leq t < \infty$ we have

$$L[f(\cdot)] = \int_0^\infty e^t e^{-st} dt = \frac{1}{s-1} \quad \text{Re}(s) > 1$$

and the function $1/(s-1)$ is analytical anywhere apart from the singular point $s = 1$.

Other important properties of the Laplace transform are those concerning integration and differentiation of a function $f(\cdot)$. In fact, the following relations hold:

$$L\left[\int_0^t f(\tau) d\tau\right] = \frac{1}{s} L[f(\cdot)]$$
$$L\left[\frac{d}{dt} f(\cdot)\right] = s L[f(\cdot)] - f(0)$$

Finally, the product of two transformed functions corresponds, in the time domain, to the operation called *convolution*, that is, if $F(\cdot)$ and $G(\cdot)$ are the Laplace transforms of two functions $f(\cdot)$ and $g(\cdot)$, the inverse transform of

$$H(\cdot) = F(\cdot) G(\cdot)$$

is

$$h(t) = \int_0^t f(t-\tau) g(\tau) d\tau \quad 0 \leq t < \infty$$

In the following table, we report some Laplace transforms $F(s)$ of functions $f(t)$:

$f(t)$	$F(s)$
imp$t$	1
step$t$	$\dfrac{1}{s}$
ramp$t$	$\dfrac{1}{s^2}$
$e^{\alpha t}$	$\dfrac{1}{s-\alpha}$
$\sin \omega t$	$\dfrac{\omega}{s^2+\omega^2}$
$\cos \omega t$	$\dfrac{s}{s^2+\omega^2}$
$f(t-\tau)$	$e^{-\tau s}F(s)$
$t^n \quad n>0$	$\dfrac{n!}{s^{n+1}}$

B.25 Z–TRANSFORM

Consider a function $f(\cdot)$ defined on the nonnegative integers, that is,

$$f(\cdot) : t \mapsto f(t) \quad t = \text{nonnegative integer}$$

The *Zeta-transform* of such a function, denoted by

$$F(\cdot) = Z[f(\cdot)]$$

is simply given by the series

$$F(\cdot) : z \mapsto F(z) = f(0) + f(1)z^{-1} + f(2)z^{-2} + \cdots \qquad \text{(B.47)}$$

Obviously, this expression makes sense if the series converges in the neighborhood of the improper point $z^{-1} = 0$, where it clearly converges. Suppose now that $f(t)$ does not increase with t more rapidly than a geometric series. Then, $|f(t)|^{1/t}$ tends to one or more positive limits, the largest being denoted by R_c, that is,

$$\lim_{t \to \infty} |f(t)|^{1/t} = R_c$$

It is easy to show that the series

$$F(z) = \sum_{i=0}^{\infty} f(t)z^{-t}$$

converges absolutely for all complex z satisfying the relationship

$$|z| > R_c$$

and, for this reason, R_c is called the *convergence radius*.

The transformation operator $f(\cdot) \mapsto F(\cdot)$ is obviously linear, since

$$Z[\alpha f_1(\cdot) + \beta f_2(\cdot)] = \alpha Z[f_1(\cdot)] + \beta Z[f_2(\cdot)]$$

Theorems completely analogous to those given for the Laplace transform can be proved for the Z-transform. By denoting $f^-(\cdot)$ as the function obtained from $f(\cdot)$ after a backward time shift, that is,

$$f^-(t) = \begin{cases} 0 & \text{for } t = 0 \\ f(t-1) & \text{for } t \geq 1 \end{cases}$$

the following holds:

$$Z[f^-(\cdot)] = z^{-1} Z[f(\cdot)]$$

while, if $f^+(\cdot)$ is the function obtained from $f(\cdot)$ after a forward time shift, that is,

$$f^+(t) = f(t+1) \quad \text{for } t \geq 0$$

the following holds:

$$Z[f^+(\cdot)] = zZ[f(\cdot)] - zf(0)$$

The simplest way to find analytical expressions of the Z-transform is to determine the sum of the series (B.47). Thus, for example, if

$$f(t) = a^t \quad t \geq 0$$

we have

$$F(z) = \sum_{t=0}^{\infty} a^t z^{-t} = \sum_{t=0}^{\infty} (az^{-1})^t = (1 - az^{-1})^{-1}$$

for $|z| > |a|$ and the same formula holds if a is replaced by a square matrix A and 1 by the identity matrix I. Other Z-transforms are reported in the table below.

$f(t)$	$F(z)$
a^t	$\dfrac{z}{z-a}$
1	$\dfrac{z}{z-1}$
t	$\dfrac{z}{(z-1)^2}$
t^2	$\dfrac{z(z+1)}{(z-1)^3}$
t^3	$\dfrac{z(z^2+4z+1)}{(z-1)^4}$

B.26 LAPLACE AND Z–TRANSFORMS AND TRANSFER FUNCTIONS

Recall that the transfer function $G(p)$ of a linear system described by an ARMA model

$$D(p)y(t) = N(p)u(t)$$

has been defined (see Section B.2) as the ratio of the two polynomials $N(p)$ and $D(p)$, that is,

$$G(p) = \frac{N(p)}{D(p)}$$

Moreover, if the system is proper, the transfer function can be computed using the formula [see (B.13)]

$$G(p) = c^T(pI - A)^{-1}b$$

Thus, the transfer function $G(p)$ is the (Laplace and Z-) transform of the impulse response. In fact, in a continuous-time system, the Laplace transform of the impulse response is

$$L[g(t)] = L[c^T e^{At} b] = c^T L[e^{At}]b = c^T(sI - A)^{-1}b$$

and therefore coincides with (B.13). An analogous check is possible for discrete-time systems.

If we take the above discussion into account, it is straightforward to see that the transfer function can also be written in the form

$$G(p) = \sum_{i=1}^{\infty} \frac{g_i}{p^i}$$

where $g_i = c^T A^{i-1} b$ are the Markov coefficients. In fact, for discrete-time systems (see Section B.21) we have

$$g(t) = \begin{cases} 0 & \text{for } t = 0 \\ g_t & \text{for } t \geq 1 \end{cases}$$

so that the Z-transform of $g(t)$ is (see Section B.25)

$$G(z) = g_1 z^{-1} + g_2 z^{-2} + g_3 z^{-3} + \cdots$$

In an analogous way, for continuous-time systems

$$g(t) = g_1 + g_2 t + g_3 \frac{t^2}{2!} + \cdots$$

so that, recalling that the Laplace transform of t^n is $(n!/s^{n+1})$, one obtains

$$G(s) = L[g(t)] = \frac{g_1}{s} + \frac{g_2}{s^2} + \frac{g_3}{s^3} + \cdots$$

Index

A
Alive sectors, 110
Age structure, 117, 125
Ageing, 117
Arrivals, 155

C
Canonical form
    control, 82
    Markov, 83
    Markov dual, 83
    reconstruction, 83
Class
    absorbing, 137
    transient, 137
Communication states, 24
Compartments, 145
Compartmental network, 147

D
D-stability, 43, 44
Demographic
    explosion, 122
    projections, 118
Departures, 155
Distribution,
    asymptotic, 158, 160
    equilibrium, 134
    probability, 132
Diversity
    topological, 151
    total, 152

Duality principle, 80

E
Eigenvalue
    dominant, 36, 54
    of Frobenius, 36, 38, 50
Eigenvector
    dominant, 36
    of Frobenius, 36, 50
Efficient sector, 112
Ehrenfest model, 136
Equilibrium
    positivity, 57
    strict positivity, 57
    reachability, 73
    stability, 57
Erlang
    Formula B, 161
    Formula C, 163
Excitability, 30, 31
Extinction, 122

F
Fertility, 117
Frobenius,
    eigenvalue, 36, 38, 50
    eigenvector, 36, 50
    canonical form, 28
Function
    nonnegative, 8
    positive, 8
    strictly positive, 8

G

Graph
    influence, 17
    paths, 21
    interconnection, 101
    Markov, 131
Growth rate, 123, 124

I

Immigration, 119
Input–output analysis, 109
Interconnection, 101
    asymptotically stable, 103
    excitable, 104
    standard, 104
    of compartments, 150, 151
    minimum phase, 104
Invasion, 31
Irreducibility, 23, 26, 54

L

Leontief, 109
Leslie model, 117
    irreducibility, 121,
    excitability, 121
    nontransparency, 121
Liapunov theorem, 15
Little formula, 159

M

Markov
    cone, 86
    chain, 131
Markovian assumptions, 131, 156
Matrix
    fundamental, 141, 142
    nonnegative, 7
    positive, 7
    strictly positive, 7
    of Leslie, 118
Mendel's law, 139
Minimum phase, 91, 95
Model
    of prices, 112
    of a queueing system, 157

N

Net maternity, 122
Net reproduction, 122

O

Observability,
    complete, 80
    almost complete, 80

P

Poisson's processes, 156
Positivity, 7, 14
    external, 8, 9

Primitivity, 27
Production dynamics, 113

Q

Queueing system, 155

R

Razor's edge, 114
Reachability, 65
    complete, 74
    almost complete, 74, 76
    of equilibria, 73
Reachability cone, 65, 74
    convexity, 66
    solidness, 66
    expansion, 67
Realization, 81
    positive, 83, 85, 86, 89
Reconstruction error, 92
Relative change, 62
Reproduction, 117

S

Selection rule, 155
Semelparous populations, 127
Services
    with unbounded queue, 162
    with bounded queue, 157, 161
    without queue, 160
Simplex, 133
Stability, 35, 40
    asymptotic, 40
    connective, 43
Stocking, 119, 128
System
    asymptotically stable, 35
    cyclic, 27
    compartmental, 145
        acyclic, 146
        cyclic, 146
    connectively stable, 43
    excitable, 30
    externally stable, 35
    unstable, 35
    irreducible, 23, 26, 50, 54
    positive, 12
    primitive, 27, 28
    reducible, 23
    marginally stable, 35
    stable, 35, 40
    standard, 59
    transparent, 31
Survival, 117

T

Trace condition, 40
Transfer constant, 91
Transfer function,

positively realizable,
Transparence, 31,32

Z
Zeros, 92, 83

PURE AND APPLIED MATHEMATICS

A Wiley-Interscience Series of Texts, Monographs, and Tracts

Founded by RICHARD COURANT
Editors Emeriti: PETER HILTON and HARRY HOCHSTADT
Editors: MYRON B. ALLEN III, DAVID A. COX, PETER LAX,
 JOHN TOLAND

ADÁMEK, HERRLICH, and STRECKER—Abstract and Concrete Catetories
ADAMOWICZ and ZBIERSKI—Logic of Mathematics
AKIVIS and GOLDBERG—Conformal Differential Geometry and Its Generalizations
ALLEN and ISAACSON—Numerical Analysis for Applied Science
*ARTIN—Geometric Algebra
AUBIN—Applied Functional Analysis, Second Edition
AZIZOV and IOKHVIDOV—Linear Operators in Spaces with an Indefinite Metric
BERG—The Fourier-Analytic Proof of Quadratic Reciprocity
BERMAN, NEUMANN, and STERN—Nonnegative Matrices in Dynamic Systems
BOYARINTSEV—Methods of Solving Singular Systems of Ordinary Differential
 Equations
BURK—Lebesgue Measure and Integration: An Introduction
*CARTER—Finite Groups of Lie Type
CASTILLO, COBO, JUBETE and PRUNEDA—Orthogonal Sets and Polar Methods in
 Linear Algebra: Applications to Matrix Calculations, Systems of Equations,
 Inequalities, and Linear Programming
CHATELIN—Eigenvalues of Matrices
CLARK—Mathematical Bioeconomics: The Optimal Management of Renewable
 Resources, Second Edition
COX—Primes of the Form $x^2 + ny^2$: Fermat, Class Field Theory, and Complex
 Multiplication
*CURTIS and REINER—Representation Theory of Finite Groups and Associative Algebras
*CURTIS and REINER—Methods of Representation Theory: With Applications to Finite
 Groups and Orders, Volume I
CURTIS and REINER—Methods of Representation Theory: With Applications to Finite
 Groups and Orders, Volume II
DINCULEANU—Vector Integration and Stochastic Integration in Banach Spaces
*DUNFORD and SCHWARTZ—Linear Operators
 Part 1—General Theory
 Part 2—Spectral Theory, Self Adjoint Operators in
  Hilbert Space
 Part 3—Spectral Operators
FARINA and RINALDI—Positive Linear Systems: Theory and Applications
FOLLAND—Real Analysis: Modern Techniques and Their Applications
FRÖLICHER and KRIEGL—Linear Spaces and Differentiation Theory
GARDINER—Teichmüller Theory and Quadratic Differentials
GREENE and KRANTZ—Function Theory of One Complex Variable
*GRIFFITHS and HARRIS—Principles of Algebraic Geometry
GRILLET—Algebra
GROVE—Groups and Characters
GUSTAFSSON, KREISS and OLIGER—Time Dependent Problems and Difference
 Methods
HANNA and ROWLAND—Fourier Series, Transforms, and Boundary Value Problems,
 Second Edition

*HENRICI—Applied and Computational Complex Analysis
        Volume 1, Power Series—Integration—Conformal Mapping—Location of Zeros
        Volume 2, Special Functions—Integral Transforms—Asymptotics—Continued Fractions
        Volume 3, Discrete Fourier Analysis, Cauchy Integrals, Construction of Conformal Maps, Univalent Functions
*HILTON and WU—A Course in Modern Algebra
*HOCHSTADT—Integral Equations
JOST—Two-Dimensional Geometric Variational Procedures
*KOBAYASHI and NOMIZU—Foundations of Differential Geometry, Volume I
*KOBAYASHI and NOMIZU—Foundations of Differential Geometry, Volume II
LAX—Linear Algebra
LOGAN—An Introduction to Nonlinear Partial Differential Equations
McCONNELL and ROBSON—Noncommutative Noetherian Rings
NAYFEH—Perturbation Methods
NAYFEH and MOOK—Nonlinear Oscillations
PANDEY—The Hilbert Transform of Schwartz Distributions and Applications
PETKOV—Geometry of Reflecting Rays and Inverse Spectral Problems
*PRENTER—Splines and Variational Methods
RAO—Measure Theory and Integration
RASSIAS and SIMSA—Finite Sums Decompositions in Mathematical Analysis
RENELT—Elliptic Systems and Quasiconformal Mappings
RIVLIN—Chebyshev Polynomials: From Approximation Theory to Algebra and Number Theory, Second Edition
ROCKAFELLAR—Network Flows and Monotropic Optimization
ROITMAN—Introduction to Modern Set Theory
*RUDIN—Fourier Analysis on Groups
SENDOV—The Averaged Moduli of Smoothness: Applications in Numerical Methods and Approximations
SENDOV and POPOV—The Averaged Moduli of Smoothness
*SIEGEL—Topics in Complex Function Theory
        Volume 1—Elliptic Functions and Uniformization Theory
        Volume 2—Automorphic Functions and Abelian Integrals
        Volume 3—Abelian Functions and Modular Functions of Several Variables
SMITH and ROMANOWSKA—Post-Modern Algebra
STAKGOLD—Green's Functions and Boundary Value Problems, Second Editon
*STOKER—Differential Geometry
*STOKER—Nonlinear Vibrations in Mechanical and Electrical Systems
*STOKER—Water Waves: The Mathematical Theory with Applications
WESSELING—An Introduction to Multigrid Methods
[†]WHITHAM—Linear and Nonlinear Waves
[†]ZAUDERER—Partial Differential Equations of Applied Mathematics, Second Edition

*Now available in a lower priced paperback edition in the Wiley Classics Library.
[†]Now available in paperback.